Advanced Series on Circuits and Systems – Vol. 1

INTERVAL METHODS FOR CIRCUIT ANALYSIS

L V Kolev
Technical University of Sofia

World Scientific
Singapore • New Jersey • London • Hong Kong

Published by

World Scientific Publishing Co. Pte. Ltd.

P O Box 128, Farrer Road, Singapore 9128

USA office: Suite 1B, 1060 Main Street, River Edge, NJ 07661

UK office: 73 Lynton Mead, Totteridge, London N20 8DH

INTERVAL METHODS FOR CIRCUIT ANALYSIS

ISBN 981-02-1413-8

Printed in Singapore by Utopia Press.

To Lydia

PREFACE

Progress in the development of a theoretical subject is, to a great extent, predetermined by the adequacy, variety and power of the mathematical models used.

Various mathematical tools have been utilised in circuit theory. They differ from one another in their complexity, generality and area of application. Thus, the topological properties of both linear and nonlinear electric circuits are best described by graph theory. Steady-state analysis of large linear circuits is carried out in the most effective way by using sparse matrix techniques. Qualitative properties of nonlinear circuits – existence, uniqueness, boundedness of steady-states and other related topics – seem to be most adequately analyzed in terms of the qualitative theory of nonlinear differential equations.

The overwhelming majority of the mathematical models now in use in circuit theory is based on the concept of real number (complex numbers can always, at least conceptually, be viewed as pairs of real numbers in a corresponding two-dimensional space). This approach is quite natural and satisfactory if the initial data about the electric circuit studied (parameters of passive elements, energy sources etc.) can be assumed to be known exactly. Since this is rarely the case, such an idealization is acceptable only when each item of the input data can be represented within resonable accuracy as a real number. As each real number can be geometrically viewed as a point on the real line, all the data relevant to the problem at hand can be visualized as a point in a space of appropriate dimensionality. Therefore, for brevity of expression, a mathematical model based on such an approach will be called a "point" model. Although intrinsically inaccurate the point model is practically the best model for tackling problems in which the input data can reasonably be assumed to be known exactly.

However, the input data are, theoretically always and practically most often, known only with some degree of uncertainty. In circuit theory the basic approach to handling such problems is to appeal to a probabilistic description of the problem and to apply a certain statistical method for its solution. This approach is associated with the necessity to introduce experimentally some distribution law describing the probability with which the point representing the input data appears in each element of a corresponding parameter space. Another possibility is to resort to some results from fuzzy sets theory. Once again, some statistical information is needed to describe the "fuzziness" of the sets involved.

An alternative approach to treating electric circuits with inaccurate data is to apply the notions and methods of the so-called interval analysis.

Interval analysis is a new and intensively developing branch of computational mathematics. It has been in existance for only three decades: the first monograph in the field by R. Moore was published in 1966. Originating from the need to control propagation of errors in computations on digital computers, interval analysis presently covers a variety of problems in applied mathematics which are difficult to solve using traditional (noninterval) approaches.

The basic concept in interval analysis is that of an interval. An interval is a bounded segment of the real line. As arithmetic operations over intervals have been introduced, intervals are also called interval numbers. Interval numbers are a generalization of real numbers: in fact, an interval number is a set of (infinitely many) real numbers. Interval numbers can be arguments of functions which are called interval functions. Broadly speaking, one of the main objectives of interval analysis is to study the properties of interval functions and to seek efficient methods for their evaluation. Based on these investigations, various interval methods have been proposed for solving numerous problems in linear and nonlinear analysis. In fact, there exists nowadays an interval counterpart for practically every significant problem and method encountered in classical mathematical analysis.

Since interval analysis deals with intervals rather than points, it is ideally suited for handling electric circuit problems whose initial data are allowed to take on values from some prescribed intervals. A mathematical model of such a problem which is based on the interval representation of the input data will be called, for brevity and in contrast to the point model mentioned earlier, an interval model. Furthermore, a method for solving a particular problem which is based on an associated interval model and appeals to appropriate interval analysis techniques will be referred to as an interval method.

It is the purpose of the present book to acquaint the reader with some applications of interval analysis in electric circuit theory. More specifically interval models and ensuing interval methods for circuit analysis are presented in detail for the following topics: linear electric circuit tolerance analysis (steady-state as well as transient analysis), linear circuit stability and nonlinear circuit analysis (including both resistive and dynamic circuits).

In order to make the book self-contained a comprehensive survey of all the necessary interval analysis notions and techniques is provided in Chapter 1. Readers familiar with interval analysis could ignore this introductory text.

Chapter 2 begins the discussion of one of the problems treated in this book by way of interval techniques, namely the tolerance analysis of linear electric circuits. Both the determenistic (worst-case) and the probabilistic statement of the tolerance analysis problem are considered. Either tolerance problem is formulated in this chapter as an associated equivalent global optimization problem. The latter problem is solved using various interval methods: zero-order method (using no derivatives of the functions involved), first-order method and second-order methods (using first- or second-order derivatives, respectively). Several algorithms implementing the above interval methods for tolerance analysis are presented and their numerical efficiency is studied.

Chapter 3 continues the discussion of the worst-case tolerance analysis of linear electric circuits. Now the mathematical model used is in the form of a specific linear system of equations with independent or dependent interval coefficients. This approach proves to be more efficient than the global optimization approach from Chapter 2 for the case of electric circuits of increased size. Exact solution to the d.c. tolerance problem and approximate solutions to the a.c. tolerance problem are thus derived.

In Chapter 4 the problem of stability analysis for linear electric circuits with interval parameters is considered. Two approaches to treating this problem are presented. According to the first one the stability of the circuit investigated is assessed by means of an associated characteristic polynomial whose coefficients are functions of the circuit interval parameters. The second approach reduces the original stability problem to assessing the stability of a corresponding interval matrix (matrix whose elements are intervals). Based on some known results on stability for circuits with exact data, necessary and sufficient conditions as well as simple sufficient conditions are thus suggested for checking the stability, instability or stability margin of linear circuits (systems) with interval parameters.

Chapter 5 deals with transient analysis of linear circuits with uncertain (interval) data. In fact, the problem herein considered is a dynamic generalisation of the static worst-case tolerance analysis problem (in both forms presented in Chapters 2 and 3). Assuming the linear circuit to be intervally stable, several methods for exact and approximate solution of the transient analysis problem are proposed.

The last chapter covers some topics relative to the analysis of nonlinear circuits with exact data. Firstly, the challenging problem of finding all operating points of nonlinear resistive circuits is considered. Two groups of interval methods for solving this problem are presented: general methods applicable for the case where the circuit is described by a system of nonlinear algebraic equations of general form and specialized methods designed for the special case where the circuit nonlinear equations are in the known hybrid representation form. Secondly, an interval method for finding all the periodic solutions (of a given period) existing in a dynamic nonlinear circuit is introduced. The latter method is based on results obtained in Chapter 5. Finally, the fundamental problem of uniqueness of the periodic solution is touched upon. A sufficient condition for uniqueness is obtained which is implemented by means of some of the results on stability of interval matrices presented in Chapter 4.

Most of the theoretical developements considered in the book are illustrated through numerical examples.

Interval methods have a number of appealing features. One of their fundamental advantages is that, unlike the point methods, they provide each output result in the form of an interval which contains the result sought, thus guaranteeing infallible bounds on the true value of the respective output variable. Using the so-called machine interval arithmetic, they automatically account for roundoff errors when implemented on a computer. Interval methods will always globally converge in a finite number of steps and the numerical accuracy achieved is basically determined by the machine accuracy of the computer used. On the other hand, programming and using interval methods may, in some cases, present some difficulties. Indeed, all the interval operations involved in the method used must be programmed individually for every problem being solved by the developer or user of the method. However, there already exist special versions of high-level algorithmic languages which permit intervals to be declared as a special data type: the commonest computations with intervals are then written in as simple a manner as for real number computations. It is to be expected that shortly all the standard numerical methods will be available in interval form as reliable and efficient software packages, thus making the use of interval methods as easy as that of the noninterval mathematical tools.

The present book is written by an electrical engineer and is intended primarily for electrical and electronics engineers applying circuit theory in their work. Most of the results presented are, however, applicable (directly or after minor modifications) to static and dynamic systems of arbitrary nature. Therefore the book may be of interest to systems analysts and control engineers as well. It will also be useful to graduate and postgraduate students interested in circuit or system theory and their applications. Some of the results herein obtained might also interest applied mathematicians who are concerned with developing and using interval computation methods.

It is hoped that the monograph will help popularize the fruitful ideas of interval analysis among electrical, electronics and control engineers and will stimulate further research in the topics covered as well as in investigating possibilities for other interval analysis applications in electrical engineering and other related disciplines.

The author has benefited from contacts (either personal or by correspondence) with leading specialists in interval analysis. To mention the names of all who have, directly or indirectly, stimulated my work would make a long list with the inevitable risk of omitting someone's important influence. However, I wish to express my thanks to my compatriot, the bulgarian mathematician Svetoslav Markov who some ten years ago lent me his library on interval analysis (a rarity in Bulgaria at that time) and thus made possible my acquaintance with the realm of interval analysis methodology.

L. KOLEV

CONTENTS

INTERVAL METHODS FOR CIRCUIT ANALYSIS

CHAPTER 1

MATHEMATICAL BACKGROUND

The present chapter introduces some of the basic notions and techniques from interval analysis needed in the sequel for presenting various uses of interval analysis in electric circuit theory and its applications. It cannot be overemphasized that the reader who is not familiar with interval analysis should master the material in this chapter before proceeding to the following chapters.

1.1. INTERVAL ARITHMETIC

1.1.1. Interval numbers

In what follows we will be dealing primarily with sets. Although the reader is assumed to be familiar with the rudiments of set theory, a definition of a set will be given here in the form which will be used in the sequel, namely

$$S = \{x: p(x)\}$$

where S is the set considered, x is an arbitrary element of the set while $p(x)$ denotes some rules which specify whether x belongs to S or not.

A fundamental notion in interval analysis is the notion of an interval [1], [2]. Let R be the set of all real numbers. By an interval X we mean a closed bounded compact subset of R:

$$X = \{x: x \in R, \quad a \leq x \leq b, \quad a,b \in R, \quad -\infty < a \leq b < \infty\} \tag{1.1a}$$

The set of all intervals will be denoted by $I(R)$. To distinguish intervals from real numbers the elements of $I(R)$ will be designated most often by capital letters while lower-case letter will be used for the elements of R (later on we will employ also lower-case letters with superscript I to explicitly denote intervals where needed to avoid ambiguity). Furthermore, if X is an interval, we will denote its lower (left) endpoint by $\underline{x}$ and its upper (right) endpoint by $\overline{x}$. Thus, alongside the full definition (1.1a) the equivalent shorter notation

$$X = [\underline{x}, \overline{x}] \tag{1.1b}$$

will also be used.

In the next section we will introduce arithmetic operations over intervals, so the elements of $I(R)$ will be also called interval numbers.

It follows from (1.1a) and (1.1b) that an interval can be treated in two different manners: we can regard it either as a set of real numbers or an ordered pair of two real numbers, the first number being $\underline{x}$ and the second number being $\overline{x}$, with $\underline{x} \leq \overline{x}$. However, it will be shown later that from a computational point of view the latter representation offers great advantages over the former since it permits operations with interval numbers to be reduced to operations involving their endpoints only, thus avoiding the more cumbersome operations with sets.

An interval $X = [\underline{x}, \overline{x}]$ is called degenerate if $\underline{x} = \overline{x}$; otherwise ($\underline{x} \neq \overline{x}$) it is referred to as nondegenerate.

The interval number is a generalisation of the real number. Indeed, in terms of interval analysis any real number x can be considered as a degenerate interval $x = [x, x]$. From this point of view the set of real numbers is contained in the set of interval numbers, i.e. $R \subset I(R)$.

We call two intervals $X = [\underline{x}, \overline{x}]$ and $Y = [\underline{y}, \overline{y}]$ equal if and only if (we shall employ the abbreviation iff) their corresponding endpoints are equal, that is, $X = Y$ iff $\underline{x} = \underline{y}$ and $\overline{x} = \overline{y}$.

The elements of $I(R)$ can be ordered in the following way: $X < Y$ iff $\overline{x} < \underline{y}$.

The intersection $X \cap Y$ of two intervals X and Y is empty, i.e. $X \cap Y = \varnothing$, if either $X < Y$ or $Y < X$. Otherwise the intersection of X and Y is again an interval:

$$X \cap Y = [\max\{\underline{x}, \underline{y}\},\ \min\{\overline{x}, \overline{y}\}]$$

If two intervals X and Y have a nonempty intersection, their union $X \cup Y$ is again an interval:

$$X \cup Y = [\min\{\underline{x}, \underline{y}\},\ \max\{\overline{x}, \overline{y}\}]$$

If $X \cap Y = \varnothing$, the union $X \cup Y$ is obviously not an interval since in this case the set $Z = X \cup Y$ is not compact (in fact, Z is formed of two distinct intervals X and Y).

A useful relation for intervals is the set inclusion:

$$X \subseteq Y \quad \text{iff} \quad \underline{y} \leq \underline{x} \quad \text{and} \quad \overline{x} \leq \overline{y}$$

We call width of an interval $X = [\underline{x}, \overline{x}]$ the real number

$$w(X) = \overline{x} - \underline{x} \tag{1.2}$$

The midpoint (or centre) of X is the real number

$$m(X) = (\underline{x} + \overline{x})/2 \tag{1.3}$$

We define the absolute value of an interval X by

$$|X| = \max(|\underline{x}|, |\overline{x}|) \tag{1.4}$$

It is easily seen that $|X| \leq |Y|$, $w(X) \leq w(Y)$ when $X \subseteq Y$.
Finally, the distance $\rho(X,Y)$ between $X,Y \in I(R)$ is defined as

$$\rho(X,Y) = \max\{|\underline{x} - \underline{y}|, |\overline{x} - \overline{y}|\} \tag{1.5}$$

A nondegenerate interval X is called symmetric if $-\underline{x} = \overline{x}$. Any (nonsymmetric) interval X can be defined either by specifying its endpoints $\underline{x}$ and $\overline{x}$, i.e. in the form (1.1a)

$$X = [\underline{x}, \overline{x}] \tag{1.6}$$

or, equivalently, as the sum of its centre and a corresponding symmetric interval, i.e.

$$X = m(X) + [-w(x)/2, w(x)/2] \tag{1.7}$$

The quantity $w(X)/2$ is usually called the radius of the interval X. Both forms (1.6) and (1.7) are used in interval analysis.

1.1.2. Interval arithmetic operations

In this subsection arithmetic operations with intervals will be introduced.
Let $X,Y \in I(R)$. The sum of X and Y, denoted by $X + Y$, is defined by the set:

$$X + Y = \{x + y: x \in X, y \in Y\} \tag{1.8}$$

It is seen that $X + Y$ is again an interval, i.e. $X + Y \in I(R)$. Indeed, from (1.8) $\underline{x} + \underline{y} \leq x + y \leq \overline{x} + \overline{y}$. Thus, we have the equivalent relation

$$X + Y = [\underline{x}, \overline{x}] + [\underline{y}, \overline{y}] = [\underline{x} + \underline{y}, \overline{x} + \overline{y}] \tag{1.9}$$

Although (1.8) and (1.9) are equivalent, formula (1.9) is by far more useful for practical applications since it permits to find the whole set $X + Y$ by computing its endpoints $\underline{x} + \underline{y}$ and $\overline{x} + \overline{y}$ using only the corresponding endpoint of X and Y.

We define the negative of an interval by the set

$$-X = \{-x: x \in X\}$$

Similarly to the previous case we have

$$-X = -[\underline{x}, \overline{x}] = [-\overline{x}, -\underline{x}]$$

For the difference of two intervals we form the set

$$X - Y = X + [-Y] = \{x - y: x \in X, y \in Y\} \tag{1.10}$$

or equivalently

$$X - Y = [\underline{x}, \overline{x}] - [\underline{y}, \overline{y}] = [\underline{x} - \overline{y}, \overline{x} - \underline{y}] \tag{1.11}$$

Obviously, $X - Y \in I(R)$.

The product $X \cdot Y$ of two intervals X and Y is defined by the set

$$X \cdot Y = \{xy\colon x \in X,\ y \in Y\} \tag{1.12}$$

It is not hard to see that $X \cdot Y$ is again an interval and

$$X \cdot Y = [\min\{\underline{x}\,\underline{y},\ \underline{x}\,\overline{y},\ \overline{x}\,\underline{y},\ \overline{x}\,\overline{y}\},$$

$$\max\{\underline{x}\,\underline{y},\ \underline{x}\,\overline{y},\ \overline{x}\,\underline{y},\ \overline{x}\,\overline{y}\}] \tag{1.13}$$

The endpoints of the product $Z = X \cdot Y = [\underline{z}, \overline{z}]$ can be computed in a cheaper way if the signs of the endpoints of X and Y are taken into account. We have the following nine cases:

$$\begin{array}{llll}
1) & \underline{z} = \underline{x}\underline{y}\,,\ \overline{z} = \overline{x}\overline{y} & \text{if} & \underline{x} \ge 0\,,\ \underline{y} \ge 0 \\
2) & \underline{z} = \underline{x}\overline{y}\,,\ \overline{z} = \overline{x}\overline{y} & \text{if} & \underline{x} < 0 < \overline{x}\,,\ \underline{y} \ge 0 \\
3) & \underline{z} = \underline{x}\overline{y}\,,\ \overline{z} = \overline{x}\underline{y} & \text{if} & \overline{x} \le 0\,,\ \underline{y} \ge 0 \\
4) & \underline{z} = \underline{x}\overline{y}\,,\ \overline{z} = \overline{x}\underline{y} & \text{if} & \underline{x} \ge 0\,,\ \underline{y} < 0 < \overline{y} \\
5) & \underline{z} = \underline{x}\overline{y}\,,\ \overline{z} = \underline{x}\underline{y} & \text{if} & \underline{x} \le 0\,,\ \underline{y} < 0 < \overline{y} \\
6) & \underline{z} = \overline{x}\underline{y}\,,\ \overline{z} = \underline{x}\overline{y} & \text{if} & \underline{x} \ge 0\,,\ \underline{y} \le 0 \\
7) & \underline{z} = \overline{x}\underline{y}\,,\ \overline{z} = \underline{x}\underline{y} & \text{if} & \underline{x} < 0 < \overline{x}\,,\ \underline{y} \le 0 \\
8) & \underline{z} = \overline{x}\overline{y}\,,\ \overline{z} = \underline{x}\underline{y} & \text{if} & \underline{x} \le 0\,,\ \underline{y} \le 0 \\
9) & \underline{z} = \min\{\underline{x}\overline{y}, \overline{x}\underline{y}\}\,,\ \overline{z} = \max\{\underline{x}\underline{y}, \overline{x}\overline{y}\} & \text{if} & \underline{x} < 0 < \overline{x}\,,\ \underline{y} < 0 < \overline{y}
\end{array} \tag{1.14}$$

It is seen that with the exception of the ninth case formulae (1.14) are twice more effective than (1.13).

If X is an interval not containing the number 0, then we can define its reciprocal as follows:

$$1/X = \{1/x : x \in X\} \tag{1.15}$$

hence

$$1/X = [1/\overline{x},\ 1/\underline{x}] \tag{1.16}$$

and again $1/X \in I(R)$, if $X > 0$ or $X < 0$.

In the general case where $0 \in X$ the set (1.15) is no more an interval. In fact, it is not hard to see that the set $1/X$ consists of two distinct (nonintersecting) unbounded subsets of the real line. We shall postpone considering this general case for section 1.1.4.

For the quotient of two intervals, we define

$$X/Y = X \cdot (1/Y) = \{x/y : x \in X, y \in Y\} \tag{1.17}$$

If $0 \notin Y$, then X/Y is again an interval whose endpoints can be computed using (1.13) or (1.14) by the formula

$$X/Y = [\underline{x}, \overline{x}]/[\underline{y}, \overline{y}] = [\underline{x}, \overline{x}] \cdot [1/\overline{y}, 1/\underline{y}] \tag{1.18}$$

Similarly to the addition given by (1.9), the formulae (1.11), (1.13) and (1.18) for subtraction, multiplication and division of two intervals X and Y permit to obtain the whole resulting set using only the endpoints of X and Y.

For brevity, we shall often drop the dot notation for the product of two intervals and simply write XY for the product of X and Y.

1.1.3. Properties of interval arithmetic

If X and Y are degenerate intervals, then equalities (1.9), (1.11), (1.13) and (1.18) reduce to the ordinary arithmetic operations over real numbers. Thus, interval arithmetic can be regarded as a generalization of real arithmetic. Therefore, it is normal to expect that the properties of interval arithmetic will be similar to those of real arithmetic, which is really the case. However, there are several striking dissimilarities that will be stressed below.

It follows from the set-theoretical definitions (1.8), (1.12) that similarly to the respective real operation, interval addition is associative and interval multiplication is commutative, that is, if $X,Y,Z \in I(R)$ then

$$X + (Y + Z) = (X + Y) + Z; \quad X + Y = Y + X$$

$$X(YZ) = (XY)Z, \quad XY = YX$$

Zero and unity in $I(R)$ are the degenerate intervals [0, 0] and [1, 1] which will be denoted by 0 and 1 respectively. In other words:

$$X + 0 = 0 + X, \quad 1 \cdot X = X \cdot 1$$

for any $X \in I(R)$.

It is important to underline that unlike real arithmetic

$$X - X \neq 0$$

and

$$X/X \neq 1$$

when $w(X) > 0$. Indeed,

$$X - X = [\underline{x}-\overline{x},\ \overline{x}-\underline{x}] = w(X)[-1,\ 1]$$

and

$$X/X = [\underline{x}/\overline{x},\ \overline{x}/\underline{x}] \text{ for } X > 0$$

or

$$X/X = [\overline{x}/\underline{x},\ \underline{x}/\overline{x}] \text{ if } X < 0.$$

Another interesting property of interval arithmetic is the fact that the distributive law

$$X(Y + Z) = XY + XZ \tag{1.19}$$

does not always hold. For example, we have $[0,\ 1](1 - 1) = 0$ whereas $[0,\ 1] - [0,\ 1] = [-1,\ 1]$. We do, however, always have the following algebraic property

$$X(Y + Z) \subseteq XY + XZ \tag{1.20}$$

Indeed, $t \in X(Y + Z)$ implies $t \in x(y + z)$ where $x \in X$, $y \in Y$, $z \in X$. On the other hand, $xy \in XY$, $xz \in XZ$ and hence $t = xy + xz \in XY + XZ$ which proves (1.20).

The property (1.20) is called subdistributivity. It is to be stressed that, as is seen from (1.20) and the above example $w(X(Y + Z)) \leq w(XY + XZ)$. Therefore, it is always advantageous to use the factored form $X(Y + Z)$ rather than the expression $XY + XZ$ since the former form leads, in general, to a narrower resultant interval.

It is proven [1] that the distributivity law (1.19) remains true in the following special cases:

1) if Y and Z are symmetric;
2) if $YZ > 0$;
3) if $0 \in X$ and sign (Y) = sign (Z)

where

$$\operatorname{sign}(A) = \begin{cases} 1, & \textit{if } A > 0 \\ 0, & \textit{if } 0 \in A \\ -1, & \textit{if } A < 0 \end{cases}$$

Another important property of interval arithmetic is inclusion monotonicity. It means that if

$$X \subseteq Z,\quad Y \subseteq W \tag{1.21}$$

then

$$X + Y \subseteq Z + W$$

$$X - Y \subseteq Z - W$$

$$XY \subseteq ZW$$

$$X/Y \subseteq Z/W \qquad (\text{if } \ 0 \notin W) \tag{1.22}$$

The inclusion monotonicity follows directly from the set-theoretical definitions of the interval arithmetic operations.

1.1.4. Alternate interval arithmetics

In this subsection, some extensions of the "ordinary" interval arithmetic introduced in the previous sections will be considered.

Interval arithmetic with nonstandard operations

This interval arithmetic has been suggested in [3]. Nonstandard subtraction $\ominus$ and division $\oplus$ are defined for $X = [a, b]$, $Y = [c, d] \in I(R)$ in the following way:

$$X \ominus Y = [\min\{ a - c \, , b - d\} \, , \max\{a - c \, , b - d\}] \tag{1.23a}$$

$$X \oplus Y = \begin{cases} [\min(a/c, \ b/d), \ \max(a/c, \ b/d)] & \text{if } X,Y > 0 \\ [\min(a/d, \ b/c), \ \max(a/d, \ b/c)] & \text{if } X,Y < 0 \\ (1/c)X \ \text{ if } \ 0 \in X, \quad Y > 0 \\ (1/d)X \ \text{ if } \ 0 \in X, \quad Y < 0 \end{cases} \tag{1.23b}$$

As is seen from (1.23b) nonstandard division is defined only when $0 \notin Y$.

The nonstandard summation $\oplus$ and multiplication $\odot$ are defined as follows:

$$X \oplus Y = A \ominus (-Y) \tag{1.24a}$$

$$X \odot Y = A \oplus (1/Y) \tag{1.24b}$$

Extended interval arithmetic [4]

In this arithmetic, intervals can be unbounded and interval division X/Y is defined even when $0 \in Y$. If $0 \notin Y$, the quotient X/Y is computed by (1.18). Otherwise

$$X/Y = \begin{cases} [b/c, \infty] & \text{if} \quad b \le 0 \quad \text{and} \quad d = 0 \\ [-\infty, b/d \cup b/c, \infty] & \text{if} \quad b \le 0 \quad \text{and} \quad c < 0 < d \\ [-\infty, b/d & \text{if} \quad b \le 0 \quad \text{and} \quad c = 0 \\ [-\infty, \infty] & \text{if} \quad a < 0 < b \\ [-\infty, a/c] & \text{if} \quad a \ge 0 \quad \text{and} \quad d = 0 \\ [-\infty, a/c \cup [a/d, \infty] & \text{if} \quad a \ge 0 \quad \text{and} \quad c < 0 < d \\ [a/d, \infty] & \text{if} \quad a \ge 0 \quad \text{and} \quad c = 0 \end{cases} \tag{1.25}$$

In this case the result is not finite. However, in our applications X/Y will be intersected with a finite interval Z. Now the result is a finite set but it can be a single interval, two intervals, or the empty interval (the reader is urged to verify geometrically the above assertion).

Machine interval arithmetic [2]

The arithmetic operations defined by (1.9), (1.11), (1.13) and (1.18) are called exact interval arithmetic operations. However, when implementing these operations on a computer we commit errors due to round-off. Therefore, we have to take special measures so that the machine computed interval result always contains the exact interval result.

When computing with interval arithmetic if a left endpoint is not machine representable, it is rounded to the nearest arithmetically smaller machine number. A right endpoint is rounded to the nearest arithmetically larger machine number. This is termed outward rounding.

It should be borne in mind that machine interval arithmetic is about five times slower than ordinary arithmetic, if no special hardware is available and outward rounding is to be implemented by high level algorithmetic languages. However, progress in computer technology makes it realistic to believe that very soon machine interval arithmetic will be comparable in speed to ordinary arithmetic.

Machine interval arithmetic (i.e. appropriate rounding) should also be used when implementing the nonstandard and extended interval arithmetic on a computer.

In the following sections and chapters we shall develop various interval methods based primarily on the ordinary and extended interval arithmetic. For simplicity of presentation, only exact interval arithmetic will be used although the actual computer implementation of these methods will, naturally, require machine interval arithmetic.

1.2. INTERVAL EXTENSIONS

1.2.1. Interval functions

In this subsection the important notion of an interval function will be introduced. For easier understanding, first an interval function of one variable will be considered; then, the generalization to interval functions of several variables will be presented.

Recall that we write f: $D \subset R \to R$ to denote a real function $y = f(x)$ of one variable x, defined in the domain D, with values in R. Geometrically, to every point $x \in D$ the function f sets in correspondence one point y (Fig. 1.1a).

Similarly, if $X = [\underline{x}, \overline{x}]$ and $Y = [\underline{y}, \overline{y}]$ are intervals, we say that Y is an interval function of X, $Y = F(X)$, if to every X in a certain domain $D \subseteq I(R)$ there corresponds one interval Y. Symbolically $F\colon D \subseteq I(R) \to I(R)$.

Capital letters, e.g. F, will be used to denote interval functions, and lower-case letters, e.g. f, to denote real functions. Geometrically, F transforms any interval X from D into a new interval Y (Fig. 1.1b).

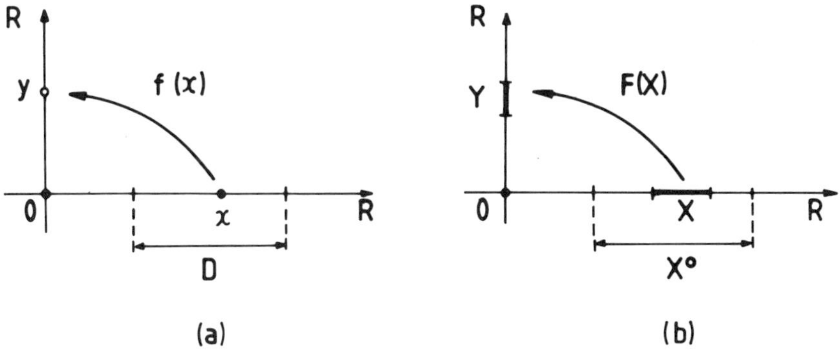

Fig. 1.1. Geometrical illustration of: (a) a real function $f(x)$, (b) an interval function $F(X)$.

The notion of an interval function of one variable can be easily extended to interval functions of several variables. To do this, we, however, need to introduce the notion of an interval vector.

Let $x = (x_1, x_2, \ldots, x_n)$ denote, as usual, an n-dimensional real vector, that is an ordered n-tuple of real numbers. For the set of all n-dimensional vectors we use the symbol R^n.

An n-dimensional interval vector is an ordered n-tuple of interval numbers $X_1, X_2, \ldots, X_n$. Loosely speaking, an interval vector is a vector whose components are intervals. We will use capital letters to denote interval vectors, i.e. we shall write $X = (X_1, X_2, \ldots, X^n)$ where X_i, $i = 1, \ldots, n$, stands for the ith component of X. Since each component

X_i is an interval, i.e. $X_i \in I(R)$, the interval vector $X \in I(R) \times I(R) \times \ldots \times I(R)$. We shall use the symbol $I(R^n)$ for the above Cartesian product of n times $I(R)$. Thus, every n-dimensional interval vector X belongs to $I(R^n)$, i.e $X \in I(R^n)$.

A two-dimensional interval vector $X = (X_1, X_2)$, where $X_1 = [\underline{x}_1, \overline{x}_1]$ and $X_2 = [\underline{x}_2, \overline{x}_2]$ are some intervals can be represented geometrically as a two-dimensional rectangle of points (x_1, x_2) such that $\underline{x}_1 \le x \le \overline{x}_1$ and $\underline{x}_2 \le x \le \overline{x}_2$ (see Fig. 1.2).

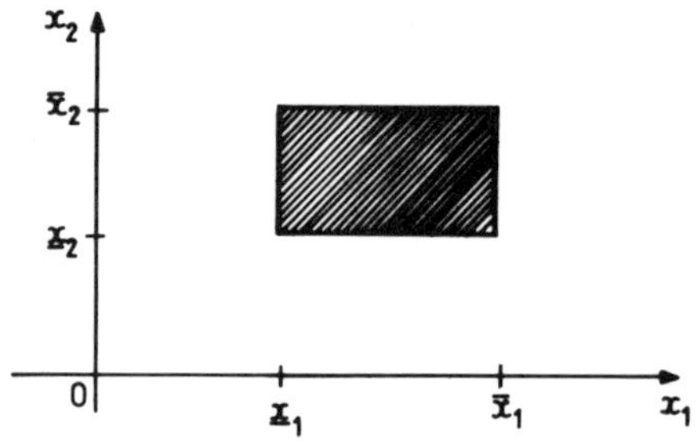

Fig. 1.2. Geometrical representation of a 2-dimensional interval vector.

An n-dimensional vector represents geometrically an n-dimensional "rectangular" region in R^n . This region will be called an n-dimensional box.

The relations equality (=), inclusion ($\subseteq$) and ordering (< or >) introduced in the previous section for interval numbers remain valid for interval vectors also if they are extended to all vector components. Thus, for any $X,Y \in I(R^n)$ the notation

$$X \subseteq Y$$

means that

$$X_i \subseteq Y_i, \quad i = \overline{1, n}$$

where X_i and Y_i are the components of X and Y, respectively, and $i = \overline{1, n}$ stands for $i = 1, 2, \ldots, n$.

Similarly, the midpoint vector (centre) $m(X)$ of an interval vector X is defined by the real vector

$$m(X) = (m(X_1), \ldots\ldots, m(X_n)) \tag{1.26}$$

The width of X is, however, given by the real number

$$w(X) = \max\{w(X_i), \ i = \overline{1, n}\} \tag{1.27}$$

Similarly to functions of one variable we write $f: D \subset R^n \to R$ to denote a real function in n real variables. The function maps each vector $x = (x_1, \ldots, x_n)$ belonging to the domain $D \subset R^n$ into a point y on the real line R.

By analogy with the definition of the real function, an interval function F: $X^0 \subset I(R^n)$ $\rightarrow I(R)$ maps an arbitrary interval vector $X \subseteq X^0$, i.e. an arbitrary box from X^0 into an interval from $I(R)$.

In interval analysis it is expedient to specify the class of "rational" functions.

By a real rational function, we mean a real function f whose values are defined by a specific finite sequence of real arithmetic operations over real numbers. Similarly, an interval function F is rational if its interval values are obtained by a specific finite sequence of interval arithmetic operations over its interval arguments. For example, consider the function F whose values are defined by

$$F(X_1, X_2) = ([1,2]X_1 + [0,1])X_2 \tag{1.28}$$

for any intervals X_1 and X_2. Here, F is a finitely represented mapping from the set of all pairs of intervals $(X_1, X_2) \in I(R^2)$ into the set of intervals.

Similarly to interval arithmetic, rational interval functions have the important property of inclusion monotonicity. Let $X,Y \in I(R^n)$ be arbitrary interval vectors and $F(X)$ be a rational interval function. The inclusion monotonicity expresses the fact that

$$X \subseteq Y \quad \text{implies} \quad F(X) \subseteq F(Y) \tag{1.29}$$

($X \subseteq Y$ if $X_i \subseteq Y_i$ for $i = \overline{1, n}$). The implication (1.29) follows from the definition of the relation $\subseteq$ between vectors and the inclusion monotonicity property (1.21), (1.22) of interval arithmetic. As an exercise the reader is advised to verify this property for the function given by (1.28).

Interval functions that are not finitely representable as interval arithmetic operations over interval numbers are not rational. (An example of such an interval function will be considered later in section 1.2.4).

If an arbitrary (not rational) function $F(X)$ satisfies (1.29) for any $X,Y \in X^0 \subset I(R^n)$ it is said to be inclusion monotonic in X^0.

It should be stressed that there is no connection between the notion of inclusion monotonicity (applicable only for interval functions) and the classical notion of monotonic (with respect to a variable or set of variables) real functions.

1.2.2. Natural interval extension

Let f be a real valued function of n real variables $x_1, \ldots, x_n$. By an interval extension of f, we mean an interval valued function $F(X_1, \ldots, X_n)$ of n interval variables $X_1, \ldots, X_n$ with the property

$$F(x_1,\ldots,x_n) = f(x_1,\ldots,x_n) \tag{1.30}$$

Thus, an interval extension of f is an interval function F: $I(R^n) \rightarrow I(R)$ which reduces to the real function f: $R^n \rightarrow R$ when all the interval arguments X_i become real (degenerate intervals). (For that reason, the corresponding function f is sometimes called the "real restriction" of F.)

Real rational functions of n variables have the so-called natural interval extensions. Given a rational expression in n real variables we can replace the real variables by corresponding interval variables and replace the real arithmetic operations by the corresponding interval arithmetic operations to obtain a rational interval function which is termed a natural extension of the rational function.

***E x a m p l e* 1.1.** Let $f(x_1, x_2) = (ax_1 + b)x_2$, where a and b are constant while x_1, $x_2 \in R$. The interval extension of this real rational function is the interval function

$$F(X_1, X_2) = (aX_1 + b)X_2$$

The value of the interval extension for some fixed constants (degenerate intervals) a and b and any given intervals $X_1, X_2 \in I(R)$ is, naturally, an interval. Thus, for $a = 2$, $b = -1$, $X_1 = [0, 1]$ and $X_2 = [-1, 2]$ we have

$$F = F([0, 1], [-1, 2]) = (2 \cdot [0, 1] - 1) \cdot [-1, 2] = [-1, 1] \cdot [-1, 2] = [-2, 2]$$

R e m a r k 1.1. For simplicity of notation we shall use one and the same symbol $F(X_1, \ldots, X_n)$ to denote both the interval extension (the corresponding interval function) and the resultant interval value of the extension after computing it for the given intervals $X_1, \ldots, X_n$.

It should be stressed that rational expressions which are identical in real arithmetic operations and, hence, represent one and the same function, may give rise to different natural interval extensions. This will be made clear by the following example.

***E x a m p l e* 1.2.** Let $f(x) = x(1-x) = x - x \cdot x$, $x \in R$. The natural extension for the first expression is

$$F_1(X) = X(1 - X);$$

for the second expression, we have

$$F_2(X) = X - X \cdot X.$$

Now, if we compute $F_1(X)$ and $F_2(X)$ for $X = [0, 1]$ we get

$$F_1[0, 1] = [0, 1] \cdot (1 - [0, 1]) = [0, 1]$$

whereas

$$F_2[0, 1] = [0, 1] - [0, 1] \cdot [0, 1] = [-1, 1].$$

Obviously, $F_1(X) \neq F_2(X)$; moreover $F_1(X) \subset F_2(X)$.

Example 1.2 shows that for polynomials the nested form (also called Horner's scheme) $A_0 + X(A_1 + X(A_2 + \dots XA_n) \dots)$ is never worse and is usually better than the sum of powers $A_0 + A_1X + A_2X{\cdot}X + \dots + A_nX{\cdot}X{\cdot}\dots{\cdot}X$ because of subdistributivity.

Henceforth, whenever we refer to the interval extension of a real function we shall assume that an expression of the function considered has already been chosen.

All natural interval extensions, being rational interval functions, have the inclusion monotonicity property (1.29).

In the general case of an arbitrary (nonrational) function of one variable $f(x)$, its interval extension is most often obtained in the following manner. First, $f(x)$ is approximated by an appropriate rational function $\tilde{f}(x)$. Then, the natural interval extension $\tilde{F}(X)$ of $\tilde{f}(x)$ is found. Finally, the interval extension $F(X)$ of the original function $f(x)$ is obtained by adding to $\tilde{F}(X)$ of an additional interval $E(X)$ which accounts for the approximation error. The extension $F(X)$ thus constructed is guaranteed to be inclusion monotonic [1].

1.2.3. Range of a function

Another important notion closely related to the interval extension of a function which will be needed in the sequel is the range of a function over a box.

Let $f: X \subset R^n \to R$ where X is a box (an interval vector). By the range $f(X)$ of f over X we mean the interval

$$f(X) = \{f(x): x \in X\} \tag{1.31}$$

Obviously, the range $f(x)$ is the union of all function values $f(x)$ for all x from X, that is, $f(X)$ is the image of the box X under f.

Finding the range of a multivariable function over a box is a fundamental problem encountered in numerous applications (all the remaining chapters of the present book will be an illustration of this assertion). The power of interval analysis approach in solving application problems derives from the following theorem due to R. Moore [1].

Theorem 1.1. If $F(X)$ is an inclusion monotonic interval extension of $f(x)$, then

$$f(X) \subseteq F(X) \tag{1.32}$$

that is, the interval extension $F(X_1, \dots, X_n)$ contains the range of $f(x_1, \dots, x_n)$ for all $x_i \in X_i$, $i = \overline{1, n}$.

Proof. By (1.30) $f(x) = F(x)$. Due to the inclusion monotonicity of F, $f(x) \in F(X)$ for each $x \in X$. Hence $f(X) = \{f(x): x \in X\} \subseteq F(X)$.

Example **1.3.** For $f(x) = x^2$ and $X = [-1, 2]$ it is easily seen that

$$f(X) = f([-1, 2]) = [0, 4]$$

On the other hand

$$F(X) = F([-1, 2]) = X \cdot X = [-1, 2] \cdot [-1, 2] = [-2, 4].$$

Hence

$$f(X) \subset F(X)$$

Example **1.4.** Let $f(x) = x(1 - x)$. The range of $f(x)$ over $X = [0, 1]$ is easily computed to be

$$f([0, 1]) = [0, 1/4].$$

From Example 1.2

$$F([0, 1]) = F_1([0, 1]) = [0, 1]$$

Once again

$$f(X) \subset F(X)$$

and the inclusion is proper, that is,

$$f(X) \neq F(X).$$

The inclusion (1.32) is one of the basic results of interval analysis. Using (1.32) we can find infallible bounds on the range of $f(x)$ over X by just computing the interval extension $F(X)$. However, the bounds thus found will, typically, be not very sharp, (as the above two examples show) especially when the box X is fairly large. Thus, one of the central problems in interval analysis is that of finding sharper bounds on $f(X)$ with a reasonable amount of computation. In section 1.2.5 we shall discuss some interval methods for computing convergent sequences of upper and lower bounds to the exact range of values.

In two special cases the range can be found in a straightforward way.

The first case refers to monotonic (in the classical sense) functions of one variable. For monotonic increasing functions $f(x)$, $x \in R$, such as $\sqrt{x}$, $\exp(x)$, $\log_e(x)$ etc. we have

$$f(X) = [f(\underline{x}), f(\overline{x})] \tag{1.33a}$$

For monotonic decreasing functions

$$f(X) = [f(\overline{x}), f(\underline{x})] \tag{1.33b}$$

If the function $f(x)$ is monotonic decreasing up to a certain point x_0 and monotonic increasing afterwards (or vice versa) and the interval X covers x_0, then the range is determined in the following way. First, $X = [\underline{x}, \overline{x}]$ is divided into two subintervals $X_1 = [\underline{x}, x_0]$ and $X_2 = [x_0, \overline{x}]$. Then formula (1.33a) is applied to X_1 and formula (1.33b) to X_2.

Finally, $f(X)$ is obtained by combining (forming the union of) the ranges $f(X_1)$ and $f(X_2)$, i.e. $f(X) = f(X_1) \cup f(X_2)$. Using this approach inclusion monotonic interval extensions have been constructed for all commonly used elementary functions [2]. We shall give here the following example. For positive integer values of k, the powers of an interval are defined by

$$X^k = \begin{cases} [\underline{x}^k, \overline{x}^k] & \text{if } X > 0 \text{ and } k \text{ is odd} \\ [\overline{x}^k, \underline{x}^k] & \text{if } X < 0 \text{ and } k \text{ is even} \\ [0, |x|^k] & \text{if } 0 \in X \text{ and } k \text{ is even} \end{cases} \tag{1.34}$$

The second special case where we can easily find the range refers to multivariable functions $f(x)$, $x \in R^n$. In this case $f(X)$ is found directly by computing the extension $F(X)$ only once.

Theorem 1.2. [2]. If $F(X)$ is any natural extension of a rational function in which each variable occurs not more than once and to the first power, then $F(X) = f(X)$ provided no division by an interval containing zero occurs.

We shall illustrated the theorem by way of the following example.

***Example* 1.5.** Consider the function

$$f(x) = \frac{x_1 x_2 x_3}{x_2 - x_3}$$

At first sight, Theorem 1.2 cannot be applied since the variables x_2 and x_3 occur twice in the function. However, we can modify the given function (by dividing the numerator and the denominator by $x_2\, x_3$) to get

$$f(x) = \frac{x_1}{\dfrac{1}{x_3} - \dfrac{1}{x_2}}$$

Now the latter expression is a candidate for applying the theorem since each variable occurs only once. If, additionally, the intervals X_2 and X_3 are such that $1/X_3 - 1/X_2$ does not contain zero then the range can be evaluated directly by the interval extension, i.e.

$$f(X) = F(X) = \frac{X_1}{\dfrac{1}{X_3} - \dfrac{1}{X_2}}$$

R e m a r k 1.2. As we shall see later (sections 1.2.5, 2.3) finding the range of a multivariable function over a box is, in the general case, a difficult problem requiring a lot of computation. Therefore, it is expedient to try to apply Theorem 1.2 whenever possible in an attempt to reduce the computational cost (further examples of such an approach will be given in section 2.1.1).

1.2.4. Excess interval

In general, the interval extension $F(X)$ is a wider interval than the range $f(X)$, as has been demonstrated in the previous examples. In order to measure the closeness of $F(X)$ to $f(X)$ we use the so-called excess interval $E(X)$ introduced as follows [2]:

$$F(X) = f(X) + E(X) \tag{1.35}$$

From (1.35)

$$w(F(X)) = w(f(X)) + w(E(X))$$

Now we are able to calculate the width of the excess interval

$$w(E(X)) = w(F(X)) - w(f(X)) \tag{1.36}$$

which is a measure of the discrepancy between $F(X)$ and $f(X)$. In the special case where $F(X) = f(X)$ it follows from (1.36) that the width of the excess and hence the excess itself is zero. (Note that the excess cannot be defined as

$$E(X) = F(X) - f(X)$$

since for $F(X) = f(X)$ the width of the excess would be

$$w(E(X)) = w[F(X) - F(X)] = w[Y - Y] \neq 0$$

which is wrong.)

The excess interval $E(X)$ always contains zero, $0 \in E(X)$. This follows from the fact that $f(X) \subseteq F(X)$.

Example **1.6.** Consider the function $f(x) = x^2$, $x \in R$, the graph of which is shown in Fig. 1.3.

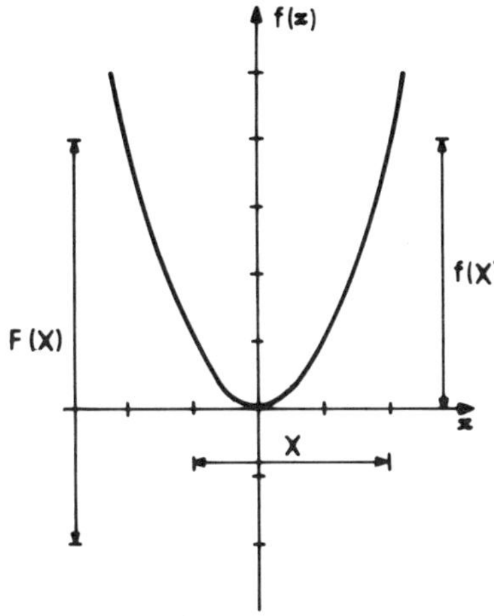

Fig. 1.3. Natural interval extension $F(X)$ and range $f(X)$ of the function $f(x) = x^2$ for $X = [-1, 2]$.

For $X = [-1, 2]$ the range $f(X)$ is seen to be $[0, 4]$. The natural interval extension $F(X) = X \cdot X$ has been calculated before as the interval $[-2, 4]$. So from (1.36)

$$w(E) = w[-2, 4] - w[0, 4] = 6 - 4 = 2$$

It is readily seen that now (1.35) is

$$[-2, 4] = [0, 4] + [-2, 0];$$

hence the excess $E(X)$ is the interval

$$E([-1, 2]) = [-2, 0]$$

which contains zero.

The following theorem [2] establishes the asymptotic behaviour of the excess for the case of rational functions.

Theorem 1.3. Let $F(X)$ be a natural interval extension of a function of n variables which is defined for $X \subseteq A$, where A is an interval vector. Then

$$w(E) = 0(w(X)) \tag{1.37}$$

(where the symbol $y = 0(x)$ means that y becomes proportional to x as x tends to zero).

Thus, the theorem essentially states that the excess width is on the same order as the width of X for narrow enough boxes.

Reduction of excess interval

Here we shall confine ourselves only to the case of functions of one variable (the multivariate case will be considered later in Chapter 2).

One way of reducing the excess interval is to partition the interval X into subintervals. Thus, if X is partitioned into two subintervals X_1 and X_2 such that

$$X = X_1 \cup X_2,$$

then clearly

$$f(X) = f(X_1) \cup f(X_2)$$

Since $f(X_1) \subseteq F(X_1)$ and $f(X_2) \subseteq F(X_2)$ we have

$$f(X) \subseteq F(X_1) \cup F(X_2)$$

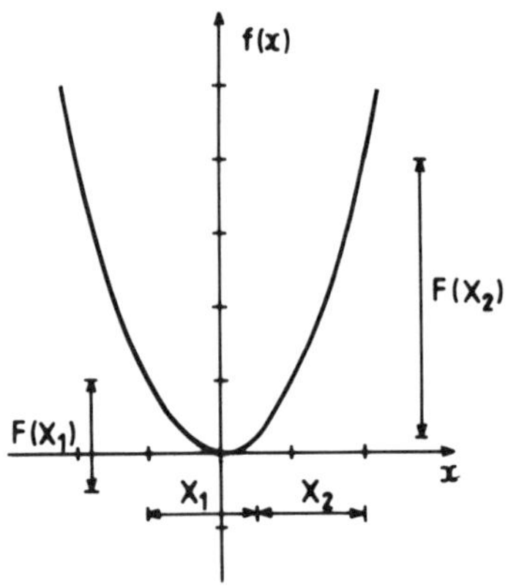

Fig. 1.4. Interval extensions $F(X_1)$ and $F(X_2)$ after halving X into X_1 and X_2.

On the other hand, by the inclusion monotonicity property $F(X_1) \subseteq F(X)$ and $F(X_2) \subseteq F(X)$ since $X_1 \subset X$ and $X_2 \subset X$.

Thus, the union

$$F(X_1) \cup F(X_2) \subseteq F(X) \tag{1.38}$$

and it is, generally, a better estimate of the range than $F(X)$.

Example **1.7.** We shall take up Example 1.6. Now divide the interval $X = [-1, 2]$ into two intervals, namely $X_1 = [-1, 0.5]$ and $X_2 = [0.5, 2]$ such that $X = X_1 \cup X_2$ (see Fig. 1.4). Interval computations lead to $F(X_1) = [-0.5, 1]$ and $F(X_2) = [0.25, 4]$. It can be easily verified that the ranges of x^2 for the smaller intervals X_1 and X_2 are, respectively,

$$f(X_1) = [0, 1]$$

and

$$f(X_2) = [0.25, 4]$$

Recall that the width of the excess E for the initial interval $X = [1, 2]$ is

$$w = w(E(X)) = w(F(X)) - w(f(X)) = 6 - 4 = 2;$$

Similarly,

$$w_1 = w(E_1(X_{1)}) = w(F(X_1)) - w(f(X_1)) = 1.5 - 1 = 0.5$$

and

$$w_2 = w(E_2(X_{2)}) = w(F(X_2)) - w(f(X_2)) = 3.75 - 3.75 = 0$$

Comparing w with w_1 and w_2 it is seen that the width of the excess interval has decreased for the narrower intervals X_1, X_2.

Moreover (as is seen from Fig. 1.4)

$$f(X) \subset F(X_1) \cup F(X_2)$$

Thus the union

$$F(X_1) \cup F(X_2) = [-0.5, 1] \cup [0.25, 4] = [-0.5, 4]$$

is seen to be a narrower interval than the interval extension

$$F(X) = [-2, 4]$$

for the interval X. Therefore, the interval $[-0.5, 4]$ can be taken as an improved estimate of the range $[0, 4]$, as compared to the interval estimate $[-2, 4]$.

This device of interval partitioning forms the basis of Moore's approach to the computation of ranges to be considered in section 1.2.5.

Monotonic functions

Another way of reducing the excess interval (theoretically to zero) which does not resort to partitioning and hence requires less computation is based on the nonstandard arithmetic operations (1.23) and (1.24). It is, however, applicable only to functions satisfying certain monotonicity conditions [6].

Let $D \in I(R)$ and let $f(x)$ be defined for $x \in D$. By $M(D)$ we will denote the set of all functions which are monotonic (in the ordinary sense) on D. If two functions $f, g \in M(D)$

are both increasing or decreasing, they will be said to satisfy monotonicity condition M_1. These functions will satisfy monotonicity condition M_2 if one of them is increasing while the other is decreasing.

Theorem 1.4. [6]. If f, g are such that $h = f + g \in M(D)$ then for all $X \in D$

$$h(X) = \begin{cases} f(X) + g(X) & \text{if } f, g \text{ satisfy } M_1 \\ f(X) \oplus g(X) & \text{if } f, g \text{ satisfy } M_2 \end{cases}$$

If f, g are such that $h = f - g \in M(D)$, then

$$h(X) = \begin{cases} f(X) \ominus g(X) & \text{if } f, g \text{ satisfy } M_1 \\ f(X) - g(X) & \text{if } f, g \text{ satisfy } M_2 \end{cases}$$

If $|f|, |g|$, $h = f \cdot g \in M(D)$, then

$$h(X) = \begin{cases} f(X) \cdot g(X) & \text{if } |f|, |g| \text{ satisfy } M_1 \\ f(X) \odot g(X) & \text{if } |f|, |g| \text{ satisfy } M_2 \end{cases}$$

If $|f|, |g| \in M(D)$, $g(x) \neq 0$ for $x \in D$ and $h = f/g \in M(D)$ then for any $X \in D$

$$h(X) = \begin{cases} f(X) / g(X) & \text{if } |f|, |g| \text{ satisfy } M_1 \\ f(X) \oslash g(X) & \text{if } |f|, |g| \text{ satisfy } M_2 \end{cases}$$

Example **1.8.** We take up the function $h(x) = x - x^2 = f(x) - g(x)$. It is seen that $h(x)$ is monotonic in $(-\infty, 0.5]$ and $[0.5, \infty)$, $g(x)$ is monotonic in $(-\infty, 0]$ and $[0, \infty)$ while $f(x)$ is monotonic in $(-\infty, \infty)$. Moreover, f and g satisfy monotonicity condition M_1 in $[0, 0.5]$ and $[0.5, \infty)$ and monotonicity condition M_2 in $(-\infty, 0]$. Hence, according to Theorem 1.4.

$$h(X) = \begin{cases} X \ominus X^2 & \text{if} \quad X \subseteq [0, 0.5] \quad \text{or} \quad X \geq 0.5 \\ X - X^2 & \text{if} \quad X \leq 0 \end{cases}$$

***Example* 1.9.** If $X \geq 0$ then $e^X = \{ e^x : x \in X \}$ is obviously given by

$$e^X = 1 + X/1! + X^2/2! + X^3/3! + \ldots$$

Let $X \leq 0$. In this case write the Taylor series for e^x in the form:

$$e^X = X + 1 + X^3/3! + X^2/2! + X^5/5! + X^4/4! + X^7/7! + \ldots$$

All partial sums of this series are monotonic increasing for $x \in (-\infty, 0]$. Thus, according to Theorem 1.4

$$e^X = X \oplus 1 + X^3/3! \oplus X^2/2! + X^5/5! \oplus X^4/4! + X^7/7! \oplus \ldots$$

for $X \leq 0$. If $0 \in X = [\underline{x}, \overline{x}]$, then $e^X = e^{[\underline{x},0]} \cup e^{[0,\overline{x}]}$.

Other examples can be found in [6].

At first glance, the range of a monotonic function could be found (without resorting to Theorem 1.4) by just evaluating the function values $f(\underline{x})$ and $f(\overline{x})$ at the endpoints $\underline{x}$ and $\overline{x}$ of the given interval X. Then, for (say) a monotonic increasing function (by formula 1.33a)

$$f(X) = [f(\underline{x}, f(\overline{x})] \tag{1.39}$$

However, the above formula ought not to be used if the computed range $\tilde{f}(X)$ must guarantee the inclusion

$$f(X) \subseteq \tilde{f}(X) \tag{1.40}$$

where $f(X)$ is the exact (ideal) range determined under the assumption that $f(\underline{x})$ and $f(\overline{x})$ are evaluated exactly. Indeed, in computing $f(\underline{x})$ and $f(\overline{x})$ on a computer we inevitably commit errors due to:

a) $\underline{x}$ and/or $\overline{x}$ may not be machine representable and their exact values have to be rounded off to the nearest machine representable numbers,

b) the real arithmetic operations involved in $f(x)$ are carried out with finite precision. Thus, if the actually computed values of $f(\underline{x})$ and $f(\overline{x})$ are $\tilde{f}(\underline{x})$ and $\tilde{f}(\overline{x})$, respectively, then by (1.39)

$$\tilde{f}(X) = [\tilde{f}(\underline{x}), \tilde{f}(\overline{x})] \tag{1.41}$$

Now, if

$$\tilde{f}(\underline{x}) > f(\underline{x}) \text{ and } \tilde{f}(\overline{x}) < f(\overline{x})$$

then, obviously, the interval $\tilde{f}(X)$ computed by (1.41) will be narrower than the exact range $f(X)$ and the requirement (1.40), important in many applications, will be violated.

One way to overcome this difficulty is to calculate $f(\underline{x})$ and $f(\overline{x})$ not as points but as intervals treating $\underline{x}$ and $\overline{x}$ as very narrow intervals (accounting for round off) and using some interval extension $F(X)$ of $f(x)$. Thus, if $\underline{x}$ is replaced by an interval X_1 and $\overline{x}$ by X_2 then we can compute

$$F(X_1) = [\underline{f}_1, \overline{f}_1]$$

and

$$F(X_2) = [\underline{f}_2, \overline{f}_2]$$

Finally, the computed range can be found as the interval

$$\tilde{f}(X) = [\underline{f}_1, \overline{f}_2] \tag{1.42}$$

Now the range defined by (1.42) is guaranteed to contain the exact range $f(X)$. However, as is not difficult to see, the inclusion is always proper, i.e.

$$f(X) \subset [\underline{f}_1, \overline{f}_2]$$

Therefore, such an approach will lead always to a nonzero excess interval.

The advantage of the approach to computing the range of monotonic functions based on Theorem 1.4 resides in the fact that it yields (whenever possible) the exact range $f(X)$. (Of course, an appropriate machine interval arithmetic must be used when implementing the nonstandard interval arithmetic on a computer).

1.2.5. Alternate forms of interval extensions

Mean – value form

The mean-value form is a particular form of interval extension which is applicable to arbitrary functions with continuous first derivatives (i.e. of class C^1).

Let $f: R^n \to R, f \in C^1$. Furthermore, $X \in I(R^n)$ (X is an interval vector) and $m = m(X)$ is its centre (midpoint vector). For any $y \in X$ the mean value theorem states that

$$f(y) = f(m) + \sum_{j=1}^{n} \frac{\partial f}{\partial x_j}(\xi)(y_j - m_j), \quad \xi \in X$$

If F_j' denotes the interval extension of $\delta f/\delta x_j = f_j'$ on X, then obviously

$$f(y) \in f(m) + \sum_{j=1}^{n} F_j'(X)(X_j - m_j) \tag{1.43}$$

for any $y \in X$. Thus, according to (1.32) the right side of (1.43) defines an interval extension of f which is called the mean value extension of f on X. It is denoted by $F_{MV}(X)$ [1], [2]:

$$F_{MV}(X) = f(m) + \sum_{j=1}^{n} F_j'(X)(X_j - m_j) \tag{1.44}$$

The mean value form is inclusion monotonic. More precisely we have the following result.

Theorem 1.5. [2]. If the functions $F_j'(X)$, $j = \overline{1, n}$, are inclusion monotonic, then $F_{MV}(X)$ is also inclusion monotonic.

Theorem 1.6. [2]. If the partial derivatives f_j' satisfy on X the Lipschits condition $|f'_j(x) - f'_j(y)| \leq L|x - y|$ where L is a constant, then $F_{MV}(X)$ approximates the range $f(X)$ apart from an excess interval E, i.e.

$$F_{MV}(X) = f(X) + E(X)$$

where $0 \in E$ and $w(E) = 0(w(X)^2)$ (the symbol $y = 0(x^2)$ means that y is proportional to x^2 as x tends to zero).

Thus, the mean–value form assures better interval extension as compared to the natural extensions (Theorem 1.3) for narrow enough intervals X. However, it should be borne in mind that the bound on E given in Theorem 1.6 is asymptotic and the natural interval extension $F(X)$ may provide sharper bounds than $F_{MV}(X)$ for larger boxes. For that reason, the intersection $F(X) \cap F_{MV}(X)$ is sometimes used in practical computations.

(Other mean-value forms will be considered in section 2.2.2).

Centred form

In [1] the so-called centred form $F_c(X)$ is introduced as a particular form of interval extension of a rational function $f(x_1, \ldots, x_n)$. To derive F_c for a particular f, we first rewrite $f(x_1, \ldots, x_n)$ as

$$f(x_1,\dots,x_n) = f(m_1,\dots,m_n) + g(y_1,\dots,y_n)$$

where m_i are the components of the centre $m = m(X)$ of the interval vector X while $y_i = x_i - m_i$. Thus, g is defined by

$$g(y_1,\dots,y_n) = f(y_1 + m_1,\dots,y_n + m_n) - f(m_1,\dots,m_n)$$

and is, therefore, dependent on m_i. We define $F_c(X)$ by

$$F_c(X) = f(m) + G(m, X-m)$$

It has been proved by E. Hansen [7] that similarly to the mean value extension the excess width of the centred form extension is on the order of $w(X)^2$. However, the bounds on $f(X)$ obtained from the centred form tend to be slightly sharper in practical computations.

A serious drawback of the centred form is the amount of work involved in deriving the function G in explicit form. For this reason, it is considerably more complicated to obtain the centred form than the mean value form. Another shortcoming of $F_c(X)$ is the fact that this form is not always inclusion monotonic. For example, let $f(x) = x(1 - x)$. It can be shown that

$$F_c(X) = f(m) + (1 - 2m)(X - m) - (X - m)^2$$

Let $X_1 = [0, 1]$ and $X_2 = [0, 0.9]$ so that $X_2 \subset X_1$. However, $F_c(X_2) = [0, 0.02925]$ whereas $F_c(X_1) = [0, 0.25]$, i.e.

$$F_c(X_2) \not\subset F_c(X_1)$$

1.2.6. Computation of the range

In the special cases where the real function considered satisfies the conditions of Theorem 1.2 or Theorem 1.4, the range $f(X)$ of $f(x)$ in X can be found exactly by computing only once the corresponding interval extension $F(X)$, using the ordinary or the nonstandard interval arithmetic, respectively.

In this section it will be shown that arbitrarily sharp upper and lower bounds on the range $f(X)$ of any Lipschitz (with bounded slopes) real function can be computed. However, as will be seen, this involves numerous evaluations of $F(X^v)$ for different subregions X^v of X and sometimes may be prohibitively expensive.

Moore's approach

To make the basic idea behind Moore's approach easier to understand first we shall consider the case where f: $X \to R$ with $X \in I(R)$.

We partition the interval X into p subinterval of equal width, i.e.

$$X = [x_0, x_1] \cup [x_1, x_2] \cup \ldots \cup [x_{p-1}, x_p],$$

$$x_0 = \underline{x}, \quad x_p = \overline{x}, \quad w(x_j, x_{j+1}) = w(X)/p$$

Let the subinterval $[x_j, x_{j+1}]$ be denoted by $X^{(j)}$, i.e. $X^{(j)} = [x_j, x_{j+1}]$. Then, obviously,

$$f(X) = \bigcup_{j=0}^{p-1} f(X^{(j)}) \subseteq \bigcup_{j=0}^{p-1} F(X^{(j)})$$

(recall Example 1.7 and Fig. 1.4 for geometrical illustration in the case where $p = 2$). Let

$$F^{(p)}(X) = \bigcup_{j=0}^{p-1} F(X^{(j)}) \tag{1.45}$$

and

$$F^{(p)}(X) = f(X) + E_p \tag{1.46}$$

Thus, we can calculate the range $f(X)$ of f in X within a preset accuracy provided we can afford to evaluate $F(X^{(j)})$ as many times as needed. What is more, the approximation $F^{(p)}(X)$ is guaranteed to contain the exact range $f(X)$ for each p.

Now we shall consider the case where X is an interval vector (box), i.e. $X \in I(R^n)$. Similarly to the previous case, we first partition the box X into subboxes $X^{(v)}$ in the following way. Each component X_i, $i = \overline{1, n}$, of X is divided into p subintervals $X_i^{(j)}$ of equal width, i.e. $w(X_i^{(j)}) = w(X_i)/p$. Then, let $X^{(v)}$ be a subbox formed by a particular combination of n subintervals $X_i^{(j)}$. Obviously, the total number of subboxes is $N = p^n$. The resulting partitioning of X

$$X = \bigcup_{v=1}^{N} X^{(v)} \tag{1.47}$$

is called uniform partitioning.

The union

$$F^{(p)}(X) = \bigcup_{v=1}^{N} F(X^{(v)}) \tag{1.48}$$

is called a refinement of $F(X)$ since $F^{(p)}(X) \subseteq F(X)$ and most often $w(F^{(p)}(X)) < w(F(X))$. This follows directly from the inclusion monotonicity property of $F(X)$. We can use any of the interval extensions introduced so far (or any others) in computing the right side of (1.45). It is, however, preferable to make use of an extension which provides narrower intervals for $F(X^{(v)})$.

An interval extension $F(X)$ is Lipschitz in X^0 if there is a constant L such that $w(F(X)) \le Lw(X)$ for every $X \subseteq X^0$. We have the following important result [2].

Theorem 1.7. Let $F(X)$ be an inclusion monotonic, Lipschitz interval extension for $X \subseteq X^0$. Then

$$F^{(p)}(X) = f(X) + E_p, \quad 0 \subseteq E_p$$

and there are constants K and K' such that

$$w(E_p) = K\, w(X)/p \tag{1.49a}$$

if the natural interval extension is used in evaluating $F^{(p)}(X)$ or

$$w(E_p) = K'\, w(X)/p^2 \tag{1.49b}$$

if $F^{(p)}(X)$ is evaluated by the mean value form (centred form).

From (1.49)

$$w(E_p) \to 0 \quad \text{as} \quad p \to \infty \tag{1.50}$$

Again, we can, theoretically, compute the range $f(X)$ within arbitrarily sharp bounds if we take p large enough. However, an evaluation of $F^{(p)}(X)$ given by (1.48) requires $N = p^n$ evaluations of $F(X^{(\nu)})$. If n is large, this would involve a prohibitive amount of computation to achieve the result (1.49) for large p. Even for $n = 2$ we have, for $p = 1000$, 10^6 evaluations to carry out.

Skelboe's approach

Skelboe [5] has introduced an algorithm which can vastly reduce the number of evaluations required to bound $f(X)$ within an accuracy as compared to Moore's algorithm. Unlike Moore's approach, where $F(X^{(\nu)})$ is evaluated for each subbox $X^{(\nu)}$ of the uniform partitioning of X, Skelboe's algorithm computes the extensions $F(X^{(\mu)})$ for a relatively small number of subboxes $X^{(\mu)}$ of varying size. These subboxes are generated dynamically in the process of sharpening the bounds on the range of the function considered. To elucidate the mechanism of Skelboe's algorithm, it will be sketched for the scalar case where $f\colon X^0 \to R$ with $X^0 \in I(R)$.

Given an interval X^0 and an inclusion monotonic interval extension $F(X)$, $X \subseteq X^0$, one seeks first a lower bound on the minimum value of f in X^0 by a procedure to be described below. The same procedure is then applied to $(-f)$ to obtain the upper bound on the maximum value of f in X^0.

To find a lower bound on $f(X^0)$ we create an ordered list of subintervals X in the following way: a subinterval X comes before another subinterval Y in the list only if $\underline{F(X)} \le \underline{F(Y)}$ where $\underline{F(X)}$ stands for the left endpoint of the interval $F(X) = [\underline{F(X)}, \overline{F(X)}]$.

Procedure 1.1. (Procedure for bounding $\underline{f(X^0)}$)

(1) Set $X = X^0$.

(2) To begin with, the list is empty

(3) Bisect X into two subintervals X' and X'' of equal width: $X = X' \cup X''$

(4) Evaluate $\underline{F(X')}$ and $\underline{F(X'')}$.

(5) Set $b = \min\{\underline{F(X')}, \underline{F(X'')}\}$
(6) Enter the subintervals X' and X'' in proper order in the list (that is, if $\underline{F(X')} \leq \underline{F(X'')}$ enter X' first and then X''; otherwise enter X'' first).
(7) Retrieve the top subinterval $X^{(p)}$ (with the lowest $\underline{F(X)}$) from the list. Set $X = X^{(p)}$ and remove $X^{(p)}$ from the list.
(8) If $w(X) > \varepsilon$ where ε is a prescribed accuracy, return to step (3). Otherwise proceed to the next step.
(9) Put $b = \underline{F(X)}$. Terminate.

Clearly on exit from Procedure 1.1 the real number b obtained is a lower bound on $\underline{f(X^0)}$. Indeed, each successive bisection tends to generate intervals of smaller width. Therefore, as the number of intervals increases b grows monotonically, thus converging to $\underline{F(X^0)}$ from below.

If Procedure 1.1 is applied to $(-f)$ then, upon termination, $-b$ is an upper bound on $\overline{f}(X^0)$. To show this, it is necessary to first consider the range of $-f(x)$ in X^0, that is, the interval

$$-f(X^0) = [-\overline{f(X^0)}, -\underline{f(X^0)}]$$

Next, form the interval extension of $-f(x)$ in each current box X:

$$-F(X) = [-\overline{F(X)}, -\underline{F(X)}]$$

Now it is seen that the lower endpoint $-\overline{F}(X)$ of $-F(X)$ tends monotonically to the lower endpoint $-\overline{f}(X^0)$ of $-f(X^0)$ from below as the bisection process proceeds. Hence $(-b)$ converges monotonically to $\overline{f}(X^0)$ from above.

Procedure 1.1 can be easily generalized to encompass the case when X^0 is an interval vector. The only difference lies in the way the current box X is partitioned into subboxes. Similarly to the scalar case, in [2] X is split only into two subboxes X' and X'' by bisecting the largest side (component) of X. In [5] the current box is divided at once into 2^n subboxes.

Skelboe's algorithm can be improved in various ways by the introduction of criteria testing certain properties of the function considered such as monotonicity (in the ordinary sense), convexity and others in X^0 [8]. These refinements will be postponed for section 2.3 of Chapter 2.

1.3. INTERVAL METHODS IN LINEAR ALGEBRA

1.3.1. Linear equations with interval coefficients

An interval matrix is a matrix whose elements are intervals. Let $R^{n\times n}$ denote as usual the set of all real (noninterval) n x n matrices. By analogy, the set of all n x n interval matrices will be denoted by $I(R^{n\times n})$. The elements of $R^{n\times n}$ will be denoted by boldface capital letters, those of $I(R^{n\times n})$ by the same symbol with a superscript I.

A real matrix A with elements a_{ij} is contained in an interval matrix A^I with elements A_{ij} if $a_{ij} \in A_{ij}$ for all $i,j = \overline{1, n}$. We write $A \in A_I$. Similarly to the interval vectors the relations =, $\subseteq$ and < (>) are valid in the case of interval matrices also if they hold for all matrix components. Thus, if A^I and B^I are interval matrices with elements A and B, respectively, then $A^I < B^I$ if $A_{ij} < B_{ij}$ for all $i,j = \overline{1, n}$.

By $m(A^I)$, we denote a real matrix with elements $m(A_{ij})$. We call $m(A^I)$ the centre of A^I. The norm $\|A^I\|$ of A^I may be defined in various ways. In this chapter, we shall use a norm introduced as follows

$$\|A^I\| = \max_i \sum_{j=1}^{n} |A_{i,j}|, \quad i = \overline{1,n} \tag{1.51}$$

where $|A_{ij}|$ is given by (1.4). We shall also need the notation $|A|$ (absolute value of a real matrix). It denotes a real matrix whose elements are defined by $|A_{ij}|$.

Consider the following system of linear algebraic equations

$$Ax = b, \quad A \in R^{n \times n}, \quad b \in R^n \tag{1.52}$$

In fact, we will be interested in the solutions of (1.52) when the elements of A and b are not known exactly, i.e., when they are intervals. Thus, we will consider the family of linear systems (1.52):

$$Ax = b, \quad A \in A^I \in I(R^{n \times n}), \quad b \in B \in I(R^n) \tag{1.53}$$

For brevity, we will write (1.53) in the form

$$A^I x = B \tag{1.54}$$

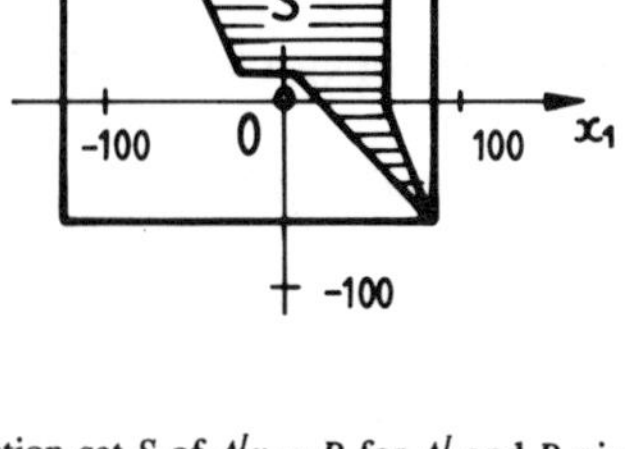

Fig. 1.5. The solution set S of $A^I x = B$ for A^I and B given by (1.56).

The solution set of (1.53) is the set

$$S = \{x: x = A^{-1}b\ ,\ \ A \in A^I,\ \ b \in B\} \tag{1.55}$$

The set S may have a very complicated structure. For example, it has been shown in [9] that if

$$A^I = \begin{bmatrix} [2,\ 3] & [0,\ 1] \\ [1,\ 2] & [2,\ 3] \end{bmatrix}\ , \quad B = \begin{bmatrix} [0,\ 120] \\ [60,\ 140] \end{bmatrix} \tag{1.56}$$

S is the nonconvex region given in Fig. 1.5.

In what follows we assume that the interval matrix A^I is regular. We call A^I regular, if each $A \in A^I$ is nonsingular.

A number of properties of S are given in [13]. We only note here that:

i) S is in general a nonconvex bounded set

ii) the intersection of S with each orthant of R^n is a convex polyhedron (see Fig. 1.5 for a geometric illustration in the case of $n = 2$).

As S is extremely difficult to find, in practice we settle for an interval vector X which contains S. However in some cases we would like to compute the smallest interval $\tilde{X}$ still containing S. The vector X is called interval solution of (1.54) while $\tilde{X}$ is referred to as the optimal interval solution of (1.54).

The methods for finding an interval solution to (1.54) are of two types:

i) iterative methods

ii) direct methods.

1.3.2. Iterative methods

There exists a variety of iterative methods for solving (1.54) (see e.g. [10]). In this section we shall only consider three such methods.

Simple iteration

Let the system (1.52) be put in the equivalent form:

$$x = Tx + b \tag{1.57}$$

where $T = E - A$, E being the identity matrix. Then, (1.54) can be written as

$$x = T^I x + B$$

Consider the interval equation

$$Z = T^I Z + B \tag{1.58}$$

A vector X^* which satisfies (1.58) is called the fixed interval vector of (1.58).

Let $\rho(|T|)$ denote the spectral radius of $|T|$. (Recall that the spectral radius of a matrix is defined by the largest modulus of its eigenvalues). The inequality $\rho(|T^I|) < 1$ means that $\rho(|T|) < 1$ holds for every $T \in T^I$.

T h e o r e m 1.8. [10]. The iteration

$$X^{k+1} = T^I X^{(k)} + B\ , \quad k \geq 0 \tag{1.59}$$

converges to the unique fixed interval X^* of (1.59) for any initial vector $X^{(0)}$ if $\rho(|T^I|) < 1$.

Let X be an interval solution to (1.57) or, equivalently, to (1.54). If we start iteration (1.59) with an arbitrary $X^{(0)}$, there is no guarantee that $X \subseteq X^*$. On the other hand, $X \subseteq X^*$ if $X \subseteq X^{(0)}$. Indeed, the inclusion monotonicity implies that

$$X^* = T^I X^* + B \subseteq T^I X^{(0)} + B = X^{(1)}$$

By induction

$$X^* \subset X^{(k)} \text{ and } X^* \subset T^I X^{(k)} + B$$

hence

$$X^* \subset (T^I X^{(k)} + B) \cap X^{(k)}, \quad k \geq 0$$

Thus, the modified iteration

$$X^{(k+1)} = (T^I X^{(k)} + B) \cap X^{(k)}, \quad k \geq 0 \tag{1.60}$$

converges to $X^* \supseteq X$ if $X^* \supseteq X^{(0)}$ for $\rho(|T^I|) < 1$. If the computations are carried out in machine interval arithmetic, then the iteration (1.60) converges in a finite number of steps since, due to outwards rounding, we will have, sooner or later, $X^{(k+1)} = X^{(k)}$ for some large enough k.

Gauss–Seidel iteration

If an initial enclosure $X^{(0)}$ for S is known a nested sequence of enclosures can be defined by the so-called Gauss–Seidel iteration with componentwise intersection [10]:

$$X^{(k+1)} = G(A^I,\ B,\ X^{(k)}), \quad k \geq 0 \tag{1.61}$$

where the vector $G^{(k)} = G(A^I, B, X^{(k)})$ has components $G_i^{(k)}$ defined by

$$Y_i = (B_i - \sum_{j=1}^{i-1} A_{ij} G^{(k)}{}_j - \sum_{j=i+1}^{n} A_{ij} X^{(k)}{}_j) / A_{ii}$$

$$G_i^{(k)} = X_i^{(k)} \cap Y_i, \quad i = \overline{1,n} \tag{1.62}$$

Clearly, the method applies only when $0 \notin A_{ii}$, $i = \overline{1, n}$, if we use ordinary interval arithmetic. It can, however, be modified for the general case on the basis of the extended interval arithmetic (see section 1.1.4). Indeed, although Y_i can be an unbounded set (or even consist of two unbounded sets), the resulting component $G_i^{(k)}$, being the intersection of $X^{(k)}$ and Y_i, is always a bounded interval (two intervals). Such a generalisation is suggested in [4].

Preconditioning

To improve the performance of Gauss–Seidel iteration the following approach called preconditioning can be used [8]. We first multiply both sides of (1.54) by a matrix Y (for instance an approximate inverse of $m(A^I)$). Let $C^I = E - YA^I$. If using (1.51) $\|C^I\| < 1$, define the sequence:

$$X^{(k+1)} = (YB + C^I X^{(k)}) \cap X^{(k)}, \quad k \geq 0 \tag{1.63a}$$

with

$$X_i^{(0)} = [-1,1] \, \|YB\| / (1 - \|C^I\|), \quad i = \overline{1,n} \tag{1.63b}$$

Then the following theorem is valid [2].

Theorem 1.9. If $\|E - YA^I\| < 1$ for some matrix Y the solution vector X of the interval system (1.54) is contained in the interval vector $X^{(k)}$ defined by (1.63) for every $k \geq 0$. Using machine interval arithmetic, the sequence $\{X^{(k)}\}$ converges to X in a finite number of steps.

1.3.3. Gauss elimination

Given the real (noninterval) system (1.52) the Gauss elimination method can be described as a transformation of the original matrix A to an upper triangular matrix $A^{(n)}$: $A \to A^{(1)} \to A^{(2)} \to \ldots \to A^{(n)}$ and a corresponding transformation of the right side: $b \to b^{(1)} \to b^{(2)} \to \ldots \to b^{(n)}$. At the first step:

$$A^{(1)} = \begin{bmatrix} 1 & a_{12}^{(1)} & . & . & . & a_{1n}^{(1)} \\ 0 & a_{22}^{(1)} & . & . & . & a_{2n}^{(1)} \\ . & . & . & . & . & . \\ 0 & a_{n2}^{(1)} & . & . & . & a_{nn}^{(1)} \end{bmatrix}, \qquad b^{(1)} = \begin{bmatrix} b_1^{(1)} \\ b_2^{(1)} \\ . \\ b_n^{(1)} \end{bmatrix}$$

where

$$a_{1j}^{(1)} = a_{1j}/a_{11}, \quad j = \overline{2,n}$$

$$a_{kj}^{(1)} = a_{kj} - a_{k1}a_{1j}^{(1)}, \quad k = \overline{2,n}, \quad j = \overline{2,n}$$

$$b_1^{(1)} = b_1/a_{11}$$

$$b_k^{(1)} = b_k - a_{k1}b_1^{(1)}, \quad k = \overline{2,n}$$

For $1 \le i \le n$, $A^{(i)}$ and $b^{(i)}$ have the form:

$$A^{(i)} = \begin{bmatrix} 1 & a_{12}^{(1)} & a_{13}^{(1)} & . & . & . & . & . & a_{1n}^{(1)} \\ 0 & 1 & a_{23}^{(2)} & . & . & . & . & . & . \\ 0 & 0 & . & . & 1 & a_{i,i+1}^{i} & . & . & a_{i,n}^{(i)} \\ 0 & 0 & . & . & 0 & a_{i+1,i+1}^{i} & . & . & a_{i+1,n}^{(i)} \\ . & . & . & . & . & . & . & . & . \\ 0 & 0 & . & . & 0 & a_{n,i+1}^{(i)} & . & . & a_{nn}^{(i)} \end{bmatrix}, \quad b^{i} = \begin{bmatrix} b_1^{(i)} \\ . \\ b_i^{(i)} \\ b_{i+1}^{(i)} \\ . \\ b_n^{(i)} \end{bmatrix}$$

The transition from the ith step to the $(i+1)$th step, $i \le n-1$, is carried out by the formulae:

$$a_{i+1,i+1}^{(i+1)} = 1$$

$$a_{k,i+1}^{(i+1)} = 0\ , \quad k = \overline{i+2,n}$$

$$a_{i+1,j}^{(i+1)} = a_{i+1,j}^{(i)} / a_{i+1,i+1}^{(i)}\ , \quad i = \overline{i+2,n}$$

$$a_{k,j}^{(i+1)} = a_{k,j}^{(i)} - a_{k,i+1}^{(i)}\, a_{i+1,j}^{(i+1)}\ , \quad k,j = \overline{i+2,n}$$

$$b_{i+1}^{(i+1)} = b_{i+1}^{(i)} / a_{i+1,i+1}^{(i)}$$

$$b_{k}^{(i+1)} = b_{k}^{(i)} - a_{k,i+1}^{(i)}\, b_{i+1}^{(i+1)}\ , \quad k = \overline{i+2,n} \tag{1.65}$$

(The above formulae involve only those elements of $A^{(i+1)}$ and $b^{(i+1)}$ which change when passing from one step to another). At the last step $A^{(n)}$ is an upper triangle matrix and the solution x is obtained by the formulae:

$$x_n = b_n^{(n)}$$

$$x_i = b_i^{(i)} - \sum_{j=i+1}^{n} a_{ij}^{(i)} x_j\ , \quad i = n-1, n-2, \ldots, 2, 1 \tag{1.66}$$

Clearly, the formulae (1.64) to (1.66) are only valid if $a_{11} \neq 0$, $a_{(i+1)}^{(i)} \neq 0$. This is always possible by using some pivoting scheme.

Now consider the interval system (1.54). By analogy with the real case, we try to transform A^I into an interval triangular matrix and B into a corresponding interval vector. We can do this by replacing in (1.64) to (1.66) the real coefficients and the real arithmetic operations by corresponding intervals and intervals arithmetic operations. If this transformation is possible, i.e. if no division by an interval containing zero occurs, then the interval solution X to (1.54) is obtained by extending (1.64) to (1.66) to interval form. Thus, we have

$$A_{1j}^{(1)} = A_{1j} / A_{11}\ , \quad j = \overline{2,n}$$

$$A_{kj}^{(1)} = A_{kj} - A_{k1} A_{1j}\ , \quad k,j = \overline{2,n}$$

$$B_1^{(1)} = B_1 / A_{11}\ , \quad B_k^{(1)} = B_k - A_{kl} B_1^{(1)}, \quad k = \overline{2,n}$$

$$A_{i+1,i+1}^{(i+1)} = 1\ , \quad A_{k,i+1}^{(i+1)} = 0, \quad k = \overline{i+2,n} \tag{1.67}$$

$$A_{i+1,j}^{(i+1)} = A_{i+1,j}^{(i)} / A_{i+1,i+1}^{(i)}, \quad j = \overline{i+2,n}$$

$$A_{kj}^{(i+1)} = A_{kj}^{(i)} - A_{k,i+1}^{(i)} A_{i+1,j}^{(i+1)}\ , \quad k,j = \overline{i+2,n}$$

$$B_{i+1}^{(i+1)} = B_{i+1}^{(i)} / A_{i+1,i+1}^{(i)}$$

$$B_k^{(i+1)} = B_k^{(i)} - A_{k,i+1}^{(i)} B_{i+1}^{(i+1)}, \quad k = \overline{i+2,n} \tag{1.68}$$

$$X_n = B_n^{(n)}$$

$$X_i = B_i^{(i)} - \sum_{j=i+1}^{n} A_{ij}^{(i)} X_j, \quad i = n-1, n-2, ..., 2, 1 \tag{1.69}$$

for the interval version of the Gauss elimination method. Obviously, it is a direct interval method for solving (1.54).

Based on the inclusion monotonicity of interval arithmetic the following theorem is readily proven.

Theorem 1.10. Let the intervals X_i be defined by formulae (1.67) to (1.69) (i.e. no division by zero occurs). Then, for any $A \in A^I$, $b \in B$ the solution $x = (x_1, \ldots, x_n)^T$ to system (1.52) does exist, it is unique and is contained in $X = (X_1, \ldots, X_n)^T$, that is,

$$x \in S \subseteq X = (X_1, \ldots, X_n)$$

In other words, det $A \neq 0$, $\forall A \in A^I$.

The inverse assertion is not always valid: the condition det $A \neq 0$, for $\forall A \in A^I$ does not imply that the interval Gauss elimination can be carried out. A convincing example is constructed in [11].

1.3.4. Computing the optimal interval solution

In many applications (see section 1.4) it is not very important to find the optimal interval solution $\tilde{X}$ to (1.54); it is sufficient to compute an interval solution X

$$X = \tilde{X} + \tilde{E}$$

which approximate $\tilde{X}$ by a relatively small excess $\tilde{E}$. In other cases (see section 3.1 from Chapter 3) what we seek is just the optimal solution $\tilde{X}$.

In this subsection we will survey some of the methods which guarantee the computation of $\tilde{X}$ (such methods will be called exact). (Other exact methods for solving (1.54) will be discussed later in section 3.3).

It is important to point out that the methods considered so far can provide very good or very bad results (with small or large excess width) depending on the particular properties of A^I and B (even breakdown of the computation process has been observed in some instances). However, if A^I and B possess some "nice" characteristics these methods can be exact [12], giving the optimal solution $\tilde{X}$. Several such cases are given below.

If A^I in (1.54) is degenerate, i.e. all $A_{ij} = a_{ij}$ are real numbers and $A^I = A$ then

$$\tilde{X} = A^{-1}B$$

A real matrix $A = (a_{ij})$ is called an M-matrix, if $a_{ij} \leq 0$ for all $i \neq j$ and one of the following equivalent conditions holds:

1) $\det A \neq 0$ and $A^{-1} \geq 0$ ($A^{-1} \geq 0$ denoting $(A^{-1})_{ij} \geq 0$)
2) $a_{ii} > 0$, $i = \overline{1, n}$, and the spectral radius of the matrix $E - D^{-1}A$ is less than unity (here D is a diagonal matrix with elements $d_{ii} = a_{ii}$)
3) all eigenvalues of A have positive real parts
4) for any vector x, $Ax \geq 0$ implies $x \geq 0$.

C o r o l l a r y. If C is an M-matrix and $C \leq A$, that is, $c_{ij} \leq a_{ij}$, then A is also a M-matrix.

An interval matrix A^I is called an M-matrix if each real matrix $A \in A^I$ is an M-matrix.

If A^I in (1.54) is an M-matrix then Gauss–Seidel iteration (1.62) yields the optimal solution $\tilde{X}$.

An interval matrix A^I is called inverse positive, if $A^{-1} \geq 0$, $\forall A \in A^I$.

In cases where the righthand side B of (1.54) satisfy one of the conditions:

$$B \geq 0\ , \quad B \leq 0 \quad \text{or} \quad 0 \in B \tag{1.70}$$

the optimal solution $\tilde{X}$ can be computed for an inverse positive matrix $A^I = [\underline{A}, \overline{A}]$ as

$$\tilde{X} = [\overline{A}^{-1}, \underline{A}^{-1}]B$$

For the more special case of M-matrices, Gauss elimination yields $\tilde{X}$ if the right-hand side B satisfies (1.70).

Several direct methods for computing the optimal solution of (1.54) in the general case where the only assumption is the regularity of A^I have been suggested by J. Rohn in a series of papers [13]–[15]. These methods are, unfortunately, very time-consuming for large n since they have a worst case complexity exponential in n. However, there are a few special cases where one of these methods is highly effective. We shall postpone the discussion of this issue for section 3.3.

1.4. SOLVING NONLINEAR EQUATIONS

1.4.1. Nonlinear equations in one variable

In this section we will consider a nonlinear function $f(x)$ with f: $A \to R$ where $A \in I(R)$. Assume that f has a continuous derivative $f'(x)$ in A, i.e. $f \in C^1$. We will be interested in solving the nonlinear equation

$$f(x) = 0\,, \quad X \in A \tag{1.71}$$

From the mean value theorem:

$$f(y) = f(x) + f'(\xi)(y - x) \tag{1.72}$$

where ξ lies between x and y. Assume y is a zero of f (i.e. y is a solution to (1.71)). Then $f(y) = 0$ and from (1.72):

$$y = x - f(x)/f'(\xi)$$

Let X be an interval containing x and y. Then $\xi \in X$ and hence $f'(\xi) \in F'(X)$. Denote

$$N(x,X) = x - f(x)/F'(X) \tag{1.73}$$

Then the following theorem due to R. Moore [1] holds.

Theorem 1.11. If a zero y of f exists in X, then, for any $x \in$ X, we have $y \in N(x, X)$.

A popular method for solving (1.71) is the Newton method. Now we shall introduce an interval generalization of this method and we shall briefly enumerate its important properties.

The interval Newton algorithm is defined as follows. Given an interval X_0 we have

$$N(x_k,X_k) = x_k - f(x_k)/F'(X_k) \tag{1.74}$$

$$X_{k+1} = X_k \cap N(x_k,X_k) \tag{1.75}$$

with $x_k \in X_k$ $(k \geq 0)$. Usually, x_k is taken to be the midpoint of X_k. The purpose of (1.75) is to discard points which are in $N(x_k, X_k)$ but not in X_k and thus produce convergence.

It has been proven that this algorithm (first proposed by Moore) is globally convergent if $0 \notin F'(X_0)$ and that its asymptotic rate of convergence is quadratic (in the sense that $w(X_{k+1}) \leq cw(X_k)^2$).

The interval Newton method has been extended by E. Hansen to allow $0 \in F'(X)$ and global convergence for this general case was proved also.

If $0 \in F'(X)$ the quotient $f(x_k)/F'(X_k)$ in (1.74) is computed using (1.25). Then X_{k+1}, as computed from (1.75), may consist of two intervals. If this is the case, one of them is stored in a list and processed later. This algorithm is called extended interval Newton algorithm. Thus, division by an interval containing zero leads to isolating all the zeros of a function from one another.

Theorem 1.12. If a zero y of f exists in X_k, then $y \in X_{k+1}$ for all $k \geq 0$.

Thus, no zero of f in X_0 is ever lost.

We assume that both f and f' have a finite number of zeros in X_0. Then the extended interval Newton algorithm will find every zero of f in X_k. Inversely, we have, as proved by Moore the following important result.

Theorem 1.13. If X_{k+1} is empty, there is no zero of f in X_k.

As a consequence, note that if the algorithm deletes all of X_0, it thus proves that there is no zero of f in X_0.

A useful property of the interval Newton algorithm is that it can prove the existence of a solution [16].

Theorem 1.14. If $N(x_k, X_k) \subset X_k$, then there exists a zero of f in X.

The condition $N(x_k, X_k) \subset X_k$ can only occur if $0 \notin F'(X)$; otherwise $N(x_k, X_k)$ is not finite. Thus, the algorithm can only prove the existence of simple zeros.

Uniqueness can also be proved.

Theorem 1.15. If $0 \notin F'(X)$, then any zero of f in X is unique.

This follows because $0 \notin F'(X)$ implies that $f(x)$ is strictly monotonic for $x \in X$.

As mentioned above, the rate of convergence to a simple zero is asymptotically quadratic. The following theorem shows the convergence of the interval algorithm is reasonably fast even for large intervals.

Theorem 1.16. [1]. Assume $0 \notin F'(X)$ and that x_k is the midpoint of X_k. Then $w(X_{k+1}) \leq w(X_k)/2$.

This result follows because either $N(x_k, X_k) \geq x_k$ or else $N(x_k, X_k) \leq x_k$.

Stopping criteria are much simpler for the interval Newton method than for the noninterval case. The following criterion is usually used. Let the user prescribe an error tolerance ε. Stop processing X_k, $(k \geq 1)$ if either $w(X_k) < \varepsilon$ or $X_k = X_{k+1}$. In the former case the zero is bounded as sharply as desired. The latter case occurs when rounding errors are such that no further refinement of X_k is possible without using higher precision arithmetic.

The extended interval Newton algorithm is a powerful tool for solving (1.71). It will find all the zeros of f in a given interval X_0 even in the presence of multiple zeros in a finite number of steps.

1.4.2. Systems of nonlinear equations

We now change to vector notation. Let $x = (x_1, \ldots, x_n)^T$ and $f(x) = (f_1(x), \ldots, f_n(x))$. We wish to solve the system of equations

$$f(x) = 0 \tag{1.76}$$

There exist various interval Newton methods for tackling this problem. They all solve iteratively a linearized version of (1.76). They differ in the choice of the linearization, and how the linearized equations are solved.

We assume $f_i(x)$, $i = \overline{1, n}$ has a continuous derivative with respect to each variable x_j, $j = \overline{1, n}$ i.e. $f \in C^1$. Denote

$$J_{ij}(x) = \frac{\partial f_i}{\partial x}(x), \quad i,j = \overline{1, n}$$

From the mean value theorem

$$f_i(y) = f_i(x) + J_{i1}(x + t(y-x))(y_1 - x_1) + \cdots + J_{in}(x + t(y-x))(y_n - x_n) \tag{1.77}$$

for some $t \in [0, 1]$. Let $X = (X_1, \ldots, X_n)$ be the interval vector

$$X = x + [0, 1](y - x)$$

and let $J(X)$ denote the interval matrix with components $J_{ij}(X)$. Then it follows from (1.77) that

$$f(y) \in f(x) + J(X)(y - x) \tag{1.78}$$

From (1.78), we see that if $x \in X$ and there is a zero y of f in X, then y is in the solution set of

$$f(x) + J(X)(y - x) = 0 \tag{1.79}$$

In fact, (1.79) is a system of linear interval equations with respect to y.

The existing interval methods for solving (1.76) differ from one another, basically, in the manner the linear interval system (1.79) is solved.

Let B be a real (i.e. noninterval) matrix computed as the approximate inverse of some matrix contained in $J(X)$. In practice B is the inverse of the centre of $J(X)$, i.e.

$$B = [J(m)]^{-1} \tag{1.80}$$

As pointed out in section1.3.2, in solving linear equations such as (1.79), it is advantageous to first premultiply (1.79) by B. Thus, we are led to consider the system

$$A(X)(y - x) = b(x) \tag{1.81}$$

where $A(X) = BJ(X)$ and $b = -Bf(x)$. Let

$$S = \{y\colon A(y - x) = b(x), \quad A \in A(X)\} \tag{1.82}$$

where A is any real matrix contained in $A(X)$. Let $N(x, X)$ denote an interval vector containing the set S of solutions of (1.81).

An interval Newton method for solving (1.76) has the following algorithm. Given $X^{(0)}$, define

$$X^{(k+1)} = X^{(k)} \bigcap N(x^{(k)}, X^{(k)}), \quad k \geq 0 \tag{1.83}$$

where $N(x^{(k)}, X^{(k)})$ is an interval solution of

$$A(X^{(k)})(y - x^{(k)}) = b(x^{(k)}) \tag{1.84}$$

and $x^{(k)} \in X^{(k)}$. Usually, $x^{(k)}$ is chosen to be the centre of $X^{(k)}$.

Since the linear interval system (1.84) is to be solved repeatedly (for different boxes $X^{(k)}$) approximate methods are used for its solution (the obtainment of the optimal solution of (1.84) for each iteration using the existing exact methods would require an unacceptable large volume of computation). In the first versions of the Newton method (1.84) was solved using the interval Gaussian elimination (see section1.3.3). Since then many other possibilities have been investigated.

A substantial improvement was made by Krawczyk [17]. He showed that it is was possible to obtain a nonsharp but adequate bound on the solution set of (1.84) by a very simple process which circumvents the necessity of solving the linear interval system (1.84).

Adding $x - y$ to both sides of (1.81), we get

$$[A(X) - E](y - x) = b(x) + x - y \tag{1.85}$$

or equivalently

$$y = b(x) + x + [E - A(X)](y - x) \tag{1.86}$$

where E is the identity matrix. Assume $y \in X$ and denote

$$K(x, X) = b(x) + x + [E - A(X)](X - x) \tag{1.87}$$

From (1.86) and (1.87), we see that

$$y \in K(x, X) \tag{1.88}$$

that is, the set of solutions to (1.81) is contained in $K(x, X)$. Thus, instead of (1.83) Krawczyk's version of the interval Newton method uses the iteration

$$X^{(k+1)} = X^{(k)} \bigcap K(x^{(k)}, X^{(k)}), \quad k \geq 0 \tag{1.89}$$

The advantage of this method resides in the fact that it avoids solving (1.84).

An alternative to Krawczyk's approach was suggested by Hansen [4], [8]. Again a nonsharp solution to (1.84) is accepted. We solve the system (1.81) in a Gauss–Seidel manner (see formulae (1.61), (1.62)), i.e. we solve the ith equation of (1.81) for the ith variable ($i = \overline{1, n}$). This is done in a "successive iteration" mode in that new information is used as soon as it is available. Given $X^{(0)}$, we compute

$$Y_i^{(k)} = x_i^{(k)} + [b_i^{(k)} - \sum_{j=1}^{i-1} A_{ij}^{(k)}(X_j^{(k+1)} - x_j^{(k+1)})$$

$$- \sum_{j=i+1}^{n} A_{ij}^{(k)}(X_j^{(k)} - x_j^{(k)})]/A_{ii}^{(k)} , \tag{1.90}$$

$$X_i^{(k+1)} = X_i^{(k)} \cap Y_i^{(k)} , \quad i = \overline{1,n}, \quad k \geq 0$$

where $A_{ij}^{(k)}$ denotes $[A(X^{(k)})]_{ij}$. Due to the inclusion monotonicity property such an approach leads to narrower intervals $Y_i^{(k)}$ and thus speeds up the convergence rate of the method.

This method is considered to be superior over Krawczyk's method.

1.4.3. Properties of the multidimensional interval newton methods

The multidimensional interval Newton methods has many of the valuable properties of the one-dimensional method described in section1.4.1. In what follows, $N(x, X)$ is either an interval solution to (1.79), (1.81) or $K(x, X)$ defined by (1.87)). We assume that $x \in X$.

From the derivation of (1.79) we see that the following theorem holds.

Theorem 1.17. If a solution y of (1.76) exists in $X^{(k)}$ then

$$y \in N(x^{(k)}, X^{(k)}), \; k \geq 0.$$

Thus, no zero of f in $X^{(0)}$ is ever lost. As a consequence, we have the following important result.

Theorem 1.18. If $X^{(k)} \cap N(x^{(k)}, X^{(k)})$ is empty, there is no zero of f in $X^{(0)}$.

Similar to the one-dimensional case, the rate of convergence for multidimensional interval methods producing a sequence of vectors $X^{(k)}$ is said to be p if

$$w(X^{(k+1)}) = 0 \; \{[(X^{(k)})]^p\} \tag{1.91}$$

(where the symbol $y = 0\{x^p\}$ means, that y is proportional to x^p for x tending to zero).

For some of the interval Newton methods proposed in the literature, it has been proven that the rate of convergence is asymptotically quadratic for simple zeros. The only question is how sharply a solution y to (1.76) is bounded in the particular method.

Interval Newton methods have reasonably good initial convergence behaviour. In the one-dimensional case, the current interval X_n is reduced to less than half its width by each Newton step if $0 \notin f'(X_n)$. A similar behaviour can occur in the multidimensional case.

At the kth step of an interval Newton method, we solve (see (1.84))

$$A(X^{(k)})z = b(x^k) \tag{1.92}$$

Assume that $x^{(k)}$ is the centre of $X^{(k)}$. Let $Z^{(k)}$ denote an interval solution to (1.92). Suppose that a component $Z_i^{(k)}$ is either nonnegative or nonpositive for some $i \in [1, n]$. Then

$$w(X_i^{(k+1)}) \le \frac{1}{2} w(X^{(k)})$$

If this occurs for m, $1 \le m \le n$, different values of i, the volume of $X^{(k)}$ is reduced by a factor of 2^{-m}.

If (1.66) has more than one solution in $X^{(0)}$ or the initial convergence rate is rather slow, it is necessary to split the current box in half and apply the method used to each half separately. Since this necessity may arise repeatedly, we have to form a list of boxes to be processed. It should be noted that the same process of generating subboxes occurs in using the method given by (1.90) if any of the intervals $A_{ii}^{(k)}$ contains zero. In this instance, extended interval arithmetic is used. A detailed discussion of an approach to generating and storing such subboxes (suggested in [4]) will be made in Chapter 6.

Just as in section 1.4.1 the most commonly used stopping criterion for the multidimensional case is to determine when either $w(X^{(k)}) < \varepsilon$ for some prescribed ε or else when $X^{(k+1)} = X^{(k)}$.

The former criterion yields a solution to within a desired tolerance. However, because of round off, it may not be possible to satisfy this condition if ε chosen is too small. The latter condition will always be satisfied eventually because machine arithmetic provides limited accuracy. It should be, however, pointed out that if $X^{(0)}$ is large, we could have $X^{(k+1)} = X^{(k)}$ while $X^{(k)}$ is still large, i.e. for $w(X^{(k)}) > \varepsilon$. If this is the case, $X^{(k)}$ should be split.

1.5. GLOBAL OPTIMIZATION

In this section, interval analysis techniques will be applied to solving the problem of global optimization. Both the unconstrained case (subsection 1.5.1) and the constrained cases (subsections 1.5.2 and 1.5.3) will be considered.

1.5.1. Unconstrained minimization

Let f: $R^n \to R$ be a twice continuously differentiable function ($f \in C^2$). Let g denote the gradient and H denote the Hessian of f. We seek the global minimum f^* of f, i.e.

$$f^* = \min f(x), \quad x \in R^n \tag{1.93}$$

We shall consider an interval method for solving (1.93) [8] which provides infallible bounds both on f^* and on the point(s) x^* for which $f(x^*) = f^*$. Instead of (1.93) we actually solve the following problem

$$f^* = \min f(x), \quad x \in X^{(0)}, \quad X^{(0)} \in I(R^n) \tag{1.94}$$

We assume that x^* is in the interior of $X^{(0)}$, so the constrained problem (1.94) is equivalent to the original unconstrained one given by (1.93). In practice, the size of $X^{(0)}$ is taken large enough (on the order of 10^6 for each component $X_i^{(0)}$ in [8]). Thus, the present method actually solves the unconstrained minimization problem provided the global solution occurs in some finite region which is enclosed by the initial box $X^{(0)}$. If $X^{(0)}$ does not contain the global minimum, we often obtain proof of this fact.

The basic strategy of the present method is to delete subboxes of $X^{(0)}$ which cannot contain the global minimum. Eventually, only a small region remains which must contain the solution.

The following techniques can be used to delete subboxes in which the global minimum cannot occur.

Searching for stationary points of f

The global solution of the unconstrained problem considered occurs where

$$g(x) = 0 \tag{1.95}$$

The most effective approach of deleting subboxes composed of nonoptimal points is to apply an interval Newton method to (1.95). Suppose we perform one interval Newton step to find the solution(s) of (1.95) in some box $X \subset X^{(0)}$. If we obtain a new box $X' \subset X$, we can actually delete that part of X which is not contained in X'. Indeed, from Theorem 1.17 any zero of g in X is also in X'. Thus, we are sure not to have deleted the global minimum after discarding all points of X not contained in X'.

If $X' \supseteq X$ or the reduction of the particular box X is insignificant, we split X in half and apply the Newton method to each part. In practice, we store one of the resulting halves (say, the right-hand one) in a list L of subboxes [8] to be processed later. Thus, we are always working on some subbox X of $X^{(0)}$.

The remaining three techniques are designed to prevent useless steps of Newton's method on stationary points that cannot be minima or on boxes that do not contain a stationary point.

Test for nonconvexity

Obviously, it is waste of time to find all the stationary points of $f(x)$, i.e. all the solutions of $g(x) = 0$. More precisely, it is superfluous to find maxima or saddle points. A necessary condition for f to have a minimum in a box X is that f be convex in X.

This will be the case if the Hessian matrix $H(x)$ is locally positive semidefinite. Thus, we can delete the whole box X if we can show that $H(x)$ is not positive definite for any $x \in X$.

There exists a simple test to verify this condition. Recall that the diagonal elements of a positive semidefinite matrix must be nonnegative. So we first evaluate $H_{ii}(X)$ (where $H_{ii}(X)$ is the interval extension of the diagonal element $h_{ii}(x)$ of the Hessian) starting with $i = 1$. If, for some $i \in \overline{1, n}$ we find $H_{ii}(X) < 0$, then, by Theorem 1.1, $h_{ii}(x) < 0$ for all $x \in X$. Hence, $H(x)$ is not positive semidefinite for any $x \in X$ and X can, therefore, be deleted.

Test for monotonicity

This test provides a simple sufficient condition that the current box X does not contain a stationary point of f.

We evaluate the interval extension $G_i(X)$ of the ith component of the gradient $g(x)$ for $i = \overline{1, n}$. If for some i, $G_i(X) > 0$ or $G_i(X) < 0$, then f is a strictly monotonic function of x_i throughout X. Hence, X can be deleted and the Newton method need not be applied.

Bounding f^*

Let x be the centre of a particular subbox X of $X^{(0)}$. We evaluate $f = f(x)$. Let $\bar{f}$ denote the currently smallest value of f found so far (that is, using the centres of all the subboxes generated so far). Obviously, we can delete any subbox X of $X^{(0)}$ for which $F(X) > \bar{f}$ ($F(X)$ is the interval extension of f in X) since this implies $f(x) > \bar{f}$ for all $x \in X$.

The bound $\bar{f}$ can be used in a more sophisticated way. Expanding f about the centre x, we have

$$f(y) \in f(x) + (y-x)^T g(x) + \frac{1}{2}(y - x)^T H(X)(y - x)$$

The set of points $y \in X$ for which

$$f(x) + (y-x)^T g(x) + \frac{1}{2}(y - x)^T X(X)(y - x) > \bar{f} \tag{1.96}$$

can be deleted (see section 2.3.4 for details).

A detailed algorithm of the present method accounting for round off errors is given in [8]. Experimental evidence shows its efficiency and robustness.

A simpler (and less efficient) method for solving (1.93) is discussed in [18].

1.5.2. Inequality constraints

In this section we consider the following constrained optimization problem:

$$f^* = \min f(x) \tag{1.97a}$$

$$p_i(x) \le 0, \quad i = \overline{1,m} \tag{1.97b}$$

We assume that f, $p_i \in C^2$.

In order to apply the efficient method from the previous section (1.97) is first equivalently transformed into a system of nonlinear equations [19]. This transformation is done on the basis of the Fritz John necessary conditions using normalized Lagrange multipliers which yields:

$$u_0 g(x) + \sum_{i=1}^{m} u_i g_i(x) = 0 \tag{1.98}$$

($g(x)$ and $g_i(x)$ being the gradient column-vector of f and p_i respectively, while u_0, u_1, . . . , u_m are scalars)

$$u_i p_i(x) = 0 \quad i = \overline{1,m} \tag{1.99}$$

$$\sum_{i=0}^{m} u_i = 1 \tag{1.100}$$

with

$$u_i \ge 0, \quad i = \overline{i,m} \tag{1.101}$$

Note that (1.98) is in fact a system of n equations

$$u_0 g^{(j)}(x) + \sum_{i=1}^{m} u_i g_i^{(j)}(x) = 0, \quad j = \overline{1,n}$$

where

$$g^{(j)} = \frac{\partial f}{\partial x_j}, \quad g_i^{(j)} = \frac{\partial p_i}{\partial x_j}$$

Thus, the set (1.98), (1.99) and (1.100) is a system of n+m+1 equations in n+m+1 unknowns: x_j, $j = \overline{1, n}$ and u_i, $i = \overline{0, m}$.

Based on (1.98) to (1.100) an interval method for solving (1.97) has been suggested in [19]. This method is essentially the same as for the unconstrained problem. We start with a box X and delete subboxes which contain the global minimum.

Let $P_i(X)$ be the interval extension of $p_i(x)$ in a box X. The box X is called certainly feasible if $\overline{P_i(X)} \le 0$ for all $i = \overline{1, m}$, and certainly infeasible if $P_i(X) > 0$ for at least one index $i \in [1, m]$. Obviously, if X is certainly feasible (infeasible), then every point $x \in X$ is feasible (infeasible), i.e. satisfying (violating) (1.97b).

If the current box X is certainly strictly feasible, i.e. if $P_i(X) < 0$ for all $i = \overline{1, m}$, then any solution part x of (1.98) to (1.100) in X is a stationary point of f. Hence we proceed exactly as in the unconstrained case (section 5.1.1).

If the current box is certainly infeasible, we delete it.

If X is neither certainly strictly feasible nor certainly infeasible, we try to reduce or delete X by the following techniques.

Newton's method

Some interval Newton method is applied to the system of equations (1.98) to (1.100). To do so, we need initial bounds on x_i and u_i. Clearly, the initial box $X^{(0)}$ must be large enough to contain the feasible region defined by (1.97a). The initial bounds on u_i may be obtained from (1.100) and (1.101) whence:

$$0 \le u_i \le 1 \ , \quad i = \overline{0, m} \tag{1.102}$$

In [19] it is shown that sharper initial bound on u_i may be provided by the Newton method itself.

Bounding f^*

Similar to the unconstrained case, we obtain and update an upper bound $\bar{f}$ on the global minimum f^*. As each subbox X of $X^{(0)}$ is generated after reducing or splitting the previous box, we examine the centre x of X. If x is certainly feasible, then $f(x)$ is a candidate for $\bar{f}$ (the smallest upper bound on f^* found so far). Thus, if $f(x) < \bar{f}$, we let $\bar{f} = f(x)$.

Any subsequent box X is deleted if $F(X) > \bar{f}$.

Solving interval inequalities

Expanding f about the centre x of X, we obtain

$$f(y) \in f(x) + (y - x)^T G(X)$$

The points for which

$$f(x) + (y - x)^T G(X) > \bar{f} \tag{1.103}$$

can be deleted.

Expanding in a similar way the constraint functions p_i about x, we get

$$p(x) + (y - x)^T G_i(X) \le 0, \quad i = \overline{1, m} \tag{1.104}$$

where $G_i(X)$ denotes the interval extension of the gradient $g_i(x)$ of $p_i(x)$.

The inequality (1.103) is rewritten as

$$\bar{f} - f(x) - (y - x)^T G(X) \le 0 \tag{1.105}$$

The set of $m + 1$ inequalities (1.104) and (1.105) is used to determine a subbox of points $y \in X$ which can be deleted. A procedure for doing this is presented in [19].

1.5.3. Equality constraints

In this case the constrained problem has the form

$$f^* = \min f(x) \tag{1.106a}$$

$$q_i(x) = 0, \quad i = \overline{1,r} \tag{1.106b}$$

The corresponding system of nonlinear equations using the normalized Lagrange multipliers is now [19]:

$$v_0 g(x) + \sum_{i=1}^{r} v_i g_i(x) = 0 \tag{1.107}$$

($g_i(x)$ being now the gradient of $q_i(x)$)

$$q_i(x) = 0, \quad i = \overline{1,r} \tag{1.108}$$

$$v_o + \sum_{i=1}^{r} v_i^2 = 1 \tag{1.109}$$

The normalization (1.109) yields the initial bounds on v_i:

$$0 \le v_0 \le 1 \tag{1.110}$$

$$-1 \le v_i \le 1 \tag{1.111}$$

It should be noted that the normalization (1.100) is not possible here because v_i, $i = \overline{1, r}$, may be negative.

A method for solving (1.106) has been suggested in [19]. Since it involves the same techniques as for the inequality constrained case, we shall only point out the slight differences in the nature of these techniques.

The interval Newton method is applied to system (1.107) to (1.109).

The equality constraints (1.106b) are replaced by

$$\begin{array}{l} q_i(x) \le 0 \\ -q_i(x) \le 0 \end{array}, \quad i = \overline{1,r} \tag{1.112}$$

Based on (1.112) we form a set of ($2r + 1$) interval inequalities of the type (1.104), (1.105) to determine a subbox of points $y \in X$ which can be deleted.

The main difference is in finding an upper bound $\overline{f}$ on f^*. Because of roundoff, we cannot, in general, claim that $q_i(x) = 0$ exactly for a given x. Hence we cannot decide whether x is feasible. To overcome this difficulty the authors of [19] prove that there exists a feasible point in a small box X'. They then bound f over X'. This provides a candidate for $\overline{f}$ (for details, see [19]).

A simple method for solving (1.106) is presented in [20]. It is based on the use of Gauss elimination applied to system (1.107) to (1.109).

Remark 1.3. The case with both equality and inequality constraints can be treated by combining the procedures of this section and section 5.1.2.

Comments

In view of the applications of interval analysis to be presented in the following chapters we have confined ourselves to a restricted number of topics from interval analysis. Moreover, these topics have been described in a rather succinct manner. For a deeper acquaintance with the scope of interval analysis the reader is referred to [1], [2], [10].

In what follows we shall briefly point out some generalizations of the subjects covered in this chapter.

Section 1.1. The interval arithmetic considered in this section deals with real intervals, that is, intervals which are closed bounded subsets of the real line R. Such an interval arithmetic is called real interval arithmetic. In interval analysis, there exists a generalization of this arithmetic which encompasses the case where the "intervals" are bounded closed subsets of the complex plane C (e.g. see [10]). Such an arithmetic dealing with complex intervals is called complex interval arithmetic. It has not been used for circuit analysis as yet.

Referring to real interval arithmetic, some implementation considerations are now due. It should be borne in mind that machine interval arithmetic is about five times slower than ordinary (real) arithmetic if no special hardware is available. First, for each arithmetic operation, we must compute two interval endpoints instead of a single number. However, hardware implementation of interval arithmetic could produce endpoints using parallel computation, thus making it comparable in speed to ordinary arithmetic. Second, and most important, to implement outward rounding by high level algorithmic languages is actually time-consuming. However, one of the recent IEEE floating point arithmetic standards specifies that the rounding direction be specifiable. Thus, directed rounding should not continue to be slow.

Another difficulty in applying interval analysis methods is programming. A dedicated compiler can simplify the programming by allowing intervals to be declared as a special data type. This is already possible, for example, in the versions Pascal-SC, extended Turbo-Pascal (TPX) and Fortran-SC. (With such a compiler, but without hardware rounding options, interval arithmetic remains about five times slower that ordinary arithmetic). Languages allowing operator overloading for user defined variable types, such as Ada and Fortran 88 eliminate the need for special compilers.

Section 1.2. Having in mind practical application of interval methods, the most important subject in this section is undoubtedly subsection 1.2.6. where some means for obtaining arbitrarily sharp bounds on the range $f(X)$ of a function $f(x_1, \ldots, x_n)$ over a box X are considered. This problem is central in interval analysis. Still it has to be underlined that there are few methods (if any) which are capable of bounding the range sufficiently sharply with a reasonable amount of computational effort when the number n of variables x_i is high enough. Obviously, the number of interval variables that can be handled depends on the particular function and the size of the box X. Thus, for Skelboe's method, it is recommended in [21] that the number of intervals subject to partitioning should not exceed ten. In section 2.3.2 this method will be improved by the introduction of interval forms others than those considered in section 1.2.5.

If $f \in C^2$ the range can be determined by the global optimization method from section 1.5.2. However, the numerical efficiency of such an approach depends heavily on how easily the Hessian matrix is evaluated in interval form.

For functions $f \in C^1$, a simple method is suggested in [18]. Its efficiency is, however, limited to simpler problems of moderate dimensionality.

When implementing any of the above methods it is essential to compute the interval extensions involved (for example, $F_j'(X)$ in the mean-value form (1.44)) as sharply as possible. In some (rather) simple cases the approach based on Theorem 1.4 can be useful to find exactly the corresponding ranges of values. In [22] a more general approach is suggested which is based on the so-called generalized interval arithmetic (making use of the representation (1.7) of the intervals involved). The interval extensions thus obtained are often much narrower as compared to ordinary arithmetic results. However, it should be noted that, for wider initial intervals, it can result in wider extension than ordinary arithmetic extensions.

Another approach which is based on the so-called interval slopes is suggested in [23]. It provides extensions which are never worse (and are usually better) than those obtained by the mean-value form (1.44) where the derivatives are computed using ordinary arithmetic.

In general, if the computational cost can be afforded, it is expedient to obtain several extensions for one and the same function (applying different methods) and to use their intersection.

Section 1.3. Linear interval equations have been intensively studied in the interval analysis literature (see the survey paper [12] and the references therein cited).

Since most of the methods presently available are not exact there are cases in which they may lead to prohibitively large overstimation. Therefore, it is essential to derive bounds on the overestimation. In [24] the solution set S of (1.53) is enclosed in a "skew interval" and several bounds on the distance (in Hausdorf's sense) between S and the skew enclosure are proposed. In [25], [26] two skew enclosures (upper and lower) are introduced. The boundary of the solution set is proven to lie between these extreme enclosures.

It should be stressed that preconditioning can be applied only for full matrices of low dimensionality: it is too time and/or space consuming for large sparse matrices. For a discussion of the available methods for solving linear interval equations with sparse matrices, see [24].

Sections 1.4 and 1.5. The interval methods sketched in these sections may seem misleadingly simple. In fact, the elaboration of operable algorithms ensuring appropriate dynamic partitioning of the initial region $X^{(0)}$ into subboxes and accounting for the outward rounding is quite a job. Detailed algorithms for such methods are given in [4], [8], [19], [80].

Interval algorithms for a system of nonlinear equations and global optimization have a number of appealing properties. Unlike their noninterval counterparts, they are capable of locating all the solutions contained in the initial box $X^{(0)}$ providing infallible bounds on each solution. Termination for both types of problems is quite reliable and occurs in a finite number of steps. At the same time, even the best noninterval algorithms can terminate prematurely with a poor or totally incorrect answer. Convergence of the interval methods is monotonic.

If the particular nonlinear system or constrained optimisation problem has no solution in the initial box $X^{(0)}$, interval methods will establish the nonsolvability of the problem in a finite number of iterations after deleting all of the generated subboxes. Making use of noninterval methods, one can never tell when it is time to stop the computation process resulting sometimes in long fruitless searches for a solution that does not exist.

On the other hand, an interval method is generally slower than the noninterval counterpart when the latter solves the problem considered. Indeed, interval methods cannot use certain shortcuts such as updating of the Jacobian or the Hessian matrix. However, hardware for interval arithmetic will narrow the efficiency gap. Presumably when the interval approach is more mature, this gap will narrow and, even disappear.

Interval methods are well suited for parallel computation. We need only separate the initial region $X^{(0)}$ into subboxes. Different processors can be applied to each subregion. This will also make interval methods more attractive as parallel computing capabilities develop.

CHAPTER 2

TOLERANCE ANALYSIS OF LINEAR ELECTRIC CIRCUITS – GLOBAL OPTIMIZATION APPROACH

In this chapter the tolerance analysis of steady-states in linear electrical circuits is studied by interval analysis techniques. The original tolerance problems are formulated as corresponding global optimization problems. These latter problems are then solved using appropriate interval methods for global constraint optimization. The main emphasis is placed on methods for solving the worst-case tolerance analysis problem.

2.1. GLOBAL OPTIMIZATION APPROACH TO SOLVING THE LINEAR CIRCUIT TOLERANCE PROBLEM

2.1.1. Deterministic statement of the tolerance analysis problem

In this section the (steady-state) tolerance analysis problem for linear electric circuit will be stated in a deterministic setting. A probabilistic approach to formulating the tolerance problem will be presented in section 2.1.3.

Let **N** be a linear lumped time-invariant electric circuit in some direct current (d.c.) or sinusoidal (a.c.) steady-state. Let x_i, $i = \overline{1, n}$ denote some "input" parameter such as resistance r, inductance L (mutual inductance M), capacitance C, voltage or current source value, etc. Furthermore, let y be some "output" characteristic such as output voltage, total impedance, etc. We assume that:

i) there is no interdependence between the input parameters x_i;

ii) the functional relationship $y = f(x)$ between the output y and the parameter vector $x = (x_1, \ldots, x_n)$ is explicity known.

E x a m p l e **2.1.** Let **N** be the resistive circuit shown in Fig. 2.1 (a circuit of such configuration is called a ladder circuit)

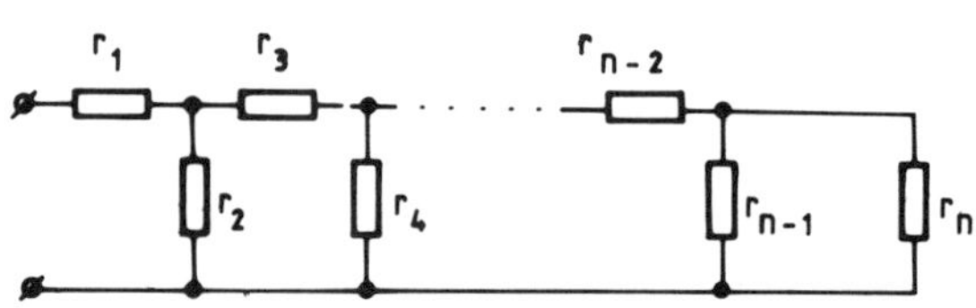

Fig. 2.1. A ladder circuit.

We are interested in the equivalent resistance r_e of the ladder circuit. It is readily seen that

$$r_e = r_1 + \cfrac{1}{\cfrac{1}{r_2} + \cfrac{1}{r_3 + \cfrac{1}{\cfrac{1}{r_4} + \cfrac{1}{r_5 + \dots}}}} \tag{2.1}$$

In this example the input parameters x_i are the resistances r_i, the output quantity y is the equivalent resistance r_e and the functional relationship between y and $x = (x_1, \dots, x_n)$ is given by formula (2.1).

***Example* 2.2.** In this example we consider an arbitrary linear a.c. circuit. We are interested in some output characteristic (e.g. voltage or current transfer function) which can be written in the form

$$H(s,x) = \frac{N(s,x_1,\dots,x_n)}{D(s,x_1,\dots,x_n)} \tag{2.2}$$

where s is the complex frequency. The function (2.2) is a real function in a complex variable and is bilinear in each of the parameters x_i (N and D are linear function with respect to x_i if all the remaining arguments are treated as constants) when they characterize network elements or controlled sources.

Most often in practice, we would like to find the modulus of (2.2) for $s = j\omega$. If ω is fixed this leads to the following real function in real variables

$$|H(j\omega,x)| = h(x_1,\dots,x_n;\omega) \tag{2.3a}$$

Now $|H(j\omega, x|$ is the output characteristic y related to the input parameters through (2.3a). If ω is variable, then it can be treated as an additional $(n+1)$th variable and (2.3a) becomes

$$|H(j\omega,x)| = h(x_1,\dots,x_n,x_{n+1}) \tag{2.3b}$$

Let $y = f(x)$ be the function which relates the input parameter vector $x = (x_1, \dots, x_n)$ to the output variable y for the particular circuit studied. In this section we assume that the input parameter x_i takes on values (owing to various causes: inperfect technology process, aging etc.) within a prescribed tolerance X_i, each X_i being an interval. Therefore, the output y, being the image of x, will also vary within a corresponding tolerance Y. Loosely speaking, the tolerance analysis problem herein considered consists in finding the output variable tolerance for the circuit investigated if the tolerances on the input parameters are known.

We shall now proceed to strictly formulating the tolerance analysis problem considered. Let $x_i \in X_i \in I(R)$ and $X = (X_1, \ldots, X_n)$ so that $x \in X$. Thus, the function f: $X \subset R^n \to R$ is defined in the n-dimensional interval vector (the box) X with values in R. Now, the tolerance problem investigated can be stated as follows.

Problem 2.1. Given the multivariate function $f(x)$ defined in a given box X, find the range $f(X)$ of f over the box X.

The tolerance problem formulated is usually referred to as the worst-case tolerance analysis problem. Whenever this will not lead to misinterpretation the shorter term of tolerance problem will, however, be used (in section 2.1.3 another type of tolerance analysis problem will be considered).

It follows from the above formulation that any interval method designed for range evaluation (e.g. from section 1.2.5) could be used to solve the worst-case tolerance analysis problem. As usual, the classical tradeoff between efficiency and cost of computations will finally guide the choice of a particular method.

In some cases (most often for d.c. circuits) the tolerance analysis can be carried out in a most effective way by applying Theorem 1.2. To illustrate this possibility consider again Example 2.1 with $n = 5$ and $r_i \in R_i$, $i = 1,\ldots,5$, when R_i (the tolerances on r_i) are some given intervals. The corresponding natural extension of (2.1) is then

$$R_e = F(R_1,\ldots,R_5) = R_1 + \cfrac{1}{\cfrac{1}{R_2} + \cfrac{1}{R_3 + \cfrac{1}{\cfrac{1}{R_4} + \cfrac{1}{R_5}}}}$$

It is seen that each interval variable R_i, $i = 1, \ldots, 5$, appears only once to the first power; moreover since $R_i > 0$ all divisions are realizable. Therefore, the above interval extension R_e satisfies all the conditions of Theorem 1.2, so that $F(R_1, \ldots, R_5)$ provides the range $f(R_1, \ldots, R_5)$. Thus, a single computation of F solves the problem of determining the tolerance on the equivalent resistance r_e.

Since application of Theorem 1.2 is a rather effective way of solving d.c. tolerance analysis problems it is worthwhile trying to extend the scope of its applicability. Unlike Example 2.1 where the function (2.1) is written in a form suitable for direct use of Theorem 1.2, in some cases we have first to transform the original expression of $f(x)$ to be able to apply the theorem. The following simple example will make the idea behind such an approach clear (see also Example 1.5.).

***Example* 2.3.** Consider a resistive circuit made up of three resistors r_1, r_2 and r_3; the resistors r_2 and r_3 connected in parallel are in series with r_1. Given the tolerance R_i on each r_i, $i = 1, 2, 3$, find the tolerance on the equivalent resistor r_e:

$$r_e = r_1 + \frac{r_2 r_3}{r_2 + r_3}$$

At first glance, it seems that Theorem 1.2 cannot be applied since the resistances r_2 and r_3 appear twice in the expression for r_e. However, this expression can be written in the following equivalent form

$$r_e = r_1 + \frac{1}{\frac{1}{r_2} + \frac{1}{r_3}}$$

Now, the natural interval extension of r_e is

$$R_e = R_1 + \frac{1}{\frac{1}{R_2} + \frac{1}{R_3}}$$

It is seen that by Theorem 1.2, $R_e = [\underline{r}, \overline{r}]$ defined as above determines the tolerance on the equivalent resistance r_e. Applying the corresponding arithmetic operations involved, we finally get

$$\underline{r} = \underline{r}_1 + \frac{1}{\frac{1}{\underline{r}_2} + \frac{1}{\underline{r}_3}}$$

$$\overline{r} = \overline{r}_1 + \frac{1}{\frac{1}{\overline{r}_2} + \frac{1}{\overline{r}_3}}$$

Based on example 2.3 the following more general result is easily obtained.

Proposition 2.1. Let **N** be a resistive one-port of series-parallel type (i.e. a one-port circuit made up only of a finite combination of series and parallel connections). The supply voltage v and the branch resistances r_i, $i = \overline{1, n}$ are given as the intervals V and R_i, respectively. Then a natural interval extension $I(V, R_1, \dots, R_n)$ for the input current $i = f(v, r_1, \dots, r_n)$ can be written in such a form that Theorem 1.2 is satisfied.

P r o o f. For each parallel connection, the equivalent resistance can be written as

$$r_p = \frac{1}{\sum_\alpha \frac{1}{r_{p\alpha}}}$$

Hence $r = f_1(r_1, \ldots, r_n)$ can always be written in such a way that each resistance occurs only once in f_1. But $i = v/r$ so that the interval extension I of $i = v/f_1$ is seen to satisfy all the conditions of Theorem 1.2. Thus, the tolerance on i can be determined directly through a single computation of I.

In some cases Theorem 1.2 can be applied to the tolerance analysis of active circuits also.

***E x a m p l e* 2.4.** [29]. Find the worst-case tolerance limits for the amplifier gain $G = V_2/V_1$ of the circuit shown in Fig.2.2. if the resistor tolerances are ±2% and the operational amplifier gain tolerance is +50% and −10%.

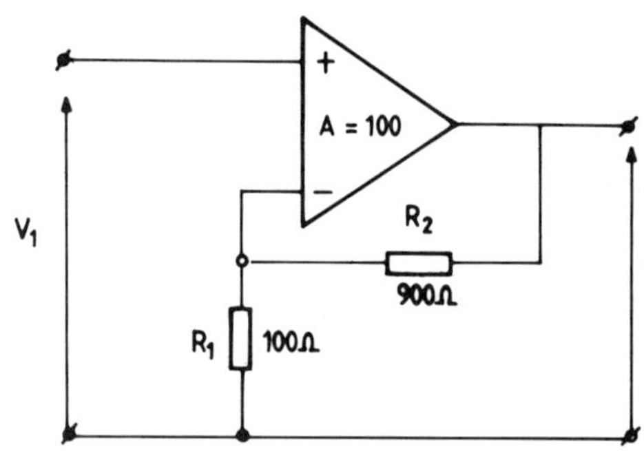

Fig. 2.2. An amplifier circuit.

It is known [29] that

$$G = \frac{1}{\frac{1}{A} + \frac{R_1}{R_1 + R_2}}$$

To apply Theorem 1.2 the expression for G is rewritten as

$$G = \frac{1}{\dfrac{1}{A} + \dfrac{1}{1 + \dfrac{R_2}{R_1}}}$$

From the nominal values and the given tolerances

$$R_1^I = [882, 918],\ R_2^I = [98, 102],\ A^I = [90, 150]$$

So the tolerance limits on G can be found as the endpoints of the interval

$$G^I = \frac{1}{\dfrac{1}{A^I} + \dfrac{1}{1 + \dfrac{R_2^I}{R_1^I}}}$$

In most practical tolerance problems Theorem 1.2 is not applicable and a single computation of the interval extension of $f(x)$ only yields an interval $F(X)$ which is wider than the tolerance $f(X)$ on the output variable y. However, by the inclusion property (1.32) the interval $F(X)$ is guaranteed to enclose $f(X)$. Thus, $F(X)$ may, in some cases, serve as an initial, rough estimate of the output variable tolerance providing infallibly outward bounds on it.

2.1.2. Equivalent formulation

Let X^0 denote the interval vector whose components are the given tolerance intervals on the input variables (we shall have to distinguish between this initial box and the current subboxes X generated in the process of solving Problem 2.1). In the general case (when Theorem 1.2 cannot be applied) the problem of computing the range $f(X^0)$ of f over X^0 may be formulated equivalently as two global optimization problems. Indeed,

$$f_L^* = \min_{x \in X^0} f(x) \tag{2.4}$$

and

$$f_U^* = \max_{x \in X^0} f(x) \tag{2.5}$$

The solution f_L^* of (2.4) provides the lower endpoint of the range $f(X^0)$ while the solution f_U^* of (2.5) provides the upper endpoint of $f(X^0)$.

The problem (2.5) can be transformed into an equivalent minimization problem

$$f_U^* = -[\min_{x \in X^0} (-f(x))] \tag{2.6}$$

Thus, the tolerance analysis problem considered reduces to two global minimization problems (2.4) and (2.6).

Problem (2.4) can be written in a detailed form as follows:

$$f_L^* = \min f(x_1,\ldots,x_n) \tag{2.7a}$$

$$x_i \le \overline{x}_i^0, \quad -x_i \le -\underline{x}_i^0, \quad i = \overline{1,n} \tag{2.7b}$$

where $\underline{x}_i^0$ and $\overline{x}_i^0$ are the left and right endpoint, respectively, of the component X_i^0 of the interval vector X^0. As is seen from (2.7) the lower endpoint f_L^* of the output variable tolerance can be determined by solving a global minimization problem with inequality constraints (2.7b). Similarly, the upper endpoint f_U^* of the output tolerance can be found by solving a corresponding minimization problem of the type (2.7), associated with (2.6).

In some tolerance problems the dimension of the arising minimization problems can be reduced. In the case of d.c. circuits the dimension reduction is based on Theorem 1.2 which is applied to part of the input variables. In the case of a.c. circuits it is made possible by using Thevenen's theorem. The following two examples will illustrate this possibility.

***Example* 2.5.** For the bridge circuit shown in Fig. 2.3(a) the tolerance on the total current i is sought if the tolerances on v and r_k , $k = 1, 2, \ldots, 6$ are given.

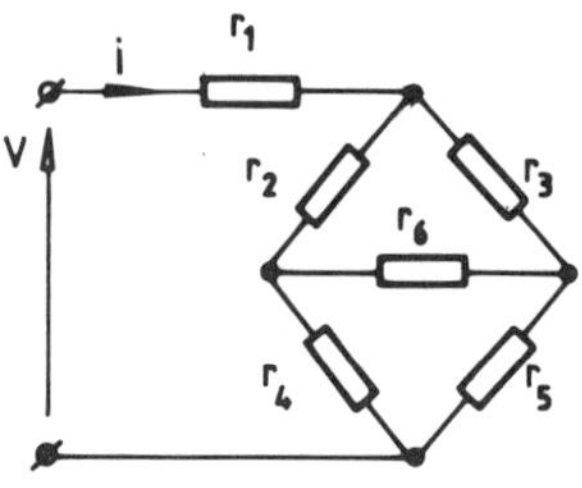

Fig. 2.3(a). A bridge circuit.

In this example Theorem 1.2 cannot be applied directly since, as is easily seen, it is impossible to find an expression for the input resistance r such that each individual resistance r_i occurs only once. Therefore, we are at first glance led to solve two minimization problems of type (2.7) with $n = 7$. However, the dimension of the arising minimization problem can be reduced to $n' = 3$ if the following approach is used. First, the delta connection formed by resistors r_2, r_3 and r_6 is transformed into an equivalent star of resistors r_{23}, r_{36} and r_{26} to obtain the equivalent circuit of Fig. 2.3(b).

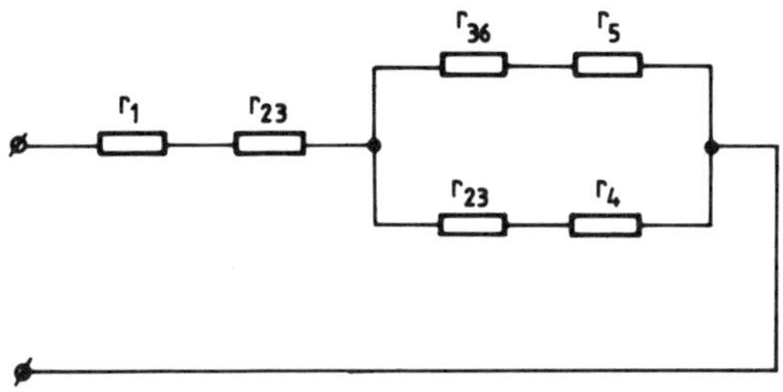

Fig. 2.3(b). Equivalent circuit.

The latter circuit is of series-parallel type; hence it is expedient to write i as the expression:

$$i = \frac{v}{r_1 + r_{23} + \dfrac{1}{\dfrac{1}{r_{36} + r_5} + \dfrac{1}{r_{26} + r_4}}} \tag{2.8}$$

In (2.8) each resistance r_{23}, r_{36} and r_{26} is a function of the branch resistances r_2, r_3 and r_6. Now fix r_2, r_3 and r_6 at arbitrary values within the corresponding intervals R_2, R_3 and R_6 and consider the following interval extension of (2.8).

$$I = \frac{V}{R_1 + r_{23} + \dfrac{1}{\dfrac{1}{r_{36} + R_5} + \dfrac{1}{r_{26} + R_4}}} \tag{2.9}$$

At this stage we can apply Theorem 1.2 to the expression (2.9) since V, R_1, R_4 and R_5 occur only once. It is easily seen that the lower endpoint $\underline{I}$ of I is given by the formula

$$\underline{I} = \frac{\underline{v}}{\overline{r}_1 + r_{23} + \dfrac{1}{\dfrac{1}{r_{36} + \overline{r}_5} + \dfrac{1}{r_{26} + \overline{r}_4}}} \tag{2.10}$$

Similarly, the upper endpoint of I is

$$\overline{I} = \frac{\overline{v}}{\underline{r}_1 + r_{23} + \dfrac{1}{\dfrac{1}{r_{36} + \underline{r}_5} + \dfrac{1}{r_{26} + \underline{r}_4}}} \tag{2.11}$$

On substituting the expressions for r_{23}, r_{36} and r_{26} into (2.10) the quantity $\underline{I}$ is seen to be a function of r_2, r_3 and r_6 only, i.e.

$$\underline{I} = f_1(r_2, r_3, r_6)$$

Now the lower endpoint i_L^* of the tolerance on i can be determined by "freeing" the input variables r_2, r_3 and r_6 and solving the following global minimization problem

$$i_L^* = \min f_1(r_2, r_3, r_6)$$

$$\underline{r}_i \le r_i \le \overline{r}_i \quad , \quad i = 2,3,6. \tag{2.12}$$

In a similar way, writing (2.11) as

$$\overline{I} = f_2(r_2, r_3, r_6)$$

the upper endpoint i_U^* of the tolerance on i can be determined by solving the global minimization problem

$$i_U^* = -\min\ (-f_2(r_2, r_3, r_6) \tag{2.13}$$

$$\underline{r}_i \le r_i \le \overline{r}_i \quad , \quad i = 2,3,6.$$

Thus, it has been shown that using Theorem 1.2 with respect to V, R_1, R_4 and R_5 has reduced the original minimization problems dimension from $n = 7$ to $n = 3$.

***E x a m p l e* 2.6.** Consider a complex a.c. circuit containing independent sources and passive elements R, L, C for which the current $\dot{I}$ through a given branch can be determined by Thevenen's theorem, namely

$$\dot{I} = \frac{\dot{V}_{oc}}{Z_e + Z}$$

where $Z = R + jX$ is the impedance of the branch with current $\dot{I}$. Now assume that R, X as well as all the other elements and sources parameters belong to some prescribed intervals. The problem is to determine the corresponding tolerance on the modulus $|\dot{I}| = I$.

We shall show that the dimension of this tolerance problem can always be reduced by one and most often by two. Indeed,

$$I = f(p,R,X) = \frac{|\dot{V}_{oc}|}{\sqrt{(R_e + R)^2 + (X_e + X)^2}}$$

where $p = (p_1, \ldots, p_n)$ is the vector of all input parameters other than R and X which are treated as $(n+1)$th and $(n+2)$th parameters, respectively. Clearly, $|\dot{V}_{0c}| = f_1(p)$, $R_e = f_2(p)$ and $X_e = f_3(p)$, that is, f_1, f_2 and f_3 are functions only of the first n parameters because they are determined when Z is removed from the original circuit. So

$$I = f(p,R,X) = \frac{f_1(p)}{\sqrt{(f_2(p) + R)^2 + (f_3(p) + X)^2}}$$

Now let $p \in P$, $R \in R^I = [\underline{R}, \overline{R}]$, $X \in X^I = [\underline{X}, \overline{X}]$. Since $f_1(P) \geq 0$, ($f_1(p) = |\dot{V}_{0c}| \geq 0$ for each $p \in P$) and $f_2(P) \geq 0$ ($f_2(p) = R_e \geq 0$ for each $p \in P$) it is easily seen from the above formula that I reaches its global minimum with respect to R for any fixed $p \in P$, $X \in X^I$ if $R = \overline{R}$ similarly $R = \underline{R}$ secures the global maximum of I for any fixed $p \in P$, $X \in X^I$. Thus, we have managed to reduce the dimension of the original global optimization problem by one. Indeed, let $\underline{I}$ denote the lower endpoint of the resultant tolerance on I. Then

$$\underline{I} = \min_{\substack{p \in P \\ X \in X^I}} f(p,\overline{R},X)$$

similarly, if $\overline{I}$ denotes the upper endpoint of the resultant tolerance on I then

$$\overline{I} = \min_{\substack{p \in P \\ X \in X^I}} f(p,\underline{R},X)$$

Now consider the effect of X on I. It is easily seen that if $X_e = f_3(p)$ and X have the same sign for any $p \in P$ and $X \in X^I$ then the effect of X on I is identical to that of R; that is, $\overline{X}$ is associated with the minimum of I while $\underline{X}$ leads to the maximum of I. So that

$$\underline{I} = \min f(p,\overline{R},\overline{X}) \ , \quad p \in P$$

$$\overline{I} = \min f(p,\underline{R},\underline{X}) \ , \quad p \in P$$

Thus, the dimension of the original optimization problems associated with the determination of the range of $I = f(p, R, X)$ when $p \in P$, $R \in R^I$, $X \in X^I$ is seen to have been reduced by two since now R and X are fixed at their respective endpoint values.

The assumption

$$X_e = f_3(p) > 0 \ , \quad p \in P$$

or

$$X_e = f_3(p) < 0 \quad , \quad p \in P$$

can be checked by the following sufficient condition

$$F_3(P) > 0$$

or

$$F_3(P) < 0$$

where $F_3(P)$ is some interval extension of $f_3(p)$.

If $X_e = f_3(p)$ and X have opposite signs and $X_e + X \geq 0$ or $X_e + X \leq 0$ for all $p \in P$ and $X \in X^I$ then, as is easily seen, the effect of X on I is reversed: $\underline{X}$ leads to the minimum of I while $\overline{X}$ leads to the maximum of I. Therefore,

$$\underline{I} = \min f(p, \overline{R}, \underline{X}) \ , \quad p \in P$$

$$\overline{I} = \min f(p, \underline{R}, \overline{X}) \ , \quad p \in P$$

The corresponding assumptions related to this latter case can be easily checked by using $F_3(p)$.

Thus, the application of Thevenen's theorem guarantees the reduction of the dimension of the problem considered by one (with respect to R). If additionally certain easily verifiable conditions on $f_3(p)$ are met, a second dimension reduction (with respect to X) is also guaranteed.

Interval methods for solving minimization problems with inequality constraints have been briefly considered in section 1.2.6 and section 1.5.2. In section 2.3 we will resume discussing this topic in a more detailed manner taking into account the specific features of problem (2.7).

2.1.3. Probabilistic statement of the tolerance problem

The deterministic problem we were considering in section 2.1.1 is termed (true) worst-case tolerance analysis in electric circuit literature because the tolerance on the output variable y accounts for the worst possible combinations of the admissible values of all input variables x_i (e.g. [21], [27], [28]).

In practice, each variable x_i takes on a particular value on an element of the real line R with certain probability. From this point of view the deterministic statement of the tolerance problem from section 2.1.1. can be reformulated as follows. Each variable x_i has zero probability to occur if $x_i \notin X_i$ and a uniform probability $P_i = 1/w(X_i)$ if $x_i \in X_i$. Given a function $y = f(x)$, we want to find the set of values for the output variable y which occur with nonzero probability.

Very often each variable x_i is, however, distributed according to some probability law which is different from the uniform probability law. If the circuit under investigation is implemented by discrete elements, then the most common case encountered in practice is the case where all variables x_i are statistically independent (uncorrelated) and satisfy the normal (Gauss) distribution law. However, if the circuit investigated is implemented by the integrated circuit technology, then the variables x_i , being as a rule normally distributed, are not statistically independent. For the sake of generality we shall consider the more complex situation when the random variables are correlated. Then, as is well known, the probability density $\varphi(x)$ is given by the formula

$$\varphi(x) = \frac{1}{\sqrt{(2\pi)^n \det C}} \exp\left[-\frac{1}{2}(x-\xi)^T C^{-1}(x-\xi)\right] \tag{2.14}$$

where C is the correlation matrix, ξ is the mean vector and the symbol T stands for transpose. Letting $\varphi(x) = a = \text{const}$, the so-called hyperellipse of equal probability can be formed. From (2.14) it is easily seen that the points pertaining to a particular hyperellipse (for a fixed constant a_0) are determined by the equation

$$(x-\xi)^T C^{-1}(x-\xi) = \gamma^2 \tag{2.15}$$

where

$$\gamma^2 = -2\ln a_o \sqrt{(2\pi)^n \det C} \tag{2.16}$$

To be more specific we shall consider the hyperellipse which is determined in the case where each variable x_i is allowed to belong to the interval $\xi_i + [-3\sigma_i, 3\sigma_i]$ (here σ_i^2 is the variance of the normally distributed component x_i). Then, it follows directly from (2.15) that in this instance $\gamma = 3$. Indeed, choosing $x_1 = \zeta_1 + 3\sigma_1$ and $x_i = \xi_i$ for all $i \neq 1$ we have

$$\left(\frac{x_1 - \xi_1}{\sigma_1}\right)^2 = \gamma^2$$

whence

$$\left(\frac{3\sigma_1}{\sigma_1}\right)^2 = \gamma^2$$

and $\gamma = 3$.

Let H denote the set of points $x \in R^n$ enclosed by the hyperellipse (2.15) when $\gamma = 3$. Clearly, H is determined by the inequality

$$(x-\xi)^T C^{-1}(x-\xi) \le \gamma^2 \tag{2.17}$$

Now we are in a position to formulate the probabilistic statement of the tolerance problem.

Problem 2.2. Given the multivariate nonlinear function $y = f(x)$, find the range of y when $x \in H$.

Similarly to the deterministic statement this problem resolves to two global optimization problems:

$$f_L^* = \min_{x \in H} f(x) \tag{2.18}$$

and

$$f_U^* = \max_{x \in H} f(x) \tag{2.19}$$

where f_L^* and f_U^* are the lower and upper endpoint, respectively, of the range for y. As we already know, problem (2.19) can be transformed into an equivalent minimization problem

$$f_U^* = -\min_{x \in H}(-f(x)) \tag{2.20}$$

Since H is determined by (2.17) it is seen that the tolerance analysis problem considered here reduces to solving twice the following inequality constraint minimization problem:

$$f^* = \alpha \min h(x) \tag{2.21a}$$

$$(x-\xi)^T C^{-1}(x-\xi) - \gamma^2 \le 0 \tag{2.21b}$$

where firstly, $\alpha = 1$, $h(x) = f(x)$, and secondly $\alpha = -1$, $h(x) = -f(x)$.

The above problem will be referred to as the basic tolerance analysis problem in probabilistic setting. However, the following more complex situation arises sometimes in practice when the parameters x_i are independent (being normally distributed).

Consider the density distribution for an arbitrary x_i shown in Fig. 2.4. There are cases when part of the components satisfying the restrictions

$$\xi_i - \omega_i < x_i < \omega_i + \zeta_i\,, \quad i = \overline{1,n} \tag{2.22a}$$

$$\frac{\omega_i}{\sigma_i} = \gamma_1 = const.\,, \quad i = \overline{1,n} \tag{2.22b}$$

are chosen to be used in high-precision devices. The remaining components satisfying the inequalities

$$\begin{aligned} \xi_i - 3\sigma_i \le x_i \le \xi_i - \omega_i, \\ \xi_i + \omega_i \le x_i \le \xi_i + 3\sigma_i \end{aligned} \tag{2.23}$$

are used separately for lower precision applications. The tolerance analysis of the former class of circuits can be once again treated using the basic problem formulation (2.21) in which C is obviously a diagonal matrix (because of variables independence) while γ should be (as is easy to verify) replaced by γ_1 from (2.22b).

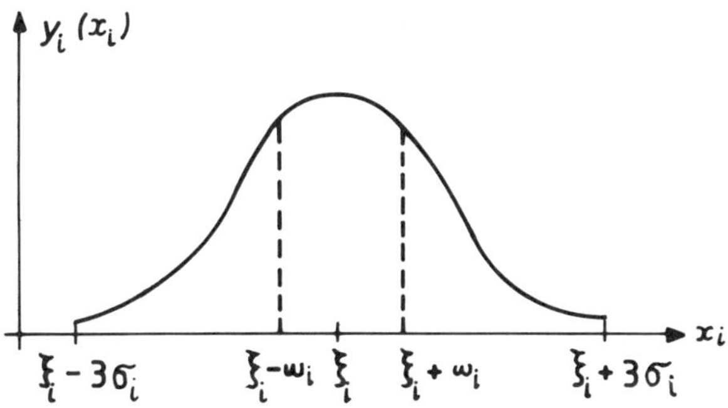

Fig. 2.4. Density distribution of a parameter x_i.

The tolerance analysis of the latter class of circuits for which inequalities (2.23) must hold can be carried out in the following manner. Conditions (2.23) may be replaced equivalently by the following inequalities

$$\xi_i - 3\sigma_i \le x_i \le \xi_i + 3\sigma_i \tag{2.24}$$

$$\xi_i - \omega_i \le x_i \le \xi_i + \omega_i \tag{2.25}$$

It is seen that now the set H' of admissible points is the set of points enclosed between the hyperellipse defined by (2.15) and the hyperellipse defined by

$$(x - \xi)^T C^{-1} (x - \xi) = \gamma_1 \tag{2.26}$$

The tolerance analysis problem considered consists (similarly to the basic problem statement) in finding the range of the given nonlinear function $y = f(x)$ when $x \in H'$. Based on the previous developement this problem may be reduced to solving twice the following inequality constraint minimization problem:

$$f^* = \alpha \min h(x) \tag{2.27a}$$

$$(x-\xi)^T C^{-1}(x-\xi) - 9 \le 0 \tag{2.27b}$$

$$-(x-\xi)^T C^{-1}(x-\xi) - \gamma_1^2 \le 0 \tag{2.27c}$$

(C is a diagonal matrix while firstly $\alpha = 1$, $h = f$ and secondly $\alpha = -1$, $h = -f$).

Problems (2.21), (2.27) can be solved (at least for tolerance problems of small size) using the method from section 1.5.2 of Chapter 1. This topic will be dealt with in detail in section 2.5 of the present chapter.

It should be stressed that the basic tolerance analysis problem in probabilistic setting related to high-precision devices can be solved approximately in the deterministic framework of the worst-case tolerance analysis formulated in section 2.1.1. In this case inequalities (2.22a) hold and, as is seen from Fig. 2.4, the actual normal distribution $\varphi_i(x_i)$ for $x_i \in [\xi_i - \omega_i, \xi_i + \omega_i]$ can be reasonably well approximated by a uniform distribution with

$$P_i(\xi_i - \omega_i \le x_i \le \xi_i + \omega_i) = \int_{\xi_i-\omega_i}^{\xi_i+\omega_i} \varphi_i(x_i)dx_i / (2\omega_i)$$

Therefore, the approximate solution of the probabilistic tolerance problem considered can be found as the exact solution of the following worst-case problem.

Problem 2.3. Determine the range of $f(x)$ over the box defined by $x_i \in [\xi_i - \omega_i, \xi_i + \omega_i]$, $i = \overline{1, n}$.

In the following two sections we shall be concerned with the exact solution of the worst-case tolerance problem as formulated in subsection 2.1.1.

2.2. MODIFIED MEAN-VALUE INTERVAL FORMS

In this section we shall consider various recent forms of interval extensions that will be used in implementing the interval methods for tolerance analysis to be described in sections 2.3, 2.4.

2.2.1. Mean-value interval forms

Let $f: X \subset R^n \to R$ be a multivariate function with continuous first-order derivatives ($f \in C^1$) defined in the interval vector (box) X.

In section 1.2.5, the mean-value form (1.44) was introduced as a possible interval extension of $f(x)$ in X. We rewrite (1.44) as

$$F_{MV}(X) = f(m) + \sum_{i=1}^{n} G_i(X)(X_i - m_i) \tag{2.28}$$

where X is an n-dimensional interval vector with components $X_i = [\underline{x}_i, \overline{x}_i]$, $i = \overline{1, n}$, m is the centre of X and $G_i(X)$ is the interval extension of the derivative $\delta f/\delta x_i$ of f. Recall that the mean-value form (2.28) is inclusion monotonic if all the functions $G_i(X)$ are inclusion monotonic, i.e. if $X \subseteq Y$ then $G_i(X) \subseteq Y_i(X)$, $i = \overline{1, n}$ implies $F_{MV}(X) \subseteq F_{MV}(Y)$.

Another mean-value interval extension called the monotonicity test form is suggested in [2]:

$$F_{MT}(X) = [f(u), f(v)] + \sum_{i \in S} G_i(X)(X_i - m_i) \tag{2.29}$$

where S is the set of integers i such that $G_i(X)$ properly contains zero and

$$(u_i\ ,\ v_i) = \begin{cases} (\underline{x}_i,\ \overline{x}_i) & if\ \ \underline{G_i(X)} \geq 0 \quad (2.30a) \\ (\overline{x}_i,\ \underline{x}_i) & if\ \ \overline{G_i(X)} \leq 0 \quad (2.30b) \\ (m_i,\ m_i) & if\ \ i \in S \quad (2.30c) \end{cases}$$

This form provides a better interval extension of $f(x)$ than (2.28). It has been used in [21] in a method designed to solve the worst-case tolerance analysis problem for linear electric circuits.

In the above mean-value forms (2.28) and (2.29) the interval extensions $G_i(X)$ are defined as follows:

$$G_i(X) = G_i(X_1, \ldots, X_n) \tag{2.31}$$

that is, in the general case G_i is dependent on all the intervals X_i, $i = \overline{1, n}$.

Another way of improving the basic mean-value form (2.28) is introduced in [30] where the interval extensions $G_i(X)$ are computed in the following manner:

$$G_i(\tilde{X}_i, \tilde{m}_i) = G_i(X_i, \ldots, X_i, m_{i+1}, \ldots, m_n) \tag{2.32}$$

and

$$\tilde{X}_i = (X_1, \ldots, X_i)\ ,\ \ \tilde{m}_i = (m_{i+1}, \ldots, m_n)$$

Since in (2.32) part of the arguments m_j ($j = i+1, \ldots, n$) are real numbers rather than intervals X_j and G_i are assumed inclusion monotonic, the inclusion $G_i(\tilde{X}_i, \tilde{m}_i) \subseteq G_i(X)$ holds. Therefore, the interval extension

$$F_{MV}(X,m) = f(m) + \sum_{i=1}^{n} G_i(\tilde{X}_i, \tilde{m}_i)(X_i - m_i) \tag{2.33}$$

is, in general, narrower than that computed by formula (2.28). It should, however, be stressed that (with the exception of the special case $i = n$ where $G_n(\tilde{X}_n,\tilde{m}_n) = G_n(X_1, \ldots, X_n)$) the monotonicity tests (2.30a) and (2.30b) are not applicable for the form (2.33).

In all of the above mean-value forms (2.28), (2.29) and (2.33) the midpoint vector m of X is used to compute the corresponding interval extension.

In the following two sections, two modifications of the mean-value forms (2.29) and (2.33) are suggested which generally result in narrower (but never wider) interval extensions. In contrast to (2.29) and (2.33) the modifications suggested appeal to two new points distinct from m to compute the interval extension.

2.2.2. Modified MT-form

Let $f \in C^1$, $f\colon X \subset R^n \rightarrow R$ and $X \in I(R^n)$. Assume that the interval extensions $G_i(X)$ of $\delta f/\delta x_i$ are inclusion monotonic. By analogy with (2.28) we write the interval extension of f in the form

$$F(X,x) = f(x) + \sum_{i=1}^{n} G_i(X)(X_i - x_i) \tag{2.34}$$

where $x = (x_1, \ldots, x_n) \in X$ is an arbitrarry fixed point. For simplicity of notation let the left endpoint $\underline{F(X)}$ of the interval $F(X)$ be denoted by $\inf F(X)$.

Definition 2.1. The expression

$$F(X,x^L) = f(x^L) + \sum_{i=1}^{n} G_i(X)(X_i - x_i^L) \tag{2.35}$$

with $x^L = (x_1^L, \ldots, x_n^L)$ will be called an optimal minoring form of (2.34) if

$$\inf F(X,x) \leq \inf F(X,x^L) \tag{2.36}$$

for any $x \in X$. The point x^L will be called lower pole of the form (2.34).

As seen from Definition 2.1 the lower pole x^L secures in X the global maximum of the lower bound of the form (2.9).

Let $G_i(X) = [a_i,b_i]$, $i = \overline{1, n}$. We will prove the following theorem.

Theorem 2.1. The point x^L is the lower pole of the form (2.34) if its components are determined as follows:

$$x_i^L = \begin{cases} \underline{x}_i & \text{if} \quad a_i \geq 0 & (2.37a) \\ \overline{x}_i & \text{if} \quad b_i \leq 0 & (2.37b) \\ (b_i \underline{x}_i - a_i \overline{x}_i)/(b_i - a_i) & \text{if} \quad i \in S & (2.37c) \end{cases}$$

where $i \in S$ if $0 \in \text{int } G_i(X)$ (i.e. $a_i < 0 < b_i$).Moreover,

$$x_i^L \in \text{int } X_i \quad \text{for} \quad i \in S \tag{2.38}$$

and

$$\begin{aligned} \inf F(X,x^L) &= f(x^L) + \sum_{i \in S} b_i(\underline{x_i} - x_i^L) \\ &= f(x^L) + \sum_{i \in S} a_i(\overline{x_i} - x_i^L) \end{aligned} \tag{2.39}$$

P r o o f. In order to prove that x^L given by (2.37) is the lower pole of the form (2.34) it suffices to prove inequality (2.36).

Using the rules for interval multiplication and (2.37) it can be easily shown that for any $x \in X$

$$\inf [a_i , b_i](X_i - x_i) \leq \inf [a_i , b_i](X_i - x_i^L) = 0 \quad , \quad i \notin S$$

$$\inf [a_i , b_i](X_i - x_i) \leq \inf [a_i , b_i](X_i - x_i^L) =$$

$$a_i(\overline{x_i} - x_i^L) = b_i(\underline{x_i} - x_i^L) = \frac{a_i b_i}{b_i - a_i} \leq 0 \quad , \quad i \in S \tag{2.40}$$

Furthermore, we introduce the subsets I_1 and I_2 of the index set $I = \overline{1, n}$ for which $x_i > x_i^L$ and $x_i \leq x_i^L$, respectively. Then based on (2.40) it is easy to verify that

$$\inf [a_i, b_i](X_i - x_i) = b_i(\underline{x_i} - x_i) \quad , \quad i \in I_1$$

$$\inf [a_i, b_i](X_i - x_i) = a_i(\overline{x_i} - x_i) \quad , \quad i \in I_2$$

Thus

$$\inf F(X,x) = f(x) + \sum_{i \in I1} b_i(\underline{x_i} - x_i) + \sum_{i \in I2} a_i(\overline{x_i} - x_i) \tag{2.41}$$

On the other hand, owing to the inclusion (1.32)

$$f(x) \in f(x^L) + \sum_{i \in I1} [a_i, b_i] (x_i - x_i^L) + \sum_{i \in I2} [a_i, b_i] (x_i - x_i^L)$$

hence

$$f(x) \le f(x^L) + \sum_{i \in I1} b_i (x_i - x_i^L) + \sum_{i \in I2} a_i (x_i - x_i^L) \tag{2.42}$$

On replacing (2.42) into (2.41) we get

$$\inf F(X,x) \le f(x^L) + \sum_{i \in I1} b_i (\underline{x_i} - x_i^L) + \sum_{i \in I2} a_i (\overline{x_i} - x_i^L) \tag{2.43}$$

Taking into account (2.41) we obtain from (2.43)

$$\inf F(X,x) \le \inf F(X,x^L)$$

which completes the first assertion of the theorem.

Formula (2.39) follows from (2.43),(2.37) and (2.40).

To prove (2.38) we note that

$$x_i^L = \frac{b_i}{b_i - a_i} \underline{x_i} + \frac{-a_i}{b_i - a_i} \overline{x_i} = \alpha_i \underline{x_i} + \beta_i \overline{x_i} \quad , \quad i \in S$$

Hence

$$\alpha_i + \beta_i = 1 \quad , \quad \alpha_i > 0 \quad , \quad \beta_i > 0$$

since $-a_i > 0$ for $i \in S$. Therefore, x_i^L being a strictly convex combination of $\underline{x_i}$ and $\overline{x_i}$ belongs to the interior of X_i.

Let sup $F(X)$ denote the right endpoint of $F(X)$.

Definition 2.2. The expression

$$F(X,x^U) = f(x^U) + \sum_{i=1}^{n} G_i(X)(X_i - x_i^U) \tag{2.44}$$

where $x^U \in X$ will be called optimal majoring form of (2.34) if for any $x \in X$

$$\sup F(X,x) \ge \sup F(X,x^U)$$

The point x^U will be called upper pole of the form (2.34).

It is seen from Definition 2.2 that the upper pole secures in X the global minimum of the upper bound of the form (2.34).

Theorem 2.2. The point x^U is the upper pole of the form (2.34) if its components are determined as follows:

$$x_i^U = \begin{cases} \overline{x}_i & \text{if } a_i \geq 0 \quad (2.45a) \\ \underline{x}_i & \text{if } b_i \leq 0 \quad (2.45b) \\ (b_i\overline{x}_i - a_i\underline{x}_i)/(b_i - a_i) & \text{if } i \in S \quad (2.45c) \end{cases}$$

where $i \in S$ if $a_i < 0 < b_i$. Furthermore, x_i^U

$$x_i^U \in \operatorname{int} X_i \quad \text{for} \quad i \in S$$

and

$$\sup F(X, x^U) = f(x^U) + \sum_{i \in S} b_i(\overline{x}_i - x_i^U) = f(x^U) + \sum_{i \in S} a_i(\underline{x}_i - x_i^U)$$

The proof of this theorem repeats with obvious modifications the proof of Theorem 2.1.

Definition 2.3. The expression

$$F_{MT}(X, x^L, x^U) = F(X, x^L) \cap F(X, x^U) = [\inf F(X, x^L), \sup F(X, x^U)] \tag{2.46}$$

will be called improved MT-form.

The improved MT-form has the following important properties proven in [31].

Theorem 2.3. The width of the improved form (2.46) is, in general, smaller (not larger) than the width of the original MT-form (2.29).

Theorem 2.4. Let $Y \subseteq X$ and $G_i(Y) \subseteq G_i(X)$, $i = \overline{1, n}$. Then

$$F_{MT}(Y, y^L, y^U) \subseteq F_{MT}(X, x^L, x^U)$$

The latter theorem states the inclusion monotonicity property of the form (2.46).

2.2.3. Modified MV-form

The approach adopted in modifying the MT-form will be now applied to the MV-form defined by (2.33). First the formula (2.33) is written as

$$F_{MV}(X,x) = f(x) + \sum_{i=1}^{n} G_i(\tilde{X}_i,\tilde{x}_i)(X_i - x_i) \tag{2.47}$$

$$G_i(\tilde{X}_i,\tilde{x}_i) = G_i(X_1,\dots,X_i,x_{i+1},\dots,x_n) \tag{2.48}$$

where $x \in X$. As it has been done above, optimal minoring and majoring forms and their corresponding poles might be introduced. On account of (2.47) these forms are:

$$F_{MV}(X,x^L) = f(x^L) + \sum_{i=1}^{n} G_i(\tilde{X}_i,\tilde{x}_i)(X_i - x_i^L) \tag{2.49}$$

$$F_{MV}(X,x^U) = f(x^U) + \sum_{i=1}^{n} G_i(\tilde{X}_i,\tilde{x}_i)(X_i - x_i^U) \tag{2.50}$$

It is seen from (2.49), (2.50) and (2.48) that the interval extensions G_i are now dependent on the corresponding components of the poles x^L and x^U of the form (2.47). For this reason, to find these poles, it is necessary to globally solve two complex optimization problems with real (point) variables.

In this section, instead of exactly finding the poles x^L and x^U of the form (2.47), it is suggested to make use of some approximations x' and x'' of x^L and x^U, respectively, which could be computed in a rather efficient way and at the same time would lead to an improvement over the form (2.33). For brevity let $G_i(\tilde{X}_i,\tilde{x}_i) = [a_i(\tilde{x}_i), b_i(\tilde{x}_i)]$. To find the components x_i of the approximation x' the following procedure is suggested.

Procedure 2.1. The components of x' are determined as

$$x_i' = \begin{cases} \underline{x}_i & \text{if } a_i(\tilde{x}_i') \geq 0 \quad (2.51a) \\ \overline{x}_i & \text{if } b_i(\tilde{x}_i') \leq 0 \quad (2.51b) \\ \dfrac{b_i(\tilde{x}_i')\underline{x}_i - a_i(\tilde{x}_i')\overline{x}_i}{b_i(\tilde{x}_i') - a_i(\tilde{x}_i')} & \text{if } i \in S \quad (2.51c) \end{cases}$$

($i \in S$ if $a_i < 0 < b_i$) in the following order. First, the component x_i' is computed and $G_{n-1}(\tilde{X}_{n-1}, x'_n)$ is calculated. Then x'_{n-1} is determined and by using x'_n and x'_{n-1} the next derivative $G_{n-2}(\tilde{X}_{n-2}, x'_{n-1}, x'_n)$ is calculated. This process continues until $i = 1$.

Using the approximations x_i' we choose for a quasi-optimal minoring form the expression:

$$F_{MV}(X,x') = f(x') + \sum_{i=1}^{n} G_i(\tilde{X}_i, \tilde{x}_i')(X_i - x_i') \tag{2.52}$$

Let J_1 and J_2 denote the subsets of the index set $I = \overline{1, n}$ for which $m_i > x_i'$ and $m_i \leq x_i'$, respectively, where m_i is the corresponding component of the midpoint m from (2.33). The following theorem proven in [31] shows that under certain conditions the use of the form $F_{MV}(X, x')$ can improve the lower bound of the form (2.47) in comparison with (2.33).

Theorem 2.5. If for x_i' computed by Procedure 2.1 the following inequalities:

$$a_i(\tilde{x}_i') \geq a_i(\tilde{m}_i), \quad i \in J_1 \tag{2.53a}$$

$$b_i(\tilde{x}_i') \leq b_i(\tilde{m}_i), \quad i \in J_2 \tag{2.53b}$$

are fulfilled, then the inequality

$$\inf F_{MV}(X, m) \leq \inf F_{MV}(X, x') \tag{2.54}$$

holds. Moreover, (2.38) and (2.39) (with x' standing for x^L) are again valid.

The following theorem [31] shows that the form

$$F_{MV}(X,x'') = f(x'') + \sum_{i=1}^{n} G_i(\tilde{X}_i, \tilde{x}_i'')(X_i - x_i'') \tag{2.55}$$

can improve the upper bound of the form (2.47) in comparison with (2.33).

Theorem 2.6. If for the point x'' whose components are determined sequentially in a simillar way as in Procedure 2.1 (using (2.45) rather than (2.51) with x_i^U replaced by x_i'') the following conditions:

$$a_i(\tilde{x}_i'') \geq a_i(\tilde{m}_i), \quad i \in J_i \tag{2.56a}$$

$$b_i(\tilde{x}_i'') \leq b_i(\tilde{m}_i), \quad i \in J_i \tag{2.56b}$$

are fulfilled, then the inequality

$$\sup F_{MV}(X,m) \geq \sup F_{MV}(X,x'')$$

holds.

Definition 2.4. The expression

$$F_{MV}(X,x',x'') = F_{MV}(X,x') \cap F_{MV}(X,x'') = \tag{2.58}$$

$$[\inf F_{MV}(X,x'),\ \sup F_{MV}(X,x'')]$$

will be called modified MV-form.

Similarly to Theorem 2.3 and 2.4 the following two theorems are easily proven.

Theorem 2.7. If the conditions (2.53) and (2.56) are fulfilled, the modified MV-form (2.58) provides, in general, a narrower extension than the MV-form (2.33).

Theorem 2.8. Assume that for $X \supseteq Y$

$$\inf G_i(\tilde{Y}_i, y_i') \geq \inf G_i(\tilde{X}_i, \tilde{x}_i') \tag{2.59}$$

and

$$\sup G_i(\tilde{Y}_i, \tilde{y}_i'') \leq \inf G_i(\tilde{X}_i, \tilde{x}_i''),\quad (i = \overline{1,n}) \tag{2.60}$$

Then

$$F_{MV}(Y,y',y'') \subseteq F_{MV}(X,x',x'') \tag{2.61}$$

Thus if the corresponding interval extensions G_i are inclusion monotonic, the modified form (2.58) is also inclusion monotonic.

The following theorem shows that the modified MV-form (2.62) may be expected to provide the narrowest interval extension of $f(x)$ in X as compared to the previous mean-value forms.

Theorem 2.9. [31]. If the conditions (2.53) and (2.56) are fulfilled, the modified MV-form (2.58) ensures the narrowest enclosure of $f(X)$ among the forms (2.28), (2.29), (2.33), (2.46) and (2.58).

If any of the conditions (2.53) and (2.56) is not fulfilled it may happen that the best result is provided by the form (2.46).

2.3. INTERVAL METHODS FOR TOLERANCE ANALYSIS

In this section several methods for tolerance analysis of linear electrical circuits will be presented which are based on the global optimization formulation of the problem.

(Another class of interval tolerance analysis methods based on an alternative approach will be introduced in Chapter 3.)

Depending on the available information about the derivatives of the function describing the tolerance problem the interval methods to be considered herein can be categorized into three groups:

a) zero-order methods;
b) first-order methods;
c) second-order methods.

2.3.1. Zero-order methods

For these methods it is assumed that the function f: $X^0 \subset R^n \to R$ involved in the global minimization problems (2.4) and (2.6) is only continuous, i.e. $f \in C$. Thus, no derivatives can be used in the methods of this group. (In practice, f may be continuously differentiable in X^0, but the derivation and computation of the partial derivatives $\partial f/\partial x_i$ is assumed to be prohibitively costly).

We present here a zero-order method for solving the global minimization problem (2.7a) with inequality constraints (2.7b) which appeals to Skelboe's algorithm. The basic idea of Skelboe's approach was elucidated in section 1.2.5 for the scalar case where f: $X^0 \subset R \to R$. The reader is strongly urged to go over this material paying special attention to Procedure 1.1. In this section we generalize Procedure 1.1 for the multivariate case where the initial region X^0 is an n-dimensional box. Since this generalized procedure will serve as a basis for the more sophisticated methods from the next sections it will be considered here in detail.

Procedure 2.2. The procedure is designed to solve within a prescribed accuracy the global minimization problem (2.7)

$$f_l^* = \min f(x_1,\ldots,x_n) \tag{2.62a}$$

$$\underline{x}_i^0 \le x_i \le \overline{x}^0, \quad i = \overline{1,n} \tag{2.62b}$$

when $f \in C$. It is based on the following ideas.

1. From the initial box X^0 with sides $X_i^0 = [\underline{x}_i^0, \overline{x}_i^0]$, we generate a list L of subboxes whose union contains the global minimum. The elements in the list are generated in pairs: each pair is the result of a bisection in a single coordinate direction of some previous subbox.

2. We bisect a subbox X in the first direction in which X has maximum width. Let this direction be along the coordinate with index i_0 and let m_{i_0} denote the centre of the side X_{i_0}. Thus, we have generated two new subboxes:

$$X^1 = [X_1,\ldots,X_{i_0-1}, [\underline{x}_{i_0}, m_{i_0}], X_{i_0+1},\ldots,X_n] \tag{2.63a}$$

and

$$X^2 = [X_1,\ldots,X_{i_0-1},[m_{i_0}\ \overline{x}_{i_0}], X_{i_0+1},\ldots,X_n] \tag{2.63b}$$

whose union is equal to X, i.e, $X = X^1 \cup X^2$.

3. For every current pair X^1 and X^2, we compute $\underline{F(X^1)}$, $\underline{F(X^2)}$, $f(m(X^1))$ and $f(m(X^2))$.

4. Let $\overline{f}$ be the currently smallest value of f found so far among $f(m(X^1))$ and $f(m(X^2))$.

5. We do not list a newly generated box X at all if $\underline{F(X)} > \overline{f}$ since in that case X cannot contain the global minimum. This is called the midpoint test.

6. Every admissible subbox (not deleted by the midpoint test) is entered into L in order of increasing lower bounds $\underline{F(X)}$ so that the first (top) element in the list L always corresponds to the least current lower bound on the global minimum f_L^*.

The procedure for bounding f_L^* includes the following steps.

Step 1. Choose an accuracy $\varepsilon > 0$.

Step 2. Set $X = X^0$.

Step 3. Initially, the list L is empty.

Step 4. Bisect X into X^1 and X^2 using (2.63).

Step 5. Let $m^1 = m(X^1)$ and $m^2 = m(X^2)$. Compute $f^1 = f(m^1)$ and $f^2 = f(m^2)$ and update the upper bound $\overline{f}$ on the global minimum f_L^* that is, if $f^1 < \overline{f}$ or $\overline{f}^2 < f$, set $\overline{f} = f^1$ or $\overline{f} = f^2$, respectively.

Step 6. Compute $\underline{F(X^1)}$ and $\underline{F(X^2)}$.

Step 7. Apply the midpoint test to X^1 and X^2.

Step 8. Enter the admissible subboxes X^1 and X^2 into L in the proper order.

Step 9. Retrieve the top element X' from L (with the lowest $\underline{F(X)}$). Rename X' as X and remove X' from L.

Step 10. Set $b = \underline{F(X)}$.

Step 11. If $\overline{f} - b > \varepsilon$, return to step 4. Otherwise go to the next step.

Step 12. Terminate. The global minimum f_L^* has been bounded by the interval $[b, \overline{f}]$ that is,

$$f_L^* \in [b,\overline{f}]$$

If the above procedure is applied to $(-f)$, then, upon termination, f_U^* is bounded by the corresponding interval $[-\overline{f}, -b]$.

Clearly, when $\varepsilon > 0$ Procedure 2.2 will converge to bounds on $f_L^*(f_U^*)$ in a finite number of computational steps provided the list L is large enough to store all the admissible subboxes generated. It should, however, be pointed out that the convergence rate of this zero-order method is relatively slow and the list L may become prohibitively long.

To improve the numerical efficiency of the interval tolerance analysis methods, in the next section we incorporate the interval forms from section 2.2 into Procedure 2.2 to obtain first-order methods for tolerance analysis.

2.3.2. First-order methods

We now assume that $f \in C^1$ (f is continuously differentiable in X^0).

The methods of this group are, again, based on the ideas used in Procedure 2.2. However, numerous additional possibilities arise owing to the fact that $f \in C^1$.

First, we can use one of the mean-value forms presented in section 2.2 in computing $\underline{F(X)}$ for the current box X which would result in an improved convergence rate of Procedure 2.2. Second, we can introduce some of the techniques mentioned in section 1.5 and designed to delete subboxes or parts of boxes in which the global minimum cannot occur.

Monotonicity test

This test allows us to make use of the monotonicity of f with respect to some variables.

Suppose, for example, that the ith component $G_i(X)$ of the interval extension of the gradient $g(x)$ of $f(x)$ is positive, i.e.:

$$G_i(X) > 0 \tag{2.64}$$

Due to the basic property (1.32), it follows from (2.64) that

$$g_i(x) > 0 \quad \text{for} \quad \forall x \in X$$

Hence, $f(x)$ cannot have a stationary point in X. If X does not contain points of the boundary of the initial box X^0 we can delete the whole current box X.

If $G_i(X)$ is only non-negative or X contains boundary points of X^0 the monotonicity test is less effective. If

$$G_i(X) \geq 0$$

we cannot delete all of X. Indeed, the smallest value of $f(x)$ in X may occur, in this case, for $x_i = \underline{x}_i$ and we have to retain all points with $x_j \in X_j$ $j \neq i$ and $x_i = \underline{x}_i$. Thus, we have only managed to reduce the demensionality of X by one.

Also, we have to retain all the points $x = (x_1, \ldots, x_n)$ with $x_i = \underline{x}_i$ even when (2.64) holds if $\underline{x}_i$ belongs to the boundary of X^0 since the global minimum in X may occur on some facet of X^0.

Similar results occur if $G_i(X) \leq 0$.

To use the monotonicity test we evaluate $G_i(X_1, \ldots, X_n)$ for $i = \overline{1, n}$ and reduce the dimensionality of X for any value of i for which $G_i(X) \leq 0$ or $G_i(X) \geq 0$. Of course, we delete all of X if possible.

Sometimes, the monotonicity test may reduce X in every direction. If so, only a single point, say $\tilde{x}$, remains. In this case, we evaluate $f(\tilde{x})$. If $f(\tilde{x}) > \bar{f}$ we can eliminate $\tilde{x}$ too. If $f(\tilde{x}) \le \bar{f}$, we reset $\bar{f}$ equal to $f(\tilde{x})$ and store $\tilde{x}$ for future reference.

Using the bound $\bar{f}$

Based on the midpoint test from Procedure 2.2 we can delete a whole box X if $\underline{F(X)} > \bar{f}$. When the midpoint test is not applicable we still try to make the most of the current bound $\bar{f}$ on f_L^*.

Taking into account that $f \in C^1$ the bound $\bar{f}$ can be used in an attempt to reduce the size of the current subbox X. Indeed, since $\bar{f}$ is an upper bound on f_L^*, we can delete points $y' \in X$ for which

$$f(y') > \bar{f} \tag{2.65}$$

Actually, we would like to retain the complementary set S of points y for which (2.65) is not satisfied. However, S may have a very complex structure; therefore we will only find an interval enclosure Y of S. To do this, we first expand $f(y)$ about a point $x \in X$ in the form (usually $x = m(X)$)

$$f(y) = f(X) + (y-x)^T g(\xi)$$

Since $\xi \in X$, $g(x)$ can be replaced by its interval extension $G(X)$. Thus, we can use the interval inequality

$$f(x) + (y - x)^T G(X) \le \bar{f} \tag{2.66}$$

to find an interval enclosure Y of the set of points y for which $f(y) \le \bar{f}$.

Denote $e = \bar{f} - f(x)$ and $\tilde{y} = y - x$. Then (2.66) becomes

$$\tilde{y}^T G(X) \le e \tag{2.67}$$

We first try to reduce X in the x_1 direction. With this in mind, we first rewrite (2.67) as

$$G_1(X)\tilde{y}_1 + \sum_{i=2}^{n} G_i(X)(y_i - x_i) \le e \tag{2.68}$$

and replace y_i by X_i to obtain

$$G_1(X)\tilde{y}_1 + \sum_{i=2}^{n} G_i(X_i - x_i) \le e \tag{2.69}$$

Now we solve (2.69) for $\tilde{y}_1$ as described below. Denote

$$A = \sum_{i=2}^{n} G_i(X)(X_i - x_i) - e = [a_1, a_2]$$

$$B = G_1(X) = [b_1, b_2]\ , \quad t = \tilde{y}_1$$

Then the solution set T of the inequality

$$A + Bt \le 0 \tag{2.70}$$

is determined as follows [8]:

$$T = \begin{cases} [-a_1 / b_2, \infty] & \text{if } a_1 \le 0\ ,\ b_2 < 0 \\ [-a_1 / b_1, \infty] & \text{if } a_1 > 0\ ,\ b_2 \le 0 \\ [-\infty, \infty] & \text{if } a_1 \le 0\ ,\ b_1 \le 0 \le b_2 \\ [-\infty, -a_1 / b_2] \cup [-a_1 / b_1, \infty] & \text{if } a_1 > 0\ ,\ b_1 < 0 < b_2 \\ [-\infty, -a_1 / b_1] & \text{if } a_1 \le 0\ ,\ b_1 > 0 \\ [-\infty, -a_1 / b_2] & \text{if } a_1 > 0\ ,\ b_1 \ge 0 \\ \text{empty set} & \text{if } a_1 > 0\ ,\ b_1 = b_2 = 0 \end{cases} \tag{2.71}$$

Recall that $\tilde{y}_1 = y_1 - x_1$. Now the set Y_1 of points y that satisfy (2.68) is $Y_1 = x_1 + T$. Since we are only interested in points with $y_1 \in X_1$ we compute Y_1 as

$$Y_1 = (x_1 + T) \cap X_1 \tag{2.72}$$

Thus, although T may be unbounded, the intersection is bounded. As seen from (2.71) and (2.72) the resulting set Y_1 may consist of one or two intervals.

For the sake of argument, suppose Y_1 is a single interval. We can then try to reduce X_2 the same way we (hopefully) reduced X_1 to get Y_1. We again rewrite (2.69) using Y_1 rather than X_1:

$$G_2(X)\tilde{y}_2 + G_1(X)(Y_1 - x_1) + \sum_{i=3}^{n} G_i(X)(X_i - x_i) - e \le 0 \tag{2.73}$$

Based on (2.70) and (2.71) we solve (2.73) for y_2 to obtain Y_2 as

$$Y_2 = (x_2 + T) \cap X_2$$

The above computation process continues in a similar way until all n components Y_i of Y are determined.

Now consider the case where Y_1 consists of two disjoint subintervals Y'_1 and Y''_1. Let the gap between Y'_1 and Y''_1 be denoted by Y_1^g. The best policy would be to delete Y_1^g and to recompute (2.73) twice with Y'_1 and Y''_1 separately when trying to reduce X_2. Note that this would involve reevaluating $G_1(X)$ for the new components Y'_2 and Y''_2. Since each of the resulting intervals Y'_2 and Y''_2 may again consist of two subintervals such an optimal approach will, however, result after several steps (with $i > 2$) in too many subintervals in each coordinate direction to be treated separately. Hence the number of newly generated boxes to handle separately may grow very fast (up to 2^n at each iteration). In an attempt to keep the computational scheme as simple as possible the following simpler strategy is usually applies in practice [8].

Procedure 2.3. Whenever we encounter a set Y_i consisting of two disjoint subintervals Y'_i and Y''_i we first store the gap Y_i^g for further inspection. Then we use the whole interval

$$Y_i = Y'_i \cup Y_i^g \cup Y''_i$$

when attempting to reduce the subsequent components X_j, $j > i$, of the current box. Moreover, we use the old component $G_i(X)$ rather than the improved extensions G_i computed by means of the narrower component Y_i. At the end when $i = n$, we remove the largest gap $Y_{i_0}^g$ with coordinate index i_o. Thus, only two new subboxes Y' and Y'' have been generated: one whose ith component is Y'_{i_0} and one whose ith component is Y''_{i_0} (the remaining components of the two subboxes are the same for all $i \neq i$).

It should be noted that the efficiency of the approach described to reduce X decreases for problems with higher dimensionality n. This assertion follows from the fact that the width of A from (2.70) tends to grow with n increasing.

2.3.3. Iterative algorithms

Based on the interval mean-value forms from section 2.2 and the techniques from subsections 2.3.1 and 2.3.2 several algorithms will be considered here which solve iteratively (2.4) (or equivalently (2.6)) when $f \in C^1$.

First, we present the following basic algorithm which is common to all interval mean-values forms section 2.2.

Basic algorithm

The initial box X^o is iteratively subdivided into subboxes which are entered into a list L. At each iteration of the algorithm, the subbox X with the lowest $\underline{F(X)}$ is extracted from L and the following steps are done.

1). Let $x \in X$ (initially $X = X^0$) be a fixed point in X. Compute $f(x)$ and put $\overline{f} = f(x)$ at the first iteration. For the next iterations do not update $\overline{f}$ unless $f(x) < \overline{f}$.

2). Compute $\underline{F(X)}$ using one of the forms (2.28), (2.29), (2.33), (2.35) or (2.52), respectively. If $\underline{F(X)} > \overline{f}$ the region X is deleted (not entered into the list L) since it only contains such points $z \in X$ where $f(z) > \bar{f}$; go to Step 9. Otherwise, proceed to next step.

3). If $\bar{f}$ - $\underline{F(X)} \leq \varepsilon$ where ε is a prescribed accuracy, the algorithm is terminated; otherwise go to the next step.

Remark 2.1. In order to guarantee the accuracy ε (accounting for the roundoff errors in representing the point x and in evaluating $f(x)$) the value of $f(x)$ should be computed using machine interval arithmetic as a corresponding interval $[f_L(x), f_R(x)]$. The left endpoint $f_L(x)$ of that interval is then used for $f(x)$ in computing $\underline{F(X)}$ from step 2; the right endpoint $f_R(x)$ is, on the other hand, used to update the current upper bound $\bar{f}$ on the global minimum $f_L{}^*$. This, however, was not implemented in the present version of the algorithm since the accuracy used in the examples considered below was relatively low and therefore the roundoff effect was negligible.

4). In this step an attempt is made to reduce the size of the current box X using the inequality (2.66) and the subsequent formulae through (2.73) At the end of this step, a new set $Y \subseteq X$ with components Y_i is obtained.

5). If Y_i is a single interval for all $i = \overline{1, n}$, then the set Y is also a single box. Otherwise we apply Procedure 2.3 to generate only two new boxes Y' and Y''.

As a result only one new box Y or two new boxes Y' and Y'' can be generated after the attempt to reduce the size of the current box X.

Remark 2.2. Obviously, a more effective scheme for reducing X is possible when several intervals Y_i^g appear. For example, the second largest interval among Y_i^g can also be retained which will result in the creation of 4 subboxes.

6). If Y is not smaller than X (or the reduction is negligible (e.g. $w(Y) \geq 0.9\, w(X)$) the current box X (Y respectively) is split into two subboxes X' and X'' in the direction of its widest component X_k at its midpoint m_k.

7). The new box Y (or boxes Y' and Y'') generated in Step 5 is (are) renamed as X (X' and X''). Using X (or X' and X'') we compute $f(x)$ and update $\bar{f}$ if $f(x) < \bar{f}$.

8). Repeat Step 2 for the new box (boxes).

9). A current box X with the lowest $\underline{F(X)}$ is chosen from L and the algorithm continues from Step 3.

To compare the relative numerical efficiency of the interval forms of section 2.2 the following five versions of the basic algorithm will be considered [31].

A1). In this algorithm $x = m = m(X)$ and inf $F(X)$ is computed by the interval mean-value form (2.28).

A2). In this algorithm, the monotonicity test form (2.29) is used as an interval extension $F(X)$. Now x is determined by (2.30) and inf $F(X)$ is computed by (2.29). Additionally, the following simple monotonicity test is applied: if $G_i(X) > 0$ or $G_i(X) < 0$ is valid at least once and X does not contain points belonging to the boundary of X^o, then X is deleted.

A3). Here $x = m$ and inf $F(X)$ is computed by (2.33). Recall that the monotonicity test is not applicable for the interval form (2.33).

A4). In this algorithm, use is made of the optimal minoring form (2.35). The point x is now the optimal lower pole x^L calculated by (2.37) while inf $F(X)$ is evaluated by (2.39). The monotonicity test from A2 is also applied.

A5). Here inf $F(X)$ is evaluated using modified MV-form (2.56) and is computed according to Procedure 2.1.

The above algorithms have been tested on the following examples.

***Example* 2.7.** The problem to be solved is (2.62) with the so-called "three-hump" function

$$f(x_1,x_2) = 2x_1^2 - 1.06x_1^4 + (1/6)x_1^6 - x_1x_2 + x_2^2$$

which has two local minimum at approximately (±1.75, ±0.87), two saddle points at approximately (±1.07, ±0.535) and one global minimum at (0,0). The initial box X^o with components $X_1^0 = X_2^0 = [-2, 4]$ contains all these points. The results are given in Table 2.1 where N_i stands for the number of iterations needed to reach the global minimum f_L* within the desired accuracy ε (each iteration comprising steps 3 to 9 from the respective modified version of the basic algorithm).

Table 2.1

Algorithm	A1	A2	A3	A4	A5		
ε	10^{-3}	10^{-3}	10^{-3}	10^{-3}	10^{-3}	10^{-5}	10^{-6}
N_i	216	136	236	81	73	94	104

***Example* 2.8.** For this example

$$f(x) = f_1(x)f_2(x_2)f_3(x_3)f_4(x_4)f_5(x_5) \tag{2.74}$$

where

$$\begin{aligned} f_1(x_1) &= 0.01x_1(x_1 + 13)(x_1 - 15) \\ f_2(x_2) &= 0.01(x_2 + 15)(x_2 + 1)(x_2 - 8) \\ f_3(x_3) &= 0.01(x_3 + 9)(x_3 - 2)x_3 - 9) \\ f_4(x_4) &= 0.01(x_4 + 11)(x_4 + 5)(x_4 - 9) \\ f_5(x_5) &= 0.01(x_5 + 9)(x_5 - 9)(x_5 - 10) \end{aligned}$$

and X^0 is defined by $X_1^0 = [8, 9]$, $X_2^0 = [-10, -9]$, $X_3^0 = [-5, -4]$, $X_4^0 = [3, 4]$, $X_5^0 = [-3, -2]$. The problem is to bound the range $f(X^0)$ of $f(x)$ in X^0. First, we solved (2.4).

Table 2.2

Algorithm	A1	A2	A3	A4	A5
N_i	289	49	216	25	37
l_m	246	40	165	21	28
$\underline{F}_L^*$	23067.380	23067.370	23067.380	23067.370	23067.370

The results given in Table 2.2 are obtained for $\varepsilon = 10^{-3}$; therein l_m stands for the maximum length of the list L reached during computation and $\underline{F}_L^*$ denotes the lower bound of f_L^*.

It should be mentioned that for the some example the algorithm of [18] has yields $\underline{F}_L^* = 22110.018$.

To find an upper bound $\overline{F}_U^*$ of $f(X^0)$ we solve problem (2.6) with $\varepsilon = 10^{-2}$. The data corresponding to algorithms A4 and A5 are given in Table 2.3; the other algorithms are not given in the table since their convergence was much slower.

It is worthwhile nothing that for the same example the algorithm from [18] has given $\overline{F}_U^* = 25426.918$.

Thus, the range $f(X^0)$ of the function (2.74) in X^0 is bounded by the interval [23067.370, 24416.040].

Table 2.3

Algorithm	A4	A5
N_i	1944	1745
l_m	1143	555
$\overline{F}_U^*$	24416.040	24416.040

Example **2.9.** We take up the example of a second-order bandpass active RC-filter (Fig. 2.5) considered in [21].

The nominal design has a center frequency $\omega = 1$ and $Q = 10$. The corresponding resistor values are shown in the diagram of Fig. 2.5, and the remaining components are:

$$C_1 = 1\mu F,\ C_2 = 1\mu F,\ \mu = 2.9.$$

The voltage transfer function is

$$H(s) = \frac{2\mu C_2 s}{2C_1C_2s^2 + (C_1 + 5C_2 - 2\mu C_2)s + 2}$$

where the resistors have been assigned numerical values.

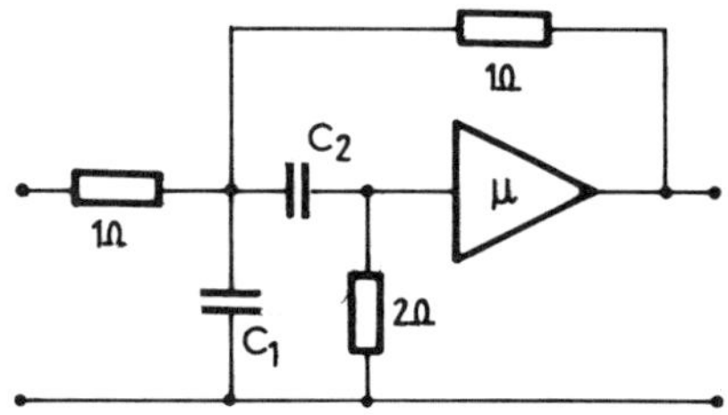

Fig. 2.5. Second-order bandpass active *RC* filter.

A tolerance analysis of the amplitude $|H(j\omega)|$ was performed, and in addition to tolerances on C_1, C_2 and μ, a temperature dependence was considered. A simple linear model was used, but higher order temperature coefficients could have been used without any complications. The models are as follows:

$$C_1 = C_{1o}(1 + T),\ C_2 = C_{2o}(1 + T),\ \mu = \mu_o(1 + rT).$$

Figure 2.6 shows the result of an analysis performed in [21] with the parameters

$$C_{1o} = C_{2o} = [0.98,\ 1.02],\ \ \mu_o = [2.871,\ 2.929]$$

$$T = [-0.0025,\ 0.0075], \qquad r = 2.$$

The range of the amplitude function is computed at the frequencies $\omega_\nu = 10^{0.1\nu-1}$, $\nu = 0, 1, \ldots, 20$.

In [31], the range of $|H(j\omega)|$ for $\omega = 1$ was determined using the algorithms A1 to A5. In actual computation, first the range of $f(x) = |H(j\omega)|^2$ was found in X^o whose components are given by C_{1o}, C_{2o}, μ_o and T. Then the range limits of $|H(j\omega)|$ were determined by taking the square root of the range limits of $f(x)$. The corresponding results for N_i and l_m in obtaining lower and upper bounds on $|H|$ with $\varepsilon = 10^{-3}$ are given in Tables 2.4 and 2.5.

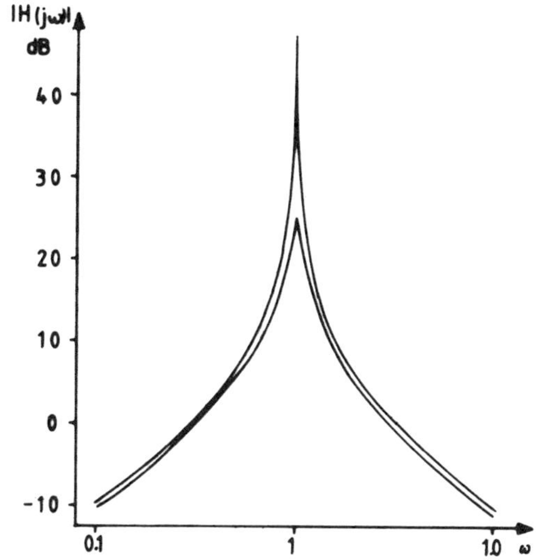

Fig. 2.6. Worst-case frequency response of a bandpass filter.

Table 2.4

Algorithm	A1	A2	A3	A4	A5
N_i	33	3	33	2	1
l_m	28	4	28	2	1

Table 2.5

Algorithm	A1	A2	A3	A4	A5
N_i	126	72	94	44	1
l_m	71	68	44	45	1

It is seen from the data of Tables 2.1 to 2.5 that the implementation of the modified interval forms (2.35) and (2.52) in the basic algorithm results in improved convergence of algorithms A4 and A5 as compared with that of algorithms A1, A2 and A3 using the interval form (2.28), (2.29) and (2.33), respectively.

It should also be mentioned that the first-order method for tolerance analysis as implemented by algorithms A4 and A5 is vastly superior over the zero-method from section 2.3.1.

Example **2.10.** We consider a problem related to the diagnosis of electrical machines by analysing their mechanical vibrations. The functional relation between the vibration velocity V of a particular machine and four basis "input" parameters x_i is given by [86]:

$$V = \sqrt{\frac{1}{2}\sum_{i=1}^{8} V_i^2}$$

where

$$\begin{aligned}
V_i &= 2\pi f_i v_i^* \delta_i \\
f_i &= a_i\,, \qquad\qquad i = 1, 2, 6, 7, 8. \\
f_i &= a_i(b_i + x_i), \quad i = 3, 4, 5. \\
v_1^* &= a_1^*(b_1 + x_1)x_2 \\
v_2^* &= a_2^*(b_2 + x_1)x_3 \\
v_i^* &= c_i(d_i + x_1)x_4, \quad i = 3, 4, 5. \\
v_6^* &= v_3^*\,,\ v_7^* = v_4^*,\ v_5^* = v_8^* \\
\delta_i &= 1/\{[1 - f_i/f_c)^2]^2 + 2.5 \times 10^{-3}(f_i/f_c)2\}^{1/2}
\end{aligned}$$

(a_i, a_i^*, b_i, c_i, d_i and f_c are constants related to the construction of the particular machine studied).

The problem is to find the range of $V(x_1, x_2, x_3, x_4)$ for $x_i \in X_i$, where X_i are given intervals. Using algorithm A5 three cases were solved with fixed intervals

$$X_3 = [0.63, 2.5],\ \ X_4 = [0.5, 1.25]$$

and variable intervals X_1 and X_2:

a) $X_1 = [-3, 3]$, $X_2 = [0.4, 6.3]$
b) $X_1 = [-2.25, 2.25]$, $X_2 = [1, 2.5]$
c) $X_1 = [-2, 2]$, $X_2 = [1.25, 2]$

The corresponding results for the tolerance interval on V are:

a) $V = [1.779, 11.292]$
b) $V = [2.039, 7.205]$
c) $V = [2.976, 6.794]$

The intervals obtained were used for diagnosis of the machine investigated [86].

2.3.4. Second-order methods

In this section it is assumed that the function f from (2.4), (2.6) is twice continuously differentiable in X^o, i.e. $f \in C^2$.

As the minimization problem (2.62) to be solved involves inequality constraints we could use the techniques of section 1.5.2 to elaborate various algorithms for solving the tolerance analysis problem. The most complete algorithm would include the following stages:

(a) test for nonconvexity of f in a current box X which does not contain points of the boundary of X^o;

(b) test for monotonicity of f in X;

(c) transformation of the constrained problem (2.62) into an equivalent problem (1.98) to (1.101);

(d) solution of (1.98) to (1.101) by some version of the Newton method for the corresponding current box;

(e) reduction of X using the upper bound $\overline{f}$ on f_L^*.

In this section we present a (relatively) simple algorithm for solving (2.62) which is based on the algorithm A1 from the previous section and the technique (e) from the above list. This second-order algorithm will be referred to as Algorithm 2.7. In fact, the only difference of the present second-order algorithm A7 from the first-order algorithm A1 concerns Steps 1 and 4 of A1 (as numbered in the basic algorithm).

Since now $f \in C^2$ the set of points y' in a current box X for which $f(y') > \overline{f}$ can be defined as follows [8]. First, we expand f as

$$f(y) = f(x) + (y-x)^T g(x) + \frac{1}{2}(y-x)^T h(x,y,\zeta)(y-x) \qquad (2.75)$$

where $g(x)$ is gradient of $f(x)$ and quantity $h(x,y,\zeta)$ is the Hessian matrix to be defined presently. For reasons related to the use of interval analysis $h(x,y,\zeta)$ will be expressed as a lower triangular matrix instead of a symmetric one so that there are fewer terms in the quadric form (2.75). Thus, the element h_{ij} of h is defined as [8]:

$$h_{ij} = \begin{cases} \delta^2 f/\delta x_i^2 & \text{for} \quad i = j\,,\ i = \overline{1,n} \\ 2\delta^2 f/\delta x_i \delta x_j & \text{for} \quad j < i\,,\ i = \overline{1,n},\ j = \overline{1,i-1} \\ 0 & \text{otherwise} \end{cases}$$

The arguments of h_{ij} depend on i and j. If we expand f sequentially in one of the its variables at a time, we can obtain the following results illustrating the case $n = 3$.

$$h(x,y\xi) = \begin{bmatrix} h_{11}(\xi_{11},x_2,x_3) & 0 & 0 \\ h_{21}(\xi_{21},x_2,x_3) & h_{22}(y_1,\xi_{22},x_3) & 0 \\ h_{31}(\xi_{31},x_2,x_3) & h_{32}(y_1,\xi_{22},x_3) & h_{33}(y_1,y_2,\xi_{33}) \end{bmatrix}$$

Assume $x_i \in X_i$ and $y_i \in X_i$ for $i = \overline{1, n}$. Then from (2.75) $\zeta_{ij} \in X_i$ for each $j = \overline{1, n}$. For general n, the arguments of h_{ij} are

$$(y_1, \ldots, y_{i-1}, \zeta_{ij}, x_{i+1}, \ldots, x_n)$$

Let x be a fixed point in X. Then for any point $y \in X$

$$h(x, y, \xi) \in H(x, X, X)$$

where $H(x, X, X)$ is the interval extension of $h(x, y, \zeta)$ for $y \in X$ and $\zeta \in X$. In the sequel, we shall shorten the notation and use $h(\xi)$ to denote $h(x, y, \xi)$ and $H(X)$ to denote $H(x, X, X)$.

The purpose of the particular Taylor expression is to obtain real (noninterval) quantities for as many arguments of the elements of $H(X)$ as possible. The standard Taylor expansion would have intervals for all elements of $H(X)$ which would lead to poorer results.

In a similar way as in section 2.3.2 we seek to find an interval enclosure Y of the set S of points $y \in X$ for which

$$f(x) + (y-x)^T g(x) + \frac{1}{2}(y-x)^T H(\xi)(y-x) \leq \bar{f}$$

If

$$e = \bar{f} - f(x)$$

and

$$\tilde{y} = y - x$$

then

$$\tilde{y}^T g(x) + \frac{1}{2}\tilde{y}^T H(\xi)\tilde{y} \leq e \tag{2.76}$$

We shall use this relation to reduce X in one dimension at a time to yield a subbox (two subboxes) Y resulting from deleting points $y' \in X$ where $f(y') > \bar{f}$. We shall illustrate the process for $n = 2$. Now (2.76) becomes

$$\tilde{y}_1 g_1(x) + \tilde{y}_2 g_2(x) + \frac{1}{2}[\tilde{y}_1^2 h_{11}(\xi) + \tilde{y}_1\tilde{y}_2 h_{21}(\xi) + \tilde{y}_2 h_{22}(\xi)] \le e \tag{2.77}$$

We first try to reduce X_1. Thus, we solve (2.77) for acceptable values of y_1. After collecting terms in y_1, we replace y_2 by X_2 (in the higher dimensional case we would also replace y_i by X_i for $i = \overline{3, n}$). Since $\xi \in X$ we also replace ξ by X to obtain

$$\tilde{y}_1[g_1 + \frac{1}{2}\tilde{X}_2 h_{21}(X)] + \frac{1}{2}\tilde{y}_1^2 h_{11}(X) + \tilde{X}_2 g_2 + \frac{1}{2}\tilde{y}_1^2 h_{22}(X) - \varepsilon \le 0 \tag{2.78}$$

where $\tilde{X}_2 = X_2 - x_2$. We solve this quadratic for the interval or intervals of points y_1 as described below. Call the resulting set Z_1. Since we are only interested in points $y_1 \in X_1$ we compute the desired set Y_1 as $Y_1 = Z_1 \cap X_1$.

Next, we try to reduce X_2. For simplicity, suppose Y_1 is a single interval. We again rewrite (2.76). This time we replace $\tilde{y}_1$ by Y_1 and, as before, ξ by X. (We could obtain sharper results by replacing some elements of ξ by Y_1 rather than X_1 but this would require reevaluation of the elements of H.) We get

$$\tilde{y}_2[g_2 + \frac{1}{2}\tilde{Y}_1 h_{21}(X)] + \frac{1}{2}\tilde{y}_2^2 h_{22}(X) + \tilde{Y}_1 g_1 + \frac{1}{2}\tilde{Y}_1^2 h_{11}(X) - e \le 0 \tag{2.79}$$

where $\tilde{Y}_1 = Y_1 - x_1$.

If the solution set Y_2 is strictly contained in X_2 we could replace X_2 by Y_2 in (2.76) and solve for a new Y_1. This has not been done in practice. Instead, we start over with the new box Y in place of X as soon as we have tried to reduce all X_i, $i = \overline{1, n}$.

The quadratic equations (2.78) or (2.79) have the general form

$$A + Bt + Ct^2 \le 0 \tag{2.80}$$

where A, B and C are intervals. We seek values of t satisfying (2.80). Let $A = [a_1, a_2]$, $B = [b_1, b_2]$ and $C = [c_1, c_2]$. Compute the discriminants

$$\Delta_1 = b_1^2 - 4a_1c_1$$

$$\Delta_2 = b_2^2 - 4a_1c_1 \tag{2.81}$$

For $i = 1, 2$ calculate

$$R_i^{\pm} = (-b_i \pm \Delta_i)/(2c_1) \tag{2.82}$$

$$S_i^{\pm} = 2a_1/(-b_i \pm \Delta_i^{1/2}) \tag{2.83}$$

The solution set of (2.80) is then determined as follows [8].

For $b_1 > 0$ and $c_1 > 0$:

$$T = \begin{cases} \varnothing\,(\text{emptyset}) & \text{if } \Delta_2 < 0 \\ [R_2^-, S_2^-] & \text{if } a_1 > 0,\ \Delta_2 \geq 0 \\ [R_2^-, S_1^-] & \text{if } a_1 \leq 0 \end{cases} \tag{2.84}$$

For $b_2 <$ and $c_1 > 0$:

$$T = \begin{cases} \varnothing & \text{if } \Delta_1 < 0 \\ [S_1^+, R_1^+] & \text{if } a_1 > 0,\ \Delta_1 \geq 0 \\ [S_2^+, R_1^+] & \text{if } a_1 \leq 0 \end{cases} \tag{2.85}$$

For $b_1 \leq 0 \leq b_2$ and $c_1 > 0$:

$$T = \begin{cases} \varnothing & \text{if } \max(\Delta_1, \Delta_2) < 0 \\ [R_2^-, S_2^-] & \text{if } |b_1| < b_2 \text{ and } \min(\Delta_1, \Delta_2) \leq 0 \leq \max(\Delta_1, \Delta_2) \\ [S_1^+, R_1^+] & \text{if } |b_1| > b_2 \text{ and } \min(\Delta_1, \Delta_2) \leq 0 \leq \max(\Delta_1, \Delta_2) \\ [R_2^-, S_2^-] \cup [S_1^+, R_1^+] & \text{if } a_1 > 0 \text{ and } \min(\Delta_1, \Delta_2) > 0 \\ R_1[R_2^-, R_1^+] & \text{if } a_1 \leq 0 \end{cases} \tag{2.86}$$

For $b_1 > 0$ and $c_1 < 0$:

$$T = \begin{cases} [-\infty, S_2^-] \cup [R_1^-, \infty] & \text{if } a_1 > 0 \\ [-\infty, S_1^-] \cup [R_1^-, \infty] & \text{if } a_1 \leq 0 \leq \Delta_1 \\ [-\infty, \infty] & \text{if } \Delta_1 < 0 \end{cases} \tag{2.87}$$

For $b_2 < 0$ and $c_1 < 0$:

$$T = \begin{cases} [-\infty, R_2^+] \cup [S_1^+, \infty] & \text{if } a_1 > 0 \\ [-\infty, R_2^+] \cup [S_2^+, \infty] & \text{if } a_1 \leq 0 \leq \Delta_2 \\ [-\infty, \infty] & \text{if } \Delta_2 < 0 \end{cases} \tag{2.88}$$

For $b_1 \leq 0 \leq b_2$ and $c_1 < 0$:

$$T = \begin{cases} [\infty,\ S_2^-] \cup [S_1^+,\ \infty] & \text{if} \quad a_1 \geq 0 \\ [-\infty,\ \infty] & \text{if} \quad a_1 < 0 \end{cases} \tag{2.89}$$

The modifications of algorithm A1 needed in the present section's algorithm concern Steps 1 and 4.

Step 1 of the present algorithm includes additionally:

Compute the gradient $g(m)$ and the interval extensions $H_{ij}(X_1, \ldots .X_j, m_{j+1}, \ldots .m_n)$ of the components h_{ij} of the Hessian h.

Step 4 is now implemented using formulae (2.76) to (2.89).

Remark 2.3. It is seen from (2.76) to (2.89) that in some cases T and hence Y_i may consist of two disjoint subintervals. Whenever this occurs we proceed in exactly the same way as explained in Step 5 of the basic algorithm from section 2.3.3.

To test the numerical efficiency of the algorithm A7 as compared to the first-order algorithm A1 from the previous section the following simple example will be considered.

***Example* 2.10.** The function to minimize is the function from Example 2.7. Since the second-order algorithm is designed to handle problems with relatively large tolerances on the input parameters the initial box X^0 was chosen to be defined as in Example 2.7 by $X_1^0 = X_2^0 = [-2, 4]$.

Using the same accuracy $\varepsilon = 10^{-3}$ the second-order algorithm yielded the interval

$$X_L^* = ([-0.01366,\ 0.00480],\ \ [-0.02742,\ 0.02112])$$

for the global solution $x_L^* = 0$ and the interval

$$[\underline{F}_L^*,\ \bar{f}] = [-0.00095,\ 0.00003]$$

for the global minimum f_L^* in 101 iterations (an iteration comprises Steps 3 to 9 of algorithm A1 modified according to the present second-order method version).The corresponding intervals

$$X_L^* = ([-0.90771,\ 0.01563],\ [-0.03125,\ 0.01563])$$

and

$$[F_L^*,\ \bar{f}] = [-0.00073,\ 0.00012]$$

here obtained by A1 in 216 iterations as shown in Table 2.1.

It is felt that the superiority of the second-order method over the first-order method will be enhanced if more sophisticated techniques such as nonconvexity test, monotonicity test and interval Newton method are used. Some such techniques will be presented in section 2.4.4.

2.4. IMPROVEMENT OF THE NUMERICAL EFFICIENCY

In this section, several devices that are designed to improve the numerical efficiency of the interval methods considered in the previous section will be introduced.

2.4.1. Alternative problem statement

In section 2.1.1 the worst-case tolerance analysis problem was formulated as Problem 2.1, namely: find the range $f(X^0)$ of a given multivariate function over X^0. In this section, we shall reformulate the tolerance problem considered as follows.

Problem 2.4. Given a multivariate function f: $X^0 \subset I(R^n) \to R$, check that the range $f(X^0) = [\underline{f}, \overline{f}]$ of f over X_0 is contained in a prescribed interval $Y = [\underline{Y}, \overline{Y}]$, that is,

$$f(X^0) \subseteq Y \tag{2.90}$$

Problem 2.4 is a more realistic formulation of the tolerance problem as compared to Problem 2.1. Indeed, in practice we compute the range $f(X^0)$ in order to verify whether the inclusion (2.90) holds (filter tolerance analysis – for a fixed frequency – provides numerous examples of the above problem).

The inclusion (2.90) is equivalent to the following two inequalities:

$$\underline{f} \geq \underline{Y} \tag{2.91}$$

$$\overline{f} \leq \overline{Y} \tag{2.92}$$

As in solving Problem 2.1 we shall consider here only the verification of inequality (2.91) since the inequality (2.92) can be transformed (multiplying it by -1) to the type of (2.91). Thus, we are led to consider the following equivalent problem.

Problem 2.5. Given a twice continuously differentiable function $f(x)$: $X^0 \subset R^n \to R$ ($f \in C^2$) check whether

$$f(x) \geq \underline{Y} \tag{2.93a}$$

$$x \in X^0 \tag{2.93b}$$

(The requirement $f \in C^2$ is needed to permit the use of second-order interval methods in solving Problem 2.5).

Naturally, Problem 2.5 can be solved by appealing to the global minimization problem (2.4) from section 2.1.2. For convenience the latter problem will be rewritten here (as Problem 2.6).

P r o b l e m 2.6. Find the global minimum f^* of $f(x)$ in X^0.

Indeed, if

$$f^* \geq \underline{Y} \tag{2.94}$$

then, obviously, (2.93) is fulfilled.

Conceptually, Problems 2.5 and 2.6 are identical. However, from the point of view of numerical efficiency the former problem can be, generally, solved in a much more effective manner than the latter one. This possibility results from the fact that to check (2.93) using a first- or second-order interval method one does not need to find the global minimum f^* (except for the cases where $f^* = \underline{Y}$). Indeed, both first- and second-order interval methods generate series of bounds: a lower bound series

$$\underline{f}^0 \leq \underline{f}^1 \leq ,\ldots, \leq \underline{f}^\rho \leq \underline{f}^{\rho+1} \leq ,\ldots, = f^* \tag{2.95}$$

which converges monotonically to f^* from below (here $\underline{f}^\rho = \underline{F}(X^\rho)$ and X^ρ is the current subbox having the lowest $\underline{F}(X)$ among all subboxes X stored in the list L) and an upper bound one

$$\bar{f}^0 \geq \bar{f}^1 \geq ,\ldots, \geq \bar{f}^\rho \geq \bar{f}^{\rho+1} \geq ,\ldots, = f^* \tag{2.96}$$

which converges monotonically to f^* from above (now $\bar{f}^\rho$ is the lowest function value, found up to the current ρth iteration). Obviously, the computation process can be terminated before f^* is reached (within the present accuracy) whenever

$$\underline{f}^\rho \geq \underline{Y} \tag{2.97}$$

for the first time even if

$$\bar{f}^\rho - \underline{f}^\rho > \varepsilon \tag{2.98}$$

Thus, the use of the stopping criterion (2.97) rather than the former one

$$\bar{f}^\rho - \underline{f}^\rho \leq \varepsilon \tag{2.99}$$

will, in general, lead to a reduced number of iterations as compared to that needed to solve Problem 2.6. Moreover, the fulfillment of (2.97) guarantees that Problem 2.5 has a solution and hence, the tolerance analysis requirement (2.91) is satisfied.

If at some current iteration ρ the following inequality

$$\bar{f}^{\rho} < \underline{Y} \tag{2.100}$$

is fulfilled, again there is no point in continuing the computation process since in this case the upper bound series (2.96) tends to a global minimum $f^* < \underline{Y}$. Thus, the fulfillment of condition (2.100) guarantees that Problem 2.5 has no solution or, equivalently, the tolerance analysis requirement (2.91) is not satisfied.

Formulation of the tolerance analysis problem as Problem 2.5 offers yet another computational advantage over Problem 2.6 when $f(x)$ is the modulus of the frequency response of an a.c. circuit. In this case

$$f(x) = |H(j\omega)| = \frac{|N(j\omega)|}{|D(j\omega)|} \tag{2.101}$$

where $N(j\omega)$ and $D(j\omega)$ are the numerator and denominator of $H(j\omega)$ while x is the input parameter vector and may, in the general case, include the frequency ω. Thus,

$$f(x) = \frac{\sqrt{a_1^2(x) + a_2^2(x)}}{\sqrt{b_1^2 + b_2^2(x)}} \tag{2.102}$$

where a_1, b_1 and a_2, b_2 are the real and imaginary parts of N and D, respectively. If the corresponding problem (2.7), (2.102) is now solved using first- or second-order method we have to find the first and second derivatives of $f(x)$ as defined by (2.102) with respect to each component x_i of x in explicit form. This alone is not an easy task, especially for the second-order derivatives of $f(x)$ for circuits of higher dimension. Furthermore, the interval extension of each derivative must be evaluated for each of the emerging subboxes X. It is not hard to see that when using (2.102) the first derivatives lead to expressions more complex than (2.102); this remains true to much greater an extent, when comparing the complexity of the second-order derivative expressions with that of their first-order counterparts. Thus, as experimental evidence shows, application of second-order methods to the original tolerance Problem 2.6 leads to execution times that are larger than those of first-order methods. We shall now show that an appropriate modification of condition (2.93a) will overcome the above difficulties.

'Similarly to Example 2.9, either side of (2.93a) is squared to get, using (2.102), the following problem.

P r o b l e m 2.5a. Check whether

$$\varphi(x) = a_1^2(x) + a_2^2(x) - \underline{Y}^2[b_1^2(x) + b_2^2(x)] \geq 0 \tag{2.103a}$$

$$x \in X^0 \tag{2.103b}$$

Obviously, problem (2.103) is equivalent to problem (2.93). However, in contrast to (2.102) the function $\varphi(x)$ has the merit that its derivative expressions become simpler and simpler as the derivative order increses. This results from the fact that the functions a_1, a_2, b_1 and b_2 are, in the overwhelming majority of cases, polynomials in the components x_i of x. Thus application of first- and second-order interval methods (or even higher order methods) to Problem (2.103) should be expected to give better results than when the equivalent problem formulation (2.93), (2.102) is used.

2.4.2. First-order method algorithmic improvements

In this section, several improvements of Algorithm A5 of the first-order interval method for tolerance analysis from section 2.3.2 will be presented. They refer to the case where the tolerance analysis problem considered has been formulated in the form (2.103). For notation simplicity henceforth the symbol $f(x)$ will be used to denote the function $\varphi(x)$ from (2.103a).

A. Improved monotonicity test

The improved monotonicity test is applied sequentially to each component $G_i(X)$ of the interval extension for the gradient of $f(x)$. It takes into account the fact that the ith side X_i^0 of the initial box X^0 may be subdivided dynamically (as the algorithm proceeds) into several distinct subintervals X_i^j, $j = \overline{0, J_i}$. Indeed, whenever Procedure 2.3 happens to be applied for the first time a side, say, X_i^0 will be split into two disjoint intervals X_i^1 and X_i^2. Subsequent use of the procedure may further split X_i^1 or X_i^2 (or both) into a pair of disjoint intervals. The maximum number of such disjoint subintervals X_i^j along the ith coordinate is denoted by J_i. It should be realized that each subinterval X_i^j gives rise to a subbox X^j which does not touch any other subbox. The new monotonicity test accounts for the presence of distinct subboxes X^j when processing the current subbox X. More specifically whenever

$$0 \notin \text{int}\, G_i(X) \tag{2.104}$$

(where int stands for interior) we compute the lowerpole x_i^L by (2.37a) or (2.37b). Then for each $j \in \overline{1, J_i}$ we check the following condition:

a) if

$$x_i^L \in \text{int } X_i^0 \tag{2.105}$$

for some j then, the whole current box X is deleted;

b) otherwise

$$x_i^L = \underline{x}_i^j \quad \text{or} \quad x_i^L = \overline{x}_i^j \tag{2.106a}$$

if

$$G_i(X) \geq 0 \quad \text{or} \quad G_i(X) \leq 0 \tag{2.106b}$$

and the dimension of X has been reduced by one.

The explanation of the above monotonicity test is as follows. In both cases x_i^L is either $\underline{x}_i$ or $\overline{x}_i$ (one of the endpoints of the ith side X_i of the current box X) because of (2.104).

First, consider case a). For the sake of argument, let $x_i^L = \underline{x}_i$ and $\overline{x}_i = \overline{x}_i^j$. Due to (2.105)

$$\underline{x}_i^j < \underline{x}_i < \overline{x}_i^j$$

Thus, the lower endpoint $\underline{x}_i$ of X_i divides X_i^j into two adjacent subintervals X_i^L (left) and X_i (right) such that

$$X_i^L \cup X = X_i^j$$

These subintervals determine two adjacent subboxes X^L and X such that

$$X^L \cup X = X^j$$

Since X^L and X are adjacent they have a common facet $\widetilde{X}$ (an (n-1)th dimensional box) defined as

$$\widetilde{X} = (X_1,\ldots,X_{i-1},\underline{x}_i,X_{i+1},\ldots,X_n) \tag{2.107a}$$

Therefore, if the global minimum f^* is attained at a point $\widetilde{x} \in \widetilde{X}$, then it will not be missed after discarding the whole box X since x will remain in the adjacent box X^L. Thus, unlike the monotonicity test from section 2.3.2 we are now able to delete the whole current box even if:

i) $\underline{G}_i(x) = 0$ or $\overline{G}_i(X) = 0$

and

ii) X is not in the interior of X^0 .

In Case b) x_i^L is one of the endpoints $\underline{x}_i^j$ or $\overline{x}_i^j$ of the interval X_i^j. For instance, let $x_i^L = \underline{x}_i^j$. Then the global minimum f^* may be attained at a point x^* lying on the $(n-1)$-dimensional facet $\tilde{X}$ of X^j defined as follows:

$$\tilde{X} = (X_1,\ldots,X_{i-1},\underline{x}_i^j,X_{i+1},\ldots,X_n) \tag{2.107b}$$

Since $\tilde{X}^j$ is not adjacent to any other box we cannot discard the current box X. Therefore, in Case b) we have to retain for further processing the reduced box $\tilde{X}$. (Similar argument is valid when $x_i^L = \overline{x}_i^j$, the ith component of the reduced dimension box X being now $\overline{x}_i^j$.)

B. Sequential evaluation of the derivatives

Since the monotonicity test is based on (2.104) and (2.106b) it is expedient to use interval extensions of the derivatives $g_i(x) = \partial f/\partial x_i$, that lead to as narrow intervals $G_i(X)$ as possible. One way to obtain this is to evaluate $G_i(X)$ in the following sequential order (for simplicity of exposition we assume that (2.105) is not true for all i).

Procedure 2.4. The expressions for $g_i(x)$ are ordered in growing complexity (which may entail reordering of the components x_i of x). First, $G_1(X)$ is evaluated, X being the current box to process. Now suppose, for the sake of argument, that the corresponding component X_1 of X has been reduced to a point x_1^L by Condition b) of the above monotonicity test – formula (2.106). Then the next component G_2 is calculated as

$$G_2 = G_2(x_1^L,X_2,X_3,\ldots,X_n) \tag{2.108}$$

Indeed, let

$$Y = (X_2,\ldots,X_n)$$

Since $f(x)$ is monotonic with respect to x_1, the global minimum of $f(x)$ in X is attained in a facet

$$X' = (x_1^L,X_2,\ldots,X_n) = (x_1,Y)$$

Now introduce the real vector

$$x' = (x_1^L,x_2,\ldots,x_n) = (x_1^L,y)$$

where

$$y = (x_2,\ldots,x_n)$$

Since x_1^L is fixed the mean-value extension of $f(x_1^L, Y, y^L)$ can be written in the form

$$F(x_1^L, Y, y^L) = f(x_1^L, y) + \sum_{i=2}^{n} G_i(x_1^L, Y)(Y_i - y_i^L) \tag{2.109}$$

which shows the validity of (2.108).

If $G_2 = G_2(x_1^L, Y)$ does not contain zero properly, X_2 is now reduced to a point and G_3 is computed as

$$G_3 = (x_1^L, x_2^L, X_3, \ldots, X_n) \tag{2.110}$$

If $0 \in \text{int}\, G_2(x_1^L, Y)$ then the next components G_i, $(i > 1)$ are calculated sequentially with fixed $x_1 = x_1^L$. As soon as for some index $i = k < n$ Condition b) of the monotonicity test is again fulfilled and the corresponding kth interval X_k has been reduced to a point x_k^L, all the remaining components G_i, $i > k$ are now computed with $x_1 = x_1^L$ and $x_k = x_k^L$.

This process of sequential reduction of some interval components X_i to points x_i^L and computing the corresponding G_i using x_i^L rather than X_i as soon as the new information is available continues until $i = n$. Obviously, due to the inclusion monotonicity of G_i the sequential procedure herein introduced will, in general, result in a narrower interval extension of $f(x)$ in X as compared with the case of simultaneous computation of all $G_i(X)$ in determining $F(X, x^L)$ by (2.35).

A better (but more expensive computationwise) version of Procedure 2.4 is possible.

Procedure 2.5. In this version, whenever the monotonicity test b) succeeds in reducing an interval component X_i to a point x_i^L, all the remaining components G_j, $j \neq i$ are recomputed using all x_i^L (available so far) instead of the corresponding X_i and the monotonicity test is applied once again to each updated G_j. This process continues until no further reduction of intervals to points is possible.

The above approach can be implemented in several different ways. One possibility is to use Procedure 2.4. The resultant version will be referred to as Procedure 2.5a.

Procedure 2.5a. For simplicity of explanation, suppose Procedure 2.4 has reduced only one interval, say X_k, to a point x_k^L. Afterwards we apply Procedure 2.4 for a second time. Thus, we start recomputing all the components G_i (using x_k^L) with $i \geq 1$ until $i = k - 1$. If no new interval reduction is possible, the procedure is terminated. Otherwise, if a new reduction of an interval to a point occurs before $i = k$, e.g. for $k' < k$, we continue applying Procedure 2.4 (first with $i > k'$ and then with $i \geq 1$) until at some stage no further interval reduction is possible.

A further improvement of Procedure 2.5a is possible. It will be referred to as Procedure 2.5b.

Procedure 2.5b. (Optimal ordering). In this version, we start with Procedure 2.4 until $i = n$ for the first time. For simplicity, suppose that only one component X_k has been reduced to a point x_k^L. At this point all the remaining components G_i, $i \neq k$ are ordered in the following way. Introduce the real vector

$$e = (e_1,\ldots,e_{k-1},e_{k+1},\ldots,e_n)$$

whose components are defined as follows

$$e_p = \min[|\underline{G}_p|, |\overline{G}_p|]/w(G_p) \tag{2.111}$$

Then reorder the components of e in increasing order to obtain the vector e', keeping track of the correspondance between the indices of the components e_p of e and the indices of the components e_p' of e'. Then start recomputing the components G_i (using x_k^L) in the order defined by the vector e'. Such an "optimal" ordering of G_i has the following advantage as compared to Procedure 2.5a: we start recomputing that component G_{i_1} which stands the best chance to reduce the corresponding interval component X_{i_1} to a point $x_{i_1}{}^L$. Indeed, according to (2.111) and the reordering rule we start with that component G_{i_1} for which the relative distance of $|\underline{G}_i|$ or $|\overline{G}_i|$ from zero (with respect to the width of G_i) is minimum. If (hopefully) we have succeeded in reducing X_{i_1} to a point x_{i_1} we go over to recomputing the second best candidate G_{i_2} using x_k^L and $x_{i_1}{}^L$. The above process continues (as before) until no further interval reduction is possible.

C. Use of the modified MV-*form*

On exit from Procedure 2.4 or Procedure 2.5 we have as many interval components X_k of X reduced to points as the respective version of the sequential evaluation of derivatives can secure. Let the set of the remaining (interval) components X_i of X be denoted by $\tilde{X}$, and the set of their indices by $\tilde{S}$. Since we have already exhausted every possibility of reducing interval components to points the best policy now in determining the lower endpoint of the interval extension of $f(x)$ in the current box X is to appeal to the modified MV-form applying it only to the reduced interval vector $\tilde{X}$. More specifically, we apply Procedure 2.1 to the set $\tilde{X}$ (making use of the optimal ordering of its components if Procedure 2.5b has been used in technique B). Thus, the interval extension of $f(x)$ in the current box X will be defined by the following formula

$$F(X) = f(u,x') + \sum_{i\in\tilde{S}} G_i(\tilde{X}_i,\tilde{x}_i)(X_i - x_i') \tag{2.112}$$

where u is the real vector whose components are defined by the monotonicity test b) while x', x_i' and $\tilde{X}_i$ have the same meaning as in section 2.2.3 but refer to the reduced vector $\tilde{X}$. The lower endpoint of the interval extension of $f(x)$ in X is then determined by (2.112) as $\underline{F(X)}$.

Such an approach to computing $\underline{F(X)}$ based on the combined effect of the sequential evaluation of G_i and application of the modified MV-form is expected to give the best possible result (the highest $\underline{F(X)}$) as compared to all other mean-value forms introduced earlier in section 2.2.

D. Choice of bisection direction

In the implementation of the algorithms from section 2.3.2, the current box X is split into two equal halves along that coordinate i_0 in which X has maximum width (see (2.63)). Such a choice of the bisection direction is in practice most often inadequate. Indeed, consider the tolerance analysis of an active RC filter. The resistors R_i are usually in kΩ while the capacitors C are normally in nF. Since the tolerances on R_i and C_i are generally on the same order (say 10% of their nominal values) the initial box X has sides corresponding to the resistors R that are on the order of 10^{12} larger than the sides corresponding to the capacitors C. Obviously, (for reasonable accuracy ε) the above simple bisection rule will leave the narrow sides (corresponding to C_i) unchanged unless they are reduced to points by the monotonicity test.

Here the coordinate number i_0 for bisection is determined in the following manner based on the mean-value form representation

$$F(X) = m(X) + \sum_{i \in \tilde{S}} G_i(X_i - m_i) \tag{2.113}$$

($\tilde{S}$ is the set of indices corresponding to components of nonzero width). Now i_0 is the coordinate direction for which the width of the product

$$G_i(X_i - m_i) \ , \quad i \in \tilde{S}$$

is maximum. Such a bisection direction choice is expected to give better results than the former one since according to (2.113) it bisects that side X_{i_0} of X which has the greatest effect on the width of $F(X)$.

E. Use of the inequality constraint

As explained in section 2.4.1, the use of conditions (2.97) or (2.100) helps to diminish the computer time needed to establish whether the tolerance analysis problem (2.93) has a solution or not for a given tolerance box X^0. A better implementation of this approach is to use (in the case of a.c. circuits) the equivalent formulation (2.103). Now, conditions (2.97) and (2.100) take on the form

$$\underline{f}^{\rho} \geq 0 \ , \tag{2.114a}$$

$$\bar{f}^{\rho} < 0 \tag{2.114b}$$

Here, $\underline{f}^{\rho}$ and $\bar{f}^{\rho}$ are associated with the function $f(x) = \varphi(x)$ given by (2.103a); $\underline{f}^{\rho} = \underline{F(X^{\rho})}$ is computed by (2.112).

There exists yet another possibility to use the inequality constraint (2.103). Consider again the inequality (2.66) which in the present context (see formula 2.112) takes on the form

$$f(u,x') + \sum_{i\in\tilde{S}} G_i(\tilde{X}_i, \tilde{x}_i)(y_i - x_i') \le \bar{f} \tag{2.115}$$

where $\bar{f}$ is the best available upper bound on f^*. Based on the inequality (2.103a) it is easily seen that now $\bar{f}$ from (2.115) should be set equal to zero right from the start. Indeed, the procedure "using the bound $\bar{f}$" from section 2.3.2 ensures that the interval enclosure Y of all points y satisfying (2.115) will be retained for subsequent processing. Thus, setting

$$\bar{f} = 0 \tag{2.116}$$

when applying the above procedure in Step 4 of the algorithm A5 from section 2.3.3 we are sure to delete only such parts of the current box X within which $f(x) > 0$. But this is exactly what we are striving to achieve (and in as fast a manner as possible) in solving problem (2.103). Since usually $\bar{f} = 0$ is smaller that the currently available upper bound $\bar{f}$ as determined by algorithm A5 the choice (2.116) leads to an improved convergence rate as compared to the stopping criterion (2.114a).

2.4.3. Numerical examples

The techniques A to E from the previous section have been incorparated in algorithm A5. The resultant new version (using optimal ordering Procedure 2.5b) will be referred to as algorithm A6. Several examples will be now considered which will demonstrate the improved numerical efficiency of the new algorithm in comparison with algorithm A5 when solving Problem 2.4.

***Example* 2.11.** We take up Example 2.8 from section 2.3.3. However, now we shall check whether the inclusion (2.90) is satisfied for several intervals Y, the initial box X^0 being the same as in Example 2.8.

Table 2.6 summarizes some results obtained by using algorithm A6.

The last row of the table gives in fact the best results (the smallest number of iterations) when f_L^* and f_U^* are sought using the global optimization algorithm A4 and A5, respectively (see Tables 2.2 and 2.3). It is seen that the introduction of a threshold $\underline{Y}$ or $\overline{Y}$ decreases substantially the number of iterations N_i and hence the computation time t (in seconds), needed to solve Problem 2.4 in comparison to the former approach when the tolerance problem is formulated as Problem 2.1 (determination of the range of $f(x)$). It will be noted that the further away the threshold is from the range, the smaller is the number of iterations required to solve Problem 2.4. Also, in reaching f_U^* (with $\varepsilon = 10^{-3}$),

A5 required X^0 to be divided into 555 versus only 17 bisections for the present method when $\overline{Y} = 24416$ (for the remaining values of $\overline{Y}$ the number of bisections is still smaller).

Table 2.6.

Checking $f_L^* \geq \underline{Y}$			Checking $f_U^* \leq \overline{Y}$		
$\underline{Y}$	N_i	t(s)	$\overline{Y}$	N_i	t(s)
21000	0	0.10	24480	65	3.90
22000	1.5	0.94	24450	75	4.51
22500	3.5	1.93	24430	95	5.43
23060	12.5	2.69	24417	13.5	8.02
23067	12.5	2.70	24416	20.5	12.14
–	25	–	–	1745	–

Remark 2.4. In Algorithms A4 and A5, an iteration comprises computation involved in Steps 3 through 9 whereas in algorithm A6 an iteration corresponds (essentially) to calculating $F(X)$. To make the results comparable an effective number of iterations (the actual number divided by two) has been introduced for algorithm A6 which explains the presence of iterations and a half in the columns for N_i in Table 2.6.

It should be also underlined that a similar reduction of N_i occurs when the threshold $\underline{Y}$ (or $\overline{Y}$) moves away from f_L^* (or f_U^*) towards the centre of the range.

***Example* 2.12.** In this example, the active RC filter shown in Fig. 2.7 is considered.

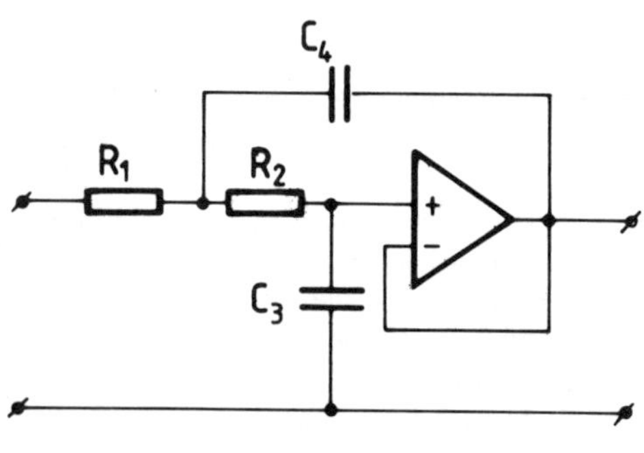

Fig. 2.7. A Sallon-Kev low-pass active filter.

The nominal values for the parameters are:

$$\omega = 1000\,s^{-1}, \quad R_1 = 100\,k\Omega, \quad R_2 = R_1, \quad C_3 = 10^{-9}\,\text{F}, \quad C_4 = 10^{-7}\,\text{F}$$

The voltage transfer function of the filter is

$$T(j\omega) = \frac{1}{1 - \omega^2 R_1 R_2 C_3 C_4 + j\omega C_3 (R_1 + R_2)}$$

First, the range of $|T(j\omega, p)|$ with $p = (R_1, R_2, C_3, C_4)$ was determined over several boxes $P = (R_1^I, R_2^I, C_3^I, C_4^I)$. For each box P, this was done by finding the range $[\underline{\varphi}, \overline{\varphi}]$ of the function

$$\varphi(p) = (1 - \omega^2 p_1 p_2 p_3 p_4)^2 + \omega^2 p_3^2 (p_1 + p_2)^2$$

in P. Obviously, the range $|T| = [\underline{T}, \overline{T}]$ of $|T(j\omega, p)|$ in P is then given by

$$\underline{T} = \frac{1}{\overline{\varphi}}, \quad \overline{T} = \frac{1}{\underline{\varphi}}$$

The range was computed for several initial boxes $P^{(0)}$ corresponding to various tolerances (tol) in percentage with respect to the nominal parameter values. The results for the lower endpoint $\underline{T}$ of the range $|T|$ obtained by algorithm A6 are given in Table 2.7.

Table 2.7

tol %	N_i	l_m	t (s)	$\underline{T}$
5	13	5	2.80	3.243
10	15	6	3.19	1.911
20	15	7	3.19	0.900

where l_m (as before) denotes the maximum length of the list L.

It should be noted that the difficulty of the tolerance problem considered (values for N_i and l_m) is not affected considerably by the size of the initial box P^0 (defined by tol). The same conclusion is also valid in determining the upper endpoint T of the range.

Remark 2.5. In Table 2.7 N_i stands for the number of iterations as defined in algorithm A6, i.e. roughly for the number of evaluations of $\underline{F(X)}$ (and the associated computation of the derivatives).

After the ranges for $|T|$ corresponding to different tolerances had been found it was possible to assign various values for the thresholds (endpoints of the interval Y covering the range). The corresponding conditions

$$1 - \underline{Y}^2\varphi(p) \geq 0, \quad p \in P$$

and

$$1 - \overline{Y}^2\varphi(p) \leq 0, \quad p \in P$$

were checked. Table 2.8 gives some results related to a 5% tolerance on the input parameters.

Table 2.8

$\underline{Y}$	2.0	2.5	3.0	3.2	3.243	1.0
t(s)	0.38	0.38	1.23	1.76	1.98	2.80

It is seen that the introduction of the thresholds reduces the computation time as compared to the case (last column of the table corresponding to $\underline{Y} = 1$) where the tolerance analysis problem considered is tackled in the former format as a global minimization problem.

2.4.4. Second-order method versions

As mentioned at the end of section 2.3.4, more sophisticated schemes than those used in Algorithm A7 can be developed to improve the numerical efficiency of the interval second-order methods for tolerance analysis. In this section, based on the results from section 2.4.2 we shall present two new second-order algorithms designed to solve tolerance analysis problems formulated in the form of Problem 2.4.

Algorithm A8.

This algorithm is based on the first-order derivative algorithm A6. Additionally, it incorporates a procedure that uses only the second derivatives $\delta^2 f/\delta x_i^2$.

Procedure 2.6. We assume that the problem to be dealt with is checking the inequality (2.91), or equivalently problem (2.103) with $f(x) = \varphi_1(x) \in C^2$. The present procedure is applied after Procedure 2.5b, i.e. only to those components X_i, $i \in \tilde{S}$, that have not been reduced to points. It is based on the following result.

Let $h_{ii}(x) = \delta^2 f/\delta x_i^2$ and $H_{ii}(X)$ be the interval extension of $h_{ii}(x)$ in X, where X denotes the current box on exit from Procedure 2.5b. We shall prove the following theorem.

Theorem 2.10. If for some $i \in \tilde{S}$

$$\overline{H_{ii}(X)} < 0 \tag{2.117}$$

then the corresponding interval $X_i = [\underline{x}_i, \overline{x}_i]$ can be reduced to two points, namely $\underline{x}_i$ and $\overline{x}_i$ when solving Problem 2.5a.

Proof. For simplicity of notation assume that $\tilde{S} = \overline{1, n}$ and $i = 1$. The function $f(x_1, \ldots, x_n)$ is then written in the form

$$f(x_i, z_2, \ldots, z_n) = f(x_1, z)$$

with

$$z = (x_2, \ldots, x_n)$$

For a fixed $z \in Z = (X_2, \ldots, X_n)$ and loose $x_1 \in X_1$ the function $y = f(x_1, z)$ can be viewed geometrically as a curve in the plane (y, x_1) (the location of the plane in R^{n+1} is determined by the fixed vector z). The derivative

$$\frac{dy}{dx_1} = \frac{\partial f(x_1, z)}{\partial x_1} = g_1(x_1, z)$$

is also a curve in the (y, x_1)-plane. The derivative

$$\frac{\partial g_1(x_1, z)}{\partial x_1} = h_{11}(x_1, z)$$

is once again a curve in the same plane.

Now let $H_{11}(X_1, Z) = H_{11}(X)$ be the interval extension of $h_{11}(x_1, z)$ for $x_1 \in X_1$, $z \in Z$. If

$$\overline{H_{11}(X)} < 0$$

then $g_1(x_1, z)$ is strictly monotonically decreasing along x_1 for each $z \in Z$. The global minimum of $g_1(x)$ in X is, therefore, somewhere in the reduced box $(\overline{x}_1, Z) = \tilde{X}$ while the global maximum of $g_1(x)$ in X is in the reduced box $(\underline{x}_1, Z) = \underset{\sim}{X}$.

Let $g_1(\overline{x}_1, Z)$ denote the range of $g_1(x_1, z)$ in $(\overline{x}_1, Z) = \tilde{X}$. If $g_1(\tilde{X}) \geq 0$ then X_1 can be reduced to $\underline{x}_1$; similarly, if $g_1(\underset{\sim}{X}) \leq 0$ then X_1 is contracted to $\overline{x}_1$. (In this case $f(x_1,z)$ is monotonically increasing or decreasing, being in either case convex along x_1.)

Now suppose that

$$\underline{g_1(\tilde{X})} < 0 < \overline{g_1(\underset{\sim}{X})}$$

In this case the equality

$$g(x_1,\ldots,x_n) = 0 \tag{2.118}$$

must hold for some $x \in X$. Eq.(2.118) defines an $(n-1)$-dimensional surface α in X. It divides X into two regions $R^{(+)}$ and $R^{(-)}$ such that

$$X = R^{(+)} \cup \alpha \cup R^{(-)}$$

($R^{(+)}$ and $R^{(-)}$ do not include α). It is easily seen that $R^{(+)}$ lies to the left of α (in the direction of x_1) and $g_1(x) > 0$ for $x \in R^{(+)}$ while $R^{(-)}$ is to the right of α and $g(x) < 0$ for $x \in R^{(-)}$. (Indeed, let $x^\alpha = (x_1^\alpha, z^\alpha)$ be a point belonging to the surface α, i.e. $g_1(x^\alpha) = 0$. Then, because of the strict monotonicity of $g_1(x)$ along the axis x_1, the component x_1^α is unique for each component z of x^α which implies the properties of $R^{(+)}$ and $R^{(-)}$). Thus, since $g_1(x) > 0$ for any $x \in R^{(+)}$ the global minimum of $f(x)$ may occur in the reduced box $(\underline{x}_1, Z)$; similarly, since $g_1(x) < 0$ for any $x \in R^{(-)}$ the global minimum of $f(x)$ may be in the box $(\overline{x}_1, Z)$. This completes the proof of Theorem 2.10.

We are now ready to present Procedure 2.6. We start computing $H_{ii}(X)$, $i \in \tilde{S}$ until Condition (2.117) is met for the first time for $i = k$. Now two boxes

$$X^1 = (\underline{x}_k, Z)$$

and

$$X^2 = (\overline{x}_k, Z)$$

are generated. At this stage we appeal to the modified monotonicity test, Condition a). It is applied sequentially to X^1 and X^2. Let $X = X^1$. Recall that X_i^j denotes the disjoint intervals along the x_i direction resulting from Procedure 2.3 (after the deleting of gaps). If for some j, $j \in \overline{1, J_k}$

$$\underline{x}_k \in \operatorname{int} X_k^j \tag{2.119}$$

then the whole box X^1 is discarded; otherwise the box X^1 is processed as in Algorithm A6 and added to the list L. The same is repeated with box X^2.

Algorithm A9.

The basis of this algorithm is again algorithm A6. However, unlike Algorithm A8 now all the second-order derivatives $h_{ij}(x) = \delta f^2/\delta x_i \delta x_j$ are used.

Comparative study of the techniques used in Algorithm A6 shows that reduction of interval components X_i of the current box X to points plays a major role in improving the efficiency of first-order tolerance analysis algorithms. That is why it is expedient to try to obtain as good bounds on the range $g_i(X)$ of the derivatives as possible. In the present algorithm this objective is achieved by computing the interval extensions $G_i(X)$ using the interval extension $H_{ij}(X)$ of the second-order derivatives. More precisely, the algorithm has the following structure.

The computational process starts as in algorithm A6. However, on exit from Procedure 2.5b the following new procedure is introduced.

Procedure 2.7. Using (2.111) we order in an optimal manner all the components X_i, $i \in \tilde{S}$, of the reduced current box $\tilde{X}$ (that have not been reduced to points). However, now additionally we keep track whether

$$G_p^* = \min [\, |\underline{G}_p|\,, |\overline{G}_p|\,] \tag{2.120}$$

is $|\underline{G}_p|$ or $|\overline{G}_p|$. To do this, we may introduce a logical variable

$$K_p = \begin{cases} \text{true} & \text{if } G_p^* = |\underline{G}_p| \\ \text{false} & \text{if } G_p^* = |\overline{G}_p| \end{cases} \tag{2.121}$$

We start with the first component of the reordered set $\tilde{X}$. If the corresponding K_p is true then we start solving the following global minimization problem

$$\underline{g}_p^* = \min g_p(\tilde{x})\,,\ \tilde{x} \in \tilde{X} \tag{2.122}$$

otherwise (if K_p is false) we start solving the global maximization problem

$$\overline{g}_p^* = \max g_p(\tilde{x})\,,\ \tilde{x} \in \tilde{X} \tag{2.123}$$

Such a policy stands the best chance of reducing the corresponding interval X_p to a point $\underline{x}_p$ or $\overline{x}_p$ depending on whether $\underline{g}_p \geq 0$ or $\overline{g}_p \leq 0$.

After converting (2.123) into an equivalent minimization problem

$$\overline{g}_p^* = -[\,(\min - g_p(\tilde{x}))\,]\,,\ \tilde{x} \in \tilde{X} \tag{2.124}$$

either problem (2.122) or (2.124) is solved by Algorithm A6. In doing so, the following points must be noted.

To be specific, consider problem (2.122) where the function to be minimized is $g_p(x)$. Now, we don't need the global minimum g_p^*: what we wish to establish is whether $\underline{g}_p^*$ is non-negative. Thus, instead of (2.202) we solve the following problem of type $\overline{2}.5$: verify whether

$$g_p(\tilde{x}) \geq 0\ ,\ \tilde{x} \in \tilde{X} \tag{2.125}$$

To check (2.125) in a most efficient way using Algorithm A6 we set the upper bound $\overline{g}_p$ on $\underline{g}_p^*$ to zero, i.e.

$$\overline{g}_p = 0 \tag{2.126}$$

(recall the technique E from section 2.4.2).

It is easily seen that the same condition (2.126) is to be used when we solve the corresponding equivalent problem (2.125) associated with the original problem (2.124).

In solving (2.125) the interval extension $G_p(\tilde{X})$ is evaluated by an expression involving the second-order derivatives

$$G_p(\tilde{X}) = g_p(\tilde{x}^L) + \sum_j H_{pj}(\tilde{X})(\tilde{X}_j - \tilde{x}_j^L) \tag{2.127}$$

Since $\tilde{X}$ may be subdivided we need a second list L' (the first list L stores information about $F(X)$).

If (2.125) has a solution then the corresponding interval X_p is reduced to a point, otherwise X_p remains an interval. In both cases we go over to the next component of the reduced box $\tilde{X}$ and the above computational process is repeated using points rather than intervals whenever possible at the next iterations. The following example will illustrate the applications of Algorithm A9.

***Example* 2.13.** We shall solve once again the problem of finding the range of the function (2.74) from Example 2.8 by means of the second-order algorithm A9. The results obtained are given in the first row of the following table.

Table 2.9

	$f_L^* = 23067.376$			$f_U^* = 24416.031$		
	N_i	l_m	t(s)	N_i	l_m	t(s)
A9	18	9	2.59	52	1	15.98
A6	25	8	2.69	52	1	13.14

For the purpose of comparison the second row of the table lists data obtained by the best first-order method A6 (used in this instance for global optimization). It is seen that for the example considered the second-order test for reducing the size of the current box cuts down the number of iterations only in the case of determining the global minimum while leaving the number unchanged in the case of the global maximum (this is because the function $f(x)$ is, generally, either convex or concave in all or almost all of its coordinates). The computation time is however not reduced (taking into account both the minimum and the maximum cases). Nevertheless, it is hoped that a better second-order method incorporating more sophisticated techniques and algorithmic improvements might lead to a faster convergence than that of the best first-order methods.

2.5. SOLVING THE PROBABILISTIC TOLERANCE ANALYSIS PROBLEM

In section 2.1.3 the tolerance analysis problem in probabilistic setting was formulated as Problem 2.2, namely: given the multivariate nonlinear function $y = f(x)$, find the range of y when $x \in H$ where H is the admissible hyperellipsoid. For reasons similar to the worst-case tolerance case (section 2.4.1) Problem 2.2 can be reformulated as follows.

Problem 2.7. Given a multivariate function f: $R^n \to R$, check that the range $f(H) = [\underline{f}, \overline{f}]$ of f over H is contained in a prescribed interval $Y = [\underline{Y}, \overline{Y}]$, i.e.

$$f(H) \subseteq Y \tag{2.128}$$

In this paragraph four methods for tackling the basic probabilistic tolerance analysis Problem 2.7 will be considered. For convenience of presentation they will be divided into two groups.

2.5.1. First group methods

The methods of this group are based on the following conjecture.

Conjecture 2.1. The endpoints of the solution of Problem 2.2 are attained at points on the boundary δH of the hyperellipsoid H.

Thus, in determining the lower solution endpoint, problem (2.18)

$$f_L^* = \min f(x) \ , \quad x \in H$$

can be replaced by the problem

$$f_L^* = \min f(x) \ , \quad x \in \partial H \tag{2.129}$$

(Problem (2.20) associated with the upper solution endpoint can be transformed in a similar way).

Let $Y = [\underline{Y}, \overline{Y}]$ be the interval which must cover the range of solutions to Problem 2.2. Then, in accordance with Problem 2.7 formulation (following exactly the same approach as in the deterministic case from section 2.4.1), Problem (2.129) should be transformed as follows

$$f(x) \geq \underline{Y}\,, \quad x \in \partial H \tag{2.130}$$

Finally, using (2.130) and (2.102) we arrive at a formulation similar to (2.103)

$$f_1(x) = a_1^2(x) + a_2^2(x) - \underline{Y}^2[b_1^2(x) + b_2^2(x)] \geq 0 \tag{2.131a}$$

$$x \in \partial H \tag{2.131b}$$

Now we are ready to present the first method of this section. It is applicable only for problems where the components x_i of x are statistically independent (the matrix C from (2.15) is diagonal). We shall show that in this case the probabilistic formulation (2.131) can can be transformed equivalently to a worst-case tolerance analysis problem.

Indeed, for a diagonal C condition (2.131b) becomes

$$\sum_{i=1}^{n} \frac{1}{\sigma_i^2}(x_i - \xi_i)^2 - \gamma^2 = 0 \tag{2.132}$$

Now, one of the variables, say x_1, can be expressed as a function of the remaining ones

$$x_1 = \xi_1 \pm \sigma_1 \sqrt{\gamma^2 - \sum_{i=2}^{n} \frac{1}{\sigma_i^2}(x_i - \xi_i)^2} \tag{2.133}$$

It can be verified that most often (for instance, when x_i are passive elements and dependent sources parameters) f from Eq. (2.131a) is a quadratic function in each x_i. Therefore, it can be written in the form

$$f_1(x_1, z) = \alpha_1(z)x_1^2 + \alpha_2(z)x_1 + \alpha_3(z) \tag{2.134}$$

where

$$z = (x_2, x_3, \ldots, x_n) \tag{2.135}$$

Using (2.135) and substituting (2.133) into (2.134) we get

$$f_1(z) = \alpha_1(z)[\xi_1 \pm \sigma_1\sqrt{\varphi(z)}\,]^2 + \alpha_2(z)[\xi_i \pm \sigma_i\sqrt{\varphi(z)}\,] + \alpha_3(z)$$

and after some manipulation

$$\pm A_1(z)\sqrt{\varphi(z)} + A_2(z) \geq 0 \tag{2.136}$$

where $\varphi(z)$ is the function under the square root in Eq. (2.133). Now, assume that $A_1(z) \neq 0$ for all $(x_1, z) \in \delta H$. Then, from (2.136)

$$\sqrt{\varphi(z)} \geq \mp \frac{A_2(z)}{A_1(z)}$$

Hence

$$A_1^2(z)\varphi(z) - A_2^2(z) \geq 0 \tag{2.137a}$$

$$z \in Z^0 \tag{2.137b}$$

where

$$Z^0 = (X_2^0, \ldots, X_n^0) \tag{2.137c}$$

The components X_i^0 are defined by the value of γ. If $\gamma = 3$ then the width of X_i^0 is $6\sigma_i$. Thus, the original tolerance problem in probabilistic statement (2.131a), (2.132) has been transformed into a deterministic problem (2.137) of a reduced (by one) dimension. The latter problem can be solved by Algorithm A6.

The second method of this group is more general: it is applicable for statistically dependent variables also (matrix C can be nondiagonal). Thus, the Condition (2.131b) is now given by (2.15) as

$$f_2(x) = (x - \xi)^T C^{-1}(x - \xi) - \gamma^2 = 0 \tag{2.138}$$

We are, therefore, led to check whether the system

$$f_1(x) \geq 0 \tag{2.139a}$$

$$f_2(x) = 0 \tag{2.139b}$$

(with (2.139a) given by (2.131a)) has a solution. If we are to use an interval method to do this, we have to introduce an initial box X^0 containing the solution(s) (if any) of (2.139). It is reasonable to define X^0 as the smallest box (interval hull) containing H.

Rather than verify (2.139), we shall check whether the system of two equations

$$f_1(x) = 0 \tag{2.140a}$$

$$f_2(x) = 0 \tag{2.140b}$$

has a solution for

$$x \in X^o \tag{2.140c}$$

It turns out that checking the validity of (2.140) is numerically an easier task than checking (2.219). Indeed, based on section 1.4.2 an interval method for verifying (2.140) can be developed. We shall, however, postpone the discussion of this method to Chapter 6, section 6.1.4. At this stage, it is important to underline that if Problem (2.140) has not a solution then the corresponding tolerance analysis problem (2.131a), (2.138) has a solution and vice versa.

Remark 2.6. If Conjecture 2.1 is proven to be valid then the above two methods will guarantee the exact (within the accuracy ε) solution of the tolerance problem considered. Otherwise it will only provide an approximate solution. It is, however, expected that even in this highly unlikely case the approximation will be rather accurate (it is difficult to imagine a circuit for which the point x_L^* securing the global minimum f_L^* in (2.129) might be far away from the boundary δH of H).

2.5.2. Second group methods

The two methods from this section do not appeal to Conjecture 2.1. Therefore, they are always guaranteed to provide the exact solution of the tolerance problem considered. Both methods appeal to formulation (2.131a). Condition (2.131b) is, however, now replaced by the general condition (2.17), i.e.

$$f_2(x) = (x-\xi)^T C^{-1}(x-\xi) - \gamma^2 \le 0 \qquad (2.141)$$

The first method verifies directly the validity of the system of inequalities

$$f_1(x) \ge 0 \qquad (2.142a)$$

$$f_2(x) \le 0 \qquad (2.142b)$$

$$x \in X^0 \qquad (2.142c)$$

It is based on a Skelboe type algorithm for seeking the global minimum of $f_1(x)$ in X^0 combined with additional rules for discarding the current box X whenever it does not satisfy (2.142a) or (2.142b). Thus, X is not entered the list L (of boxes to be processed later) if

$$\overline{F}_1 < 0 \qquad (2.143a)$$

or

$$\underline{F}_2 > 0 \qquad (2.143b)$$

The second method deals again with (2.142a), (2.142b) formulated as a global optimization problem:

$$f_1^* = \min f_1(x) \qquad (2.144a)$$

$$f_2(x) \le 0 \qquad (2.144b)$$

Obviously, if $f_1^* \geq 0$ then the original probabilistic tolerance problem considered ($f(x) \geq \underline{Y}$, $x \in H$) has a solution (and vice versa). The basic approach is to convert (2.144) into a system of nonlinear algebraic equations as done in section 1.5.2. It is readily seen that now the system of equations (1.98) to (1.100) associated with the minimization problem (2.144) is

$$u_0 g_1^{(j)}(x) + u_1 g_2^{(j)}(x) = 0 , \quad j = \overline{1,n} \tag{2.145a}$$

$$u_1 f_2(x) = 0 \tag{2.145b}$$

$$u_0 + u_1 = 1 \tag{2.145c}$$

where

$$g_i^{(j)} = \frac{\partial f_i}{\partial x_j} , \quad i = 1,2 ; \; j = \overline{1,n}$$

$$u_0 \geq 0 , \qquad u_1 \geq 0 \tag{2.145d}$$

Eliminating u_1 through (2.145c) we finally get

$$u_0 g_1^{(j)}(x) + (1 - u_0) g_2^{(j)}(x) = 0 , \quad j = \overline{1,n} \tag{2.146a}$$

Treating u_0 as an additional $(n+1)$th variable we see that we have transformed the *minimization problem (2.144) into an equivalent system (2.146) of* $(n+1)$ nonlinear algebraic equations in $(n+1)$ unknowns. The latter system can be solved by some interval Newton method from section 1.4.2.

The last point to make is to define the initial box $X^0 = (X_1, X_2, \ldots, X_n, X_{n+1})$ within which the solution of (2.146) will be sought. Obviously, the first n components of X^0 can be chosen (as in the previous section) as the components of the interval hull containing the hyperellipsoid H. The last component X_{n+1} of X^0 is defined as the interval [0, 1]. This follows directly from Eq. (2.145c) and Condition (2.145d).

2.5.3. Numerical example

In this section we shall deal again with the filter considered in Example 2.12. This time, we shall however solve the probabilistic tolerance analysis problem formulated as Problem 2.7. More precisely, the Problem (2.131) associated with the lower endpoint $\underline{Y}$

of the interval Y will be considered for various values of $\underline{Y}$. The hyperellipsoid H will be constant and will correspond to $\gamma = 3$ with axes of $6\sigma_i$. Each σ_i is defined as follows

$$\sigma_i = (\overline{x}_i^0 - \underline{x}_i^0)/6$$

where $\overline{x}_i^0$ and $\underline{x}_i^0$ are the endpoints of the interval X_i^0 associated with the ±5% worst-case tolerance analysis problem from Example 2.12.

All four methods (referred to, for convenience, as methods M1 to M4) from the preceding two subsections have been applied to solving the probabilistic tolerance problem considered in this example. We shall confine ourselves to give some results obtained by methods M2 and M3 since the remaining two methods were (at least for the problem at hand and in their present implementation) less efficient.

***Example* 2.14a.** In this example, Problem (2.131) was solved by method M2. The system (2.140) is now

$$1 - \underline{Y}[(1 - \omega^2 x_1 x_2 x_3 x_4)^2 + \omega^2 x_3^2 (x_1 + x_2)^2] = 0 \tag{2.148a}$$

$$\sum_{i=1}^{n} \frac{1}{\sigma_1^2}(\zeta_i - x_i)^2 - 9 = 0 \tag{2.148b}$$

$$x \in X^0 \tag{2.148c}$$

where X^0 is the box corresponding to ±5% tolerance on the input parameters.

Table 2.10

$\underline{Y}$	A11 a			A11 b			N_s
	N_i	l_m	t (s)	N_i	l_m	t (s)	
3	9	5	0.60	9	5	0.39	0
3.5	27	13	1.64	23	12	1.43	0
4	209	76	12.46	133	49	8.57	1
4.5	23	12	1.42	43	22	2.64	1
5	21	11	1.30	21	11	1.32	1

Method M2 is presented in detail in Chapter 6, section 6.1.4. It will be only noted here that it has been implemented by two different algorithms referred to here as algorithms A11a and A11b.

Table 2.10 gives some results regarding system (2.148) obtained by these algorithms. The notation N_s means number of solution of the system considered. Since system (2.148) has no solution for the first two rows, the corresponding probabilistic tolerance problems (2.130) associated with the lower endpoint $\underline{Y}$ of the interval Y from Problem 2.7 for the first two values of $\underline{Y}$ have a solution. Conversely, since for the last three rows system (2.148) has a solution, the corresponding tolerance problems (2.213) have no solution for the last three values of $\underline{Y}$.

Example **2.14b.** Now Problem (2.131) was solved by method M3. In order to be able to compare the efficiency of method M3 and M2 it was assumed that Conjecture 2.1 is valid for the example considered. Thus, the system to be checked for solutions in X^0 was in fact system (2.139).

An algorithm (A12) for implementing method M3 in this special case has been developed. Its structure is similar to algorithm A1 from section 2.3.3. However, A12 incorporates additionally the rules (2.143) for discarding the current box. Due to Conjecture 2.1, (2.142 b) is now an equality, so the additional discarding rule

$$\overline{F}_2(X) < 0$$

was also used.

The results obtained by algorithm A12 are given in Table 2.11.

Table 2.11

$\underline{Y}$	A12			N_s
	N_i	l_m	t (s)	
3	9	5	0.39	0
3.5	27	13	0.93	0
4	209	76	6.32	1
4.5	23	12	0.82	1
5	21	11	0.77	1

Comparing the above result with those from Table 2.10 it is seen that algorithm A12 is more efficient computationwise than algorithms A11a, A11b.

The same tolerance problem was solved by the traditional Monte-Carlo method. For $N = 2000$ where N is the number of tests the (approximate) lower endpoint $f_L = 3.854$ of the tolerance on the output variable was determined for $t = 17.30$ sec. Comparing the latter run-time with the run-times from Tables 2.10 and 2.11 it is evident that the interval methods M2 and M3 are considerably more efficient as regards execution time requirements than the traditional statistical method, especially for thresholds $\underline{Y}$ that are not quite close to the range endpoint.

Comments

Section 2.1. In the literature on tolerance analysis, it is more common to define the tolerance interval X_i on each input variable x_i not in the form (1.6)

$$X_i = [\underline{x}_i, \overline{x}_i]$$

(i.e. by specifying the endpoints of the corresponding interval X_i) but rather by the equivalent expression (1.7):

$$X_i = m(X_i) + [-w(X_i)/2, w(X_i)/2]$$

In the latter formula $m(X_i)$ is the nominal value of x_i while $w(X_i)/2$ is given in percentages of $m(X_i)$. For example, a resistance r may have a nominal value of, say, 200 Ω and – in electrical engineering jargon – a "tolerance" of ±5%; thus

$$R = 200 + [-10, 10] = [190, 200]$$

Nonsymmetrical (with respect to the nominal value) tolerances are also encountered (as in Example 2.4).

The idea of applying interval analysis techniques for tackling the worst-case tolerance problem in its global optimization setting was first suggested in [32] (see also [21] and [33]). The iterative method developed in [21] was based on the monotonicity test mean-value form representation (2.29) for the interval extension $F(X)$ of the associated function $f(x)$ relating output variable to input parameters.

The usefulness of Theorem 1.2 in worst-case tolerance analysis is illustrated by Proposition 2.1 and Examples 2.4, 2.5 and 2.6. It is felt that the scope of this theorem can be broadened. Thus, it would be of practical utility to know all possible types of resistive circuits for which the theorem is fully or partially applicable.

As shown in section 2.1.3 the basic probabilistic tolerance problem can be equated to two inequality constraint minimization problems of type (2.21). To the best of the author's knowledge such an approach has not been considered in tolerance analysis literature as yet.

Section 2.2. The idea of modifying the MT-form by choosing two distinct points x^L and x^U different from the center m in order to narrow the interval extension of the function considered was first suggested in [34] for the case of a scalar function. Later, the vector generalization of the modified MT-form was proposed in [31]. In the same paper the modified MV-form was introduced. Theorem 2.9 establishes the important result that under appropriate conditions the modified MV-form ensures the narrowest interval extension among all known mean-value forms.

The modified forms considered in section 2.2 can be further improved by making use of the so-called interval slopes (introduced in [23]) in evaluating the interval extensions of the partial derivatives involved provided the interval slopes lead to narrower intervals for derivatives than their natural extensions.

Section 2.3. Three interval methods for solving the worst-case tolerance analysis problem have been presented in this section: zero-order method (using no derivatives), first-order method and second-order method (resorting to first-order and second-order derivatives, respectively). The zero-order method was exposed only as a means for better understanding of the other two methods - its overall efficiency, as numerous examples have shown, is by far lower than that of the methods using derivatives.

The first-order method has been implemented appealing to various mean-value forms considered in section 2.2. In accordance with the theoretical predictions the numerical evidence show that the best first-order methods are those based on the modified MT- and MV-forms. It should, however, be borne in mind that their numerical efficiency might be further improved by incorporating more involved techniques: using interval slopes when evaluating the extensions of the first-order derivatives and introducing more effective schemes for reducing the current box X when several components X_i are divided into subintervals. Some alternative techniques aiming at improving the first-order interval methods for worst-case tolerance analysis are presented in section 2.4.1 and section 2.4.2.

The second-order interval method presented in section 2.3.4 has been based on a very simple algorithm (Algorithm A1 from section 2.3.3 for implementating the first-order interval method). Further improvements appealing to more sophisticated techniques such as nonconvexity tests, monotonicity tests and interval Newton method may be implemented resulting in a higher numerical efficiency. Several such approaches are briefly considered in section 2.4.4.

Section 2.4. In this section certain improvements relative to the first- and second-order methods for worst-case tolerance analysis have been presented. The first improvement suggested in subsection 2.4.1 is related to a new formulation of the tolerance analysis problem considered. Unlike section 2.1 where the tolerance analysis problem was solved by finding the global solutions of two associated minimization problems, the new formulation leads to two equivalent problems of the type 2.5a which circumvents the need to determine the respective global minimums, thus saving most often a considerable amount of computation. The second important advantage of the equivalent Problem 2.5a over the original formulation stems from the fact that the derivatives have simpler expressions as their order increases.

The equivalent Problem 2.5a checks the nonnegativeness of a function $f(x)$ in box X. A similar problem verifying the positiveness of $f(x)$ in X will occur in Chapter 4. Therefore, the efficient solution this problem is of great interest.

Five techniques for improving the numerical efficiency of the first-order tolerance analysis methods addressed to the equivalent Problem 2.5a formulation are suggested in section 2.4.2. It seems that the most important improvement is due to the more elaborate monotonicity test but further experements with a larger group of test examples are needed to determine more precisely the relative weight of each individual technique.

The best first-order algorithms are by far superior as regards computation time and accuracy over the statistical (Monte-Carlo) method in the case of circuits of moderate size.

In subsection 2.4.4 two second-order methods for worst-case tolerance analysis are suggested. Both these methods as well as the method from section 2.3.4 exploit a particular technique out of a variety of choices. Many other possibilities remain uninvestigated. For instance, rather than applying formula (2.127) involving the second-order derivatives $H_{pj}(X)$ use can be made of interval slopes [23] in evaluating the interval extensions $G_p(X)$. Such an approach would usually yield narrower (but never wider) intervals for $G_p(X)$ and thus would improve the overall efficiency of algorithm A9. The comparative study of all the options open and the elaboration of an efficient second-order method for tolerance analysis are only in their initial phase.

Section 2.5. The last section of this chapter deals with the tolerance analysis of linear circuits in probabilistic setting. Similarly to subsection 2.4.1, first an equivalent formulation of the problem considered is introduced in the form of Problem 2.7 which offers similar computational advantages over the original formulation as in the deterministic tolerance analysis case. Four interval methods for solving Problem 2.7 have been suggested.

The first two methods are based on Conjecture 2.1. Until it is confirmed these methods are theoretically only approximative although there are good reasons to believe that the conjecture may prove correct (at least for the commonest cases encountered in practice). The last two methods provide the exact solution of Problem 2.7.

The first method applicable for circuits with statistically independent parameters reduces the probabilistic problem into an equivalent worst-case tolerance problem. Thus, the whole arsenal of first- or second-order methods developed in section 2.3 and section 2.4 can be employed in solving the equivalent deterministic problem. The third method is, in fact, based on the global minimization algorithm A1 modified to take into account an additional functional constraint.

The remaining two methods appeal to solving a system of n nonlinear algebraic equations (in the case of M2) or systems of two equations (in the case of M4). These problems will be discussed in detail in Chapter 6.

It is difficult to compare the numerical efficiency of all the four methods. At the present stage the experimental data available seems to indicate that algorithm A12 yields the best results (provided that Conjecture 2.1 proves correct). Much more research is, however, needed to substantiate the validity of such a conclusion.

CHAPTER 3

LINEAR CIRCUIT TOLERANCE ANALYSIS – LINEAR INTERVAL SYSTEM APPROACH

In this chapter we continue considering the worst-case tolerance analysis of linear electrical circuits. However, unlike Chapter 2 where the tolerance analysis problem was equated to two global optimization problems here the tolerance analysis will be carried out by means of specific linear systems of equations with interval coefficients. It will be shown that such an approach results in methods for worst-case tolerance analysis which may, in some cases, be more efficient than their counterparts from Chapter 2.

3.1. PROBLEM STATEMENT FOR D.C. CIRCUITS

3.1.1. Implicit form formulation

When stating the worst-case tolerance analysis problem in section 2.1.1 it was assumed that there was only one output variable y and – what is even a more stringent stipulation – that the function $y = f(x_1, \ldots, x_n)$ relating the input parameters x_i to the output y is known explicitly. These assumptions are, in many cases, far from being realistic. Indeed, although theoretically always possible, the derivation of the function f in explicit form may prove, especially for circuits of increased size, an intractable task. Moreover, first- or second-order derivatives of f in x_i, are needed to implement the respective first- and second-order interval methods. If all these functions are to be derived repeatedly for several interval output variables $y_k = f_k(x_1, \ldots, x_n)$ the amount of preliminary analytical work needed just to formulate the problem may be prohibitively large.

In this section, it will be shown that for a large class of d.c. circuits the multiparameter - multioutput tolerance problem can be formulated (and solved) in a very efficient way as a system of linear interval equations

$$A^I y = B \tag{3.1}$$

This approach circumventing the need for explicitly deriving the functions f_k (and their derivatives) will be later (section 3.3) extended for tolerance analysis of a.c. circuits.

The formulation of the tolerance problem as a system of linear interval equations will be initially presented for a simple resistive circuit. More complex circuits will be considered in sections 3.1.3 and 3.3.3.

In this section, **N** will denote a linear resistive circuit made of uncoupled resistors and independent voltage sources. Let m be the number of branches and $(n+1)$ be the number of nodes; one of the nodes (say, the $(n+1)$ node) is grounded. The worst-case tolerance analysis problem herein considered for this class of circuits is formulated as follows.

Problem 3.1. Given the circuit **N** and the tolerances on the branch resistors and source voltages, find the tolerances on the branch currents and/or the nodal voltages.

According to the approach adopted in this chapter we have to derive a linear interval system of the form (3.1). Recall (section 1.3.1) that the elements of A^1 and B must be independent intervals.

To derive (3.1) we first have to write down an appropriate system of equations in real (noninterval) variables. To do this we make use of Kirchoff's laws. By KVL we have

$$r_\rho i_\rho - (V_{s_\rho} - V_{p_\rho}) = u_\rho, \quad \rho = \overline{1,m} \tag{3.2}$$

where r_ρ is the branch resistance, i_ρ is the branch current, s_ρ and p_ρ are the nodes of the ρth branch, $V_{s\rho}$ and $V_{p\rho}$ are the corresponding node voltages and u_ρ is the branch (independent) source voltage. By KCL we have

$$\sum_{j=1}^{m} \alpha_{kj} i_j = 0, \quad k = \overline{1,n} \tag{3.3}$$

where α_{kj} is either $+1$, -1 or 0. On introducing the (reduced) incidence matrix $\alpha = \{-\alpha_{kj}\}$ it is not hard to see that (3.2) and (3.3) can be written together as a system of $N = m + n$ linear equations in N unknowns

$$\left[\begin{array}{cccc|c} r_1 & & & & \\ & r_2 & & 0 & \\ & & \ddots & & \alpha^T \\ & 0 & & & \\ & & & r_m & \\ \hline & & \alpha & & 0 \end{array}\right] \bullet \left[\begin{array}{c} i_1 \\ i_2 \\ \vdots \\ i_m \\ \hline v_1 \\ \vdots \\ v_n \end{array}\right] = \left[\begin{array}{c} u_1 \\ u_2 \\ \vdots \\ u_m \\ \hline \\ 0 \\ \\ \end{array}\right] \tag{3.4}$$

(the symbol α^T denotes the transpose of α).

We write (3.4) in matrix form as

$$Ay = b \tag{3.5}$$

with

$$A = \begin{bmatrix} r & \alpha^T \\ \alpha & 0 \end{bmatrix}, \quad y = \begin{bmatrix} i \\ v \end{bmatrix}, \quad b = \begin{bmatrix} u \\ 0 \end{bmatrix} \tag{3.6}$$

Let each resistance r_ρ and source voltage u_ρ in the ρth branch belong to the respective interval R_ρ or U_ρ, i.e.

$$r_\rho \in R_\rho = [\underline{r}_\rho, \overline{r}_\rho] \tag{3.7}$$

$$u_\rho \in U_\rho = [\underline{u}_\rho, \overline{u}_\rho] \tag{3.8}$$

We seek the intervals of the possible values of all currents and all ungrounded node voltages. Thus we have $N = m + n$ output variables and $2m$ input parameters.

When the components of r and u vary in the intervals (3.7) and (3.8), respectively, the system (3.4) becomes an interval linear system of the type (3.1).

Indeed, let R be a diagonal interval matrix whose nonzero elements are given by (3.7) while U is an interval vector with elements defined by (3.8). Then the resulting interval linear system for tolerance analysis of the resistive circuit considered takes the form

$$\begin{bmatrix} r & \alpha^T \\ \alpha & 0 \end{bmatrix} \begin{bmatrix} i \\ v \end{bmatrix} = \begin{bmatrix} u \\ 0 \end{bmatrix}, \quad r \in R, \quad u \in U \tag{3.9}$$

which can be written symbolically as

$$\begin{bmatrix} R & \alpha^T \\ \alpha & 0 \end{bmatrix} \begin{bmatrix} i \\ v \end{bmatrix} = \begin{bmatrix} U \\ 0 \end{bmatrix} \tag{3.10}$$

Obviously, (3.10) is a system of linear interval equations of the type (3.1) with

$$A^I = \begin{bmatrix} R & \alpha^T \\ \alpha & 0 \end{bmatrix}, \quad y = \begin{bmatrix} i \\ v \end{bmatrix} \quad \text{and } B = \begin{bmatrix} U \\ 0 \end{bmatrix} \tag{3.11}$$

It is most important to underline that in the linear interval system (3.10) all interval coefficients are independent. This condition is crucial since most of the existing interval methods are capable of exactly solving only such linear interval systems.

Thus, the approach herein adopted for solving the d.c. tolerance problem considered consists in formulating and solving an associated system of linear equations with independent interval coefficients.

To contrast this approach with the global optimization approach from the previous chapter, consider again system (3.5). Having (3.5) and (3.6) in mind, we can write the vector solution y in the form

$$y = A^{-1}b = f(r,u)$$

where f is a vector function with component f_k, $k = \overline{1, N}$. So the kth component y_k of y is determined by the function

$$y_k = f_k(r,u) \tag{3.12}$$

Now let $r \in R$ and $u \in U$. Problem (3.1) might be solved by finding the range of each function $f_k(r,u)$ over the interval vector $X = (R|U)$. Since the latter problem is equivalent to two global optimization problems, we have to solve $2N$ global optimization problems in order to determine the solution of Problem 3.1. Moreover, the functions (3.12) (altogether N) must be given in explicit form. For this reason the formulation of Problem 3.1 using the global optimization approach will be referred to as explicit form formulation.

Based on this section's approach the solution of Problem 3.1 can be found as the optimal interval solution of the linear system (3.9). In this case the functions f_k relating the input variables (the components of r and u) to the output variables y_k are only given by (3.9) in implicit form. Therefore, the formulation of a tolerance analysis problem as a system of linear equations with interval coefficients will be called implicit form formulation.

3.1.2. Specific peculiarities

In this section some specific aspects associated with the implicit form formulation of the d.c. tolerance analysis problem will be considered.

It should be pointed out that only the system (3.9) is suitable for handling Problem 3.1. We shall now show that other formulations based on loop analysis or nodal analysis are not applicable since they result in a system of linear interval equations whose coefficients are not independent as required in (3.1).

We shall first consider the loop current equations formulation. For any real values $r_\rho \in R_\rho$ and $u_\rho \in U_\rho$ we have

$$\sum_{j=1}^{q} r_{sj} i'_j = e_s, \quad s = \overline{1, q} \tag{3.13}$$

where i'_j are the loop currents, q is the number of independent loops, r_{sj} is the corresponding proper or mutual loop resistance and e_s is the equivalent loop voltage.

Now consider the resistance r_{sj}. Each resistance r_{sj} is the sum of a certain number of branch resistances r_ρ, i.e.

$$r_{sj} = \sum_{\rho=1}^{m} c_{sj}^{\rho} r_{\rho} \tag{3.14}$$

where c_{sj}^{ρ} is either +1, −1 or 0. When r_ρ becomes an interval R_ρ (for simplicity we assume that u_ρ and the resulting loop voltages e_s remain constant), system (3.13) takes on the form

$$\sum_{j=1}^{q} R_{sj} i_j' = e_s, \quad s = \overline{1, q}$$

whose coefficients R_{sj} are given by

$$R_{sj} = \sum_{\rho=1}^{m} c_{sj}^{\rho} R_{\rho} \tag{3.15}$$

Obviously, R_{sj} are, generally, not independent intervals. Indeed, if $r_{\rho 0}$ belongs both to the pth and tth loops, then R_{pp} and R_{tt} are dependent since according to (3.14) and (3.15) they both depend on $R_{\rho 0}$

Next, we consider the nodal voltage equations

$$\sum_{j=1}^{n} g_{sj} v_j = \sum_{j=1}^{n} c_{sj} g_j u_j, \quad s = \overline{1, n} \tag{3.16}$$

where g_{sj} are the corresponding proper or mutual nodal conductances, g_i are the branch conductances and the constants c_{sj} are either +1 or −1 or 0. If g_j are allowed to vary in the intervals G_j (keeping, for simplicity, u_j unchanged) we have on account of (3.16)

$$\sum_{j=1}^{n} G_{sj} v_j = I_s, \quad s = \overline{1, n}$$

Here again the interval coefficients G_{sj} and I_s are not independent since some of them may depend on one or several identical conductances. If for example, some k_oth branch is incident on the pth and tth node and contains a voltage source, then obviously G_{pp}, G_{tt}, I_p and I_t are all dependent on G_{K0}.

Based on the above considerations we conclude that the only possibility of formulating the d.c. circuit tolerance analysis problem considered in the form of a linear interval system is the system (3.10)

Exactly solving (3.10) in one way or another, we are able to determine the tolerances on all the branch currents and all the nodal voltages by the corresponding components I_k, V_j of the optimal interval solution $\overline{Y}$ of (3.10).

We are, however often interested in determining the tolerances on branch voltages that are not nodal voltages. For example, let the voltage whose tolerance we would like to evaluate be across the two ungrounded nodes s and p. It should be stressed right away that in this case it is not possible to determine the tolerance V_{sp} on the branch voltage v_{sp} by means of the corresponding solution components V_s and V_p of system (3.10). Indeed, we shall show that

$$V_{sp} \neq V_s - V_p \tag{3.17}$$

To do this we assume for simplicity that the source voltages u_k, $k = \overline{1, m}$ remain constant so that

$$\begin{gathered} u_{sp}(r_1,\ldots,r_m) = v_s(r_1,\ldots,r_m) - v_p(r_1,\ldots,r_m) \\ r_i \in R_i , \qquad i = \overline{1,m} \end{gathered} \tag{3.18}$$

Since v_s and v_p are in general nonlinear functions of r_i, it is obvious from (3.18) that the range of the difference $v_s - v_p$ is not equal to the differences of the ranges of v_s and v_p which leads to (3.17).

For a similar reason V_{sp} cannot be determined either by using Ohm's law for the corresponding interval quantities since generally

$$V_{sp} \neq R_{sp} I_{sp} - U_{sp} \tag{3.19}$$

On the bases of formulae (3.17) and (3.19) the following general results are easily seen to be valid.

Proposition 3.1. The principle of superposition is not valid in the case of linear electric circuits with interval data if it is applied to equations in interval form.

The proof of this proposition is straightforward if we take into consideration the nonlinear character of the tolerance problem which is best seen from the explicit form formulation (3.12).

The next result is a corollary of Proposition 3.1.

Proposition 3.2. It is in general impossible to find first (in some way or other) the tolerances on part of the output variables and then to determine the tolerances on the remaining output variables using some linear formulae relating the output variables and the input parameters.

The following example will help clarify the implication of the last proposition.

***Example* 3.1.** Let **N** be a complex circuit of the class herein considered. Suppose (for simplicity) that only one resistor r in a given branch is allowed to take on values from a preset interval R. Find:

a) the tolerance I on the current i through r;

b) the tolerance V on the voltage v across r.

The only correct way to deal with the above problem (if we are to obtain the exact solution remaining in the framework of this chapter's approach) is to set up and solve an associated linear system with interval coefficients . We shall show that it is not possible to use the solution of a) to solve b).

Using Thevenen's theorem

$$i = \frac{v_{oc}}{r_e + r}$$

Based on Theorem 1.4 the lower endpoint $\underline{I}$ of I is

$$\underline{I} = \frac{V_{oc}}{r_e + \overline{R}}$$

On the other hand

$$v = ri$$

or in interval form

$$V = RI$$

Using this formula we might be tempted to write

$$\underline{V} = \underline{R}\,\underline{I}$$

This result is, however, incorrect. Indeed,

$$v = ri = r\frac{v_{oc}}{r_e + r} = f(r)$$

Obviously

$$\underline{V} = \min_{r \in R} f(r)$$

and $\underline{V}$ is attained for some unique $r \in R$ .Therefore, the formula

$$\underline{V} = \underline{R}\,\underline{I} = \underline{R}\frac{V_{oc}}{r_e + \overline{R}}$$

cannot be correct since it yields $\underline{V}$ as a function of two different values of r, namely $\underline{R}$ and $\overline{R}$. (The error is, of course, due to the fact that R and I have been treated as independent intervals whereas they are, in fact, dependent through r.)

Finally it should be stressed that the equivalent transformation of a current source into a voltage source is not possible in the interval case. Indeed, for a current source made of the parallel connection of i and r, the corresponding equivalent voltage source is the series connection of $u = ri$ and r, all quantities i, r and u being real numbers. Now let $i \in I$ and $r \in R$. Then $u \in U$ where $U = RI$. In the interval case the equivalent voltage source is formed of the series connection of R and $U = RI$. It is seen that while the current source is made of two independent interval quantities, the equivalent voltage source involves two dependent intervals: R and RI. That is why Problem 3.1 cannot cover circuits containing interval current sources.

3.1.3. Alternate implicit form formulations

In this section, we shall consider the implicit form formulation for several more general d.c. tolerance problems than Problem 3.1 from section 3.1.1

Problem 3.2. Given the resistive circuit **N** as defined in section 3.1.1 and the tolerances on the branch resistors and source voltages, find the tolerances on the branch currents and/or the branch voltages.

It will now be shown that an appropriate linear interval system with independent coefficients of the general form (3.1) can be found for solving Problem 3.2 Indeed, (3.2) can be rewritten as

$$r_\rho i_\rho + v_{b_\rho} = u_\rho, \quad \rho = \overline{1,m} \tag{3.20}$$

where v_{b_ρ} is the branch voltage of the ρth branch. Choosing $m - n$ independent loops we have additionally by KVL

$$\sum_{j=1}^{n} \beta_{kj} v_{b_j} = 0, \quad k = \overline{1, m-n} \tag{3.21}$$

Now let $\beta = \{\beta_{kj}\}$ be the corresponding loop matrix of the circuit considered. Using the matrices α, r (as defined before), β and the vectors i_b and v_b for the branch currents and branch voltages, respectively, Eqs.(3.20), (3.3) and (3.21) can be put in vector form:

$$ri + Ev_b = u \tag{3.22a}$$

$$\alpha i = 0 \tag{3.22b}$$

$$\beta v_b = 0 \tag{3.22c}$$

(where E is the identity matrix).

If in (3.22) r and u are replaced by their interval counterparts R and U then we arrive at a system of linear interval equations with independent coefficients:

$$\begin{bmatrix} R & E \\ \alpha & 0 \\ 0 & \beta \end{bmatrix} \begin{bmatrix} i \\ v_b \end{bmatrix} = \begin{bmatrix} U \\ 0 \end{bmatrix} \tag{3.23}$$

Now the tolerances on all currents and all branch voltages can be found by solving the interval system (3.23).

We shall now consider a d.c. circuit containing additionally independent current sources. First, we shall assume that all current sources i_{s_k}, $k = \overline{1, n}$ are known exactly. In this case the implicit form formulations of Problems 3.1 and 3.2 are not affected since the vector $i_s = (i_{s_1}, \ldots, i_{s_n})$ appearing in the right-hand side of (3.10) or (3.23) is constant. However, if the vector i_s is allowed to vary within some interval vector I_s, then Problems 3.1 and 3.2 might not be formulated as a linear interval system with independent coefficients. Indeed, suppose that some current source i_{s_0} is connected between two ungrounded nodes p and q. Then two equations of (3.3) will be of the form

$$\sum_{j=1}^{m} \alpha_{pj} i_j = I_{s_0} \tag{3.24a}$$

$$\sum_{j=1}^{m} \alpha_{qj} i_j = -I_{s_0} \tag{3.24b}$$

where I_{s_0} is the interval containing i_{s_0}. Obviously, we have two dependent interval variables on the RHS of (3.10) or (3.23). In order to obtain linear interval systems with independent coefficients we have to restrict the class of circuits having current sources.

P r o b l e m 3.3. Let **N** be a resistive circuit as defined in Problem 3.1 containing additionally current sources only in parallel branches incident on the grounded node of the circuit. The tolerance problem is: given the tolerances on the branch resistors, and voltage and current sources find the tolerances on

i) all the branch currents and/or nodal voltages;
ii) all the branch currents and/or branch voltages.

Obviously, Problem 3.3 can be formulated as a linear interval system (3.1) with independent coefficients since each current tolerance I_{s_ρ}, $\rho = \overline{1, n}$ will appear either in (3.24a) or 3.24b) (but not in both) and, hence only once in the RHS of (3.10) or (3.23). Thus, the linear interval system for Problem 3.3 (i) will be

$$\begin{bmatrix} R & \alpha^T \\ \alpha & 0 \end{bmatrix} \begin{bmatrix} i \\ v \end{bmatrix} = \begin{bmatrix} U \\ I_s \end{bmatrix} \tag{3.25}$$

while Problem 3.3(ii) will be formulated as

$$\begin{bmatrix} R & E \\ \alpha & 0 \\ 0 & \beta \end{bmatrix} \begin{bmatrix} i \\ v_b \end{bmatrix} = \begin{bmatrix} U \\ I_s \\ 0 \end{bmatrix} \tag{3.26}$$

The preceding tolerance problems deal with circuits having only independent sources. Now we shall generalize the previous results to circuits including dependent sources. Let **N** be a resistive circuit containing additionally all four types of dependent sources:

a) voltage-controlled voltage source (VCVS)

$$v_s = k_1 v_p \tag{3.27a}$$

where v_p is the independent (branch) voltage and v_s is the controlled voltage

b) current-controlled voltage source (CCVS)

$$v_s = k_2 i_p \tag{3.27b}$$

c) voltage-controlled current source (VCCS)

$$i_c = k_3 v_p \tag{3.27c}$$

d) current-controlled current source (CCCS)

$$i_c = k_4 i_p \tag{3.27d}$$

For simplicity of analysis we shall assume that v_s is a branch voltage also (if this is not the case we can always insert an artificial node between the voltage source and the series connected resistor to obtain v_s as a branch voltage).

If k_1 to k_4 are fixed constants then (as is easily seen from the previous results and (3.27)) it is always possible to derive a linear interval system of the (3.1) type with independent coefficients for each of the above tolerance problems. However, the situation becomes more complicated if k_1 to k_4 are allowed to change within some prescribed intervals K_1 to K_4. In the case of systems (3.23) or (3.26), presence of any type of controlled sources leads always to interval systems with dependent coefficients. This is clear from (3.22) (the branch current vector i and the branch voltage vector v_b occur twice) and (3.27). For a similar reason system (3.10) becomes a system with dependent coefficients if current sources of both types are included in the circuit. Indeed, (3.10) can be written as

$$Ri + \alpha^T v = U \tag{3.10a}$$

$$\alpha i = 0 \tag{3.10b}$$

and it is seen that the current vector i occurs twice which together with (3.27c), (3.27d) leads to coefficient dependence).

In a special case involving only dependent voltage sources it is possible to arrive at a linear system with independent interval coefficients.

Problem 3.4. Let the resistive circuit **N** as defined in Problem 3.3 contain additionally CCVS's (in any branch) and VCVS's only in branches incident on the grounded node with interval coefficients K_2 and K_1. Given the interval coefficients of the dependent voltage sources and the tolerances on the branch resistors and the independent sources determine the tolerances on the branch currents and/or nodal voltages.

Problem 3.4 can be formulated as a corresponding system (3.1). Indeed, the nodal voltage vector v occurs only once in (3.9) which, on account of (3.27a) and (3.27b), leads to a linear system with independent interval coefficients.

In the next section the exact solution of the d.c. tolerance Problems 3.1 to 3.4 will be obtained. A method will also be presented (subsection 2.3.4) which, under certain conditions, provides the exact solution of d.c. tolerance problems even when they are formulated as linear interval equations with dependent coefficients.

3.2. EXACT SOLUTION OF THE D.C. TOLERANCE ANALYSIS PROBLEM

3.2.1. Basic results for circuit equations with independent coefficients

In this section we shall be dealing with the interval linear system (3.1)

$$A^I y = B \ , \quad A^I \in I(R^{N\times N}) \ , \quad B \in I(R^N) \tag{3.28a}$$

Recall that (3.28a) is short notation for the following family of linear algebraic systems

$$Ay = b \ , \quad A \in A^I \ , \quad b \in B \tag{3.28b}$$

Throughout this section, we shall be using the following symbols and notations:

a_{ij} – element of the real matrix A

$|A|$ – matrix with elements $|A|_{ij} = |a_{ij}|$ (the same notation will be used for vectors)

A ≥ 0 (and similar relations) are meant componentwise (i.e. $a_{ij} \geq 0$)

A_{ij} – element of the interval matrix A^I

$\rho(A)$ – special radius of A

By analogy with the scalar case (1.7) A^I and B will be written as

$$A^I = [A_c - \Delta\ ,\ \ A_c + \Delta]\ ,\ \ \Delta \geq 0$$
$$B\ = [b_c - \delta\ ,\ \ b_c + \delta]\ ,\ \ \delta \geq 0$$

where A_c or b_c are the center (midpoint) matrix or vector of A^I or B, respectively. The real matrix Δ whose elements are

$$\Delta_{ij} = w(A_{ij})/2$$

is called radius of A^I. Similarly, the real vector δ with elements

$$\delta_i = w(B_i)/2$$

is called radius of B.

Furthermore, let

$$\begin{aligned} e &= (1,1,\ldots,1)^T \in R^N,\ \ f = -e \\ W &= \{\, w \colon w \in R^N,\ |w| = e\, \} \end{aligned} \tag{3.29}$$

It is seen that W is the set of all N-dimensional vectors whose components are either +1 or −1. Clearly, W contains 2^N distinct vectors of such form.

For each $w \in W$, T_w denotes a diagonal matrix whose diagonal is w. Thus,

$$T_e = E,\ \ T_f = -E$$

where E is the identity matrix.

For each $y \in R^N$ we assign the vector sgn y whose components are defined as

$$(\operatorname{sgn} y)_i = \begin{cases} 1 & \text{if}\ \ y_i \geq 0 \\ -1 & \text{if}\ \ y_i < 0 \end{cases} \tag{3.30}$$

Hence sgn $y \in W$. Furthermore, if

$$z = \operatorname{sgn} y \tag{3.31}$$

then, as is easily seen,

$$|y| = T_z y \tag{3.32}$$

In (3.31) and (3.32) (as well as in similar formulae encountered later) the equality is meant componentwise.

Let S be a compact bounded set in R^N. We introduce two N-dimensional vectors min S and max S such that

$$(\min S)_i = \min \{y_i\colon y \in S\}$$
$$(\max S)_i = \max \{y_i\colon y \in S\}$$
$$i = \overline{1,N}$$

Thus [min S, max S] is the narrowest interval vector containing S. This interval is called the interval hull of S.

An extreme point y^s of S is such a point that cannot be represented in the form

$$y^s = \frac{1}{2}(y^1 + y^2)$$

for two arbitrary but distinct points y^1, $y^2 \in S$.

The notation Conv S will be used for convex enclosure of S.

We now return to system (3.28). We assume A^I is regular. As before the set

$$S = \{y\colon Ay = b,\ A \in A^I,\ b \in B\} \tag{3.33}$$

denotes the solution set of system (3.28). The interval hull $\underline{Y} = [\underline{y}, \overline{y}]$ of S where $\underline{y}$ = min S, $\overline{y}$ = max S is the optimal interval solution of (3.28).

The extreme points, the convex enclosure Conv S and the interval hull Y of S are shown in Fig 3.1 for $N = 2$

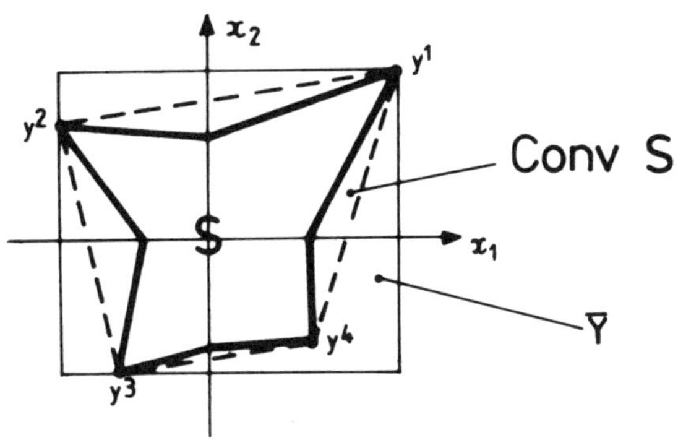

Fig. 3.1. Geometrical representation of the solution set $\underline{S}$, its extreme points y^1, y^2, y^3 and y^4, the convex hull Conv S and the interval hull $\overline{Y}$ for $N = 2$.

The solution set S has a number of properties studied in [13]–[15]. It has been shown in [13] that S can be described as

$$S = \{y: |A_c y - b_c| \le \Delta|y| + \delta\} \tag{3.34}$$

Two basic properties that follow from (3.34) will be given here.

Property 3.1. The set S may, in the general case, be not convex.

Property 3.2. The intersection of S and each orthant of R^N is a convex bounded polyhedron (see Fig 3.1 for geometrical illustration in the two-dimensional case).

It has been proven in [13] that each extreme point of Conv S satisfies the equation

$$|A_c y - b_c| = \Delta|y| + \delta \tag{3.35}$$

Let

$$w = \operatorname{sgn}(A_c y - b_c) \tag{3.36}$$

and

$$z = \operatorname{sgn} y$$

On account of (3.31), (3.32) and (3.36)

$$|A_c y - b_c| = T_w(A_c y - b_c)$$

$$|y| = T_z y$$

whence

$$T_w(A_c y - b_c) = \Delta T_z y + \delta \tag{3.37}$$

On multiplying (3.37) from the left by T_w (having in mind that $T_w^2 = E$) we get

$$A_c y - b_c = T_w \Delta T_z y + T_w \delta$$

or

$$(A_c - T_w \Delta T_z) y = b_c + T_w \delta$$

Finally

$$A_{wz} y = b_w \tag{3.38}$$

where

$$A_{wz} = A_c - T_w \Delta T_z \tag{3.39}$$

$$b_w = b_c + T_w \delta \tag{3.40}$$

Since $z = \operatorname{sgn} y$, the inequality

$$T_z y \geq 0 \tag{3.41}$$

holds for any $w \in W$. Thus, we have shown that (3.35) is equivalent to the following system

$$\begin{cases} A_{wz} y = b_w \\ T_z y \geq 0 \end{cases} \tag{3.42}$$

It is proven in [13] that for any $w \in W$ system (3.42) has a unique solution y_w which, owing to the equivalence between (3.35) and (3.42), is an extreme point of Conv S. Therefore,

$$\begin{aligned} (\underline{y_i}) &= \min \{(y_w)_i : y_w,\ w \in W\} \\ (\overline{y_i}) &= \max \{(y_w)_i : y_w,\ w \in W\} \\ & i \in \overline{1,N} \end{aligned} \tag{3.43}$$

where the set W contains, in general, 2^N vectors w. Thus the number of extreme points of Conv S is 2^N. Hence, we have to solve (3.42) 2^N times to find the optimal interval solution to (3.28).

3.2.2. General method

In this section a general method for solving the d.c. tolerance problems 3.1 to 3.4 is suggested. It consists in finding the optimal interval solution to the corresponding linear system with independent interval coefficients (3.10), (3.23), (3.25) or (3.26). The method will be presented for the case of Problem 3.1. It is based on the basic results from section 3.2.1 taking into consideration some specific features of the problem considered. On account of (3.4) to (3.8) and (3.10), in this case $N = m + n$ and

$$A_c = \begin{bmatrix} r^c & \alpha^T \\ \alpha & 0 \end{bmatrix} \tag{3.44a}$$

$$\Delta = \begin{bmatrix} \Delta r & 0 \\ 0 & 0 \end{bmatrix} \tag{3.44b}$$

$$b_c = \begin{bmatrix} u^c \\ 0 \end{bmatrix}, \quad \delta = \begin{bmatrix} \Delta u \\ 0 \end{bmatrix} \tag{3.44c}$$

where r^c is the center of the interval diagonal matrix R of the branch resistances, Δr is the radius of R; u^c and Δu are the center and the radius of the interval vector U of branch source voltages, respectively (recall that m is the number of branches and n is the number of ungrounded nodes of the circuit studied).

Due to the specific form of Δ and δ, it is seen from (3.44) and (3.38) to (3.40) that in this case the number of the extreme points of Conv S is equal to m. Indeed, the last n components of each $w \in W$ are not relevant to the solution of the tolerance problem considered since they are canceled by the last n zero components of Δ and δ. Thus, the set W from section 3.2.2 containing 2^N vectors w in the general case is reduced for the tolerance problem 3.1 to the set W' containing 2^m vectors w. Each vector $w \in W'$ has variable (+1 or −1) values for its first m components; its last n components are arbitrary and may be fixed, say, to +1.

Based on the foregoing, the d.c. tolerance problem 3.1 can be solved exactly in the following way. For each $w \in W'$ we solve system (3.42) to find the corresponding extreme point y_w. Then

$$\begin{aligned} \underline{y} &= \min\{y_w : w \in W'\} \\ \overline{y} &= \max\{y_w : w \in W'\} \end{aligned} \tag{3.45}$$

In [13] the following algorithm (called sign-accord algorithm) for solving (3.42) is suggested (see also [37]).

Sign-accord algorithm

Step 0. For a given w (now $w \in W'$) find $z = \text{sgn}\,(A_c^{-1}b_w)$

Step 1. Solve the system of linear equations.

$$A_{wz}y = b_w \tag{3.46}$$

Step 2. If

$$T_z y \geq 0$$

terminate. In this case $y_w := y$ is an extreme point (the symbol $:=$ has the usual meaning of assignment). Otherwise go to the next step.

Step 3. Find the index $k = \min\{j\colon z_j y_j < 0\}$.

Step 4. Let $z_k := -z_k$ and return to Step 1.

It is proven that the sign-accord algorithm terminates in a finite number of iterations. Very often, if A^1 is narrow enough, it actually converges in only one iteration. Indeed, z will not change if $y' = A_c^{-1}b_w$ and y_w lie in the one and the same orthant.

After all 2^m extreme points y_w have been found, the exact solution of the d.c. tolerance problem 3.1 is determined by means of (3.45).

It is clear from the foregoing that the general method is applicable, with minor modifications, to Problems 3.2 to 3.4 also. For example, $N = 2m$ for Problem 3.2; in Problem 3.3 the set W' contains 2^{m+n} vectors.

Based on the structure of the general method we have the following result.

Proposition 3.3. For any d.c. tolerance problem that can be formulated as a linear system of equations with independent interval coefficients each endpoint of the tolerance on any of the output variables is provided by a vertex of the box X of the interval input parameters (that is, by a specific combination of their endpoints).

Proof. It is seen from (3.45) that each endpoint of the tolerance on any output variable is determined by means of a corresponding extreme point y_w. On the other hand each extreme point is the image of some vertex of box X.

On account of Proposition 3.3 a "brute force" combination method for solving any of the d.c. tolerance analysis problems 3.1 to 3.4 would consist in solving 2^{n1} real systems of linear equations corresponding to 2^{n1} possible combinations of lower or upper endpoints for each input parameter with n_1 being the total number of input parameters. Under the assumption that the sign-accord method terminates in one iteration (which seems to be always the case for tolerance problems) the general method requires the solution of 2^{n2} real linear systems where n_2 is the number of elements in the set W' for the corresponding interval linear system. It is easily seen that for all problems 3.1 to 3.4 the brute-force method results in a greater amount of computation as compared to the general method since $n_1 > n_2$ always. For example, for Problem 3.3 (ii) $n_2 = m + n$ as is seen from (3.26) while $n_1 = 2m + n$. Thus, for this problem the former method is 2^m times more expensive than the latter method. Nevertheless, the general method herein suggested is rather time consuming. Thus, in the case of Problem 3.1 (assuming that the sign-accord algorithm terminates in just one iteration) it requires the ordinary (noninterval) system (3.46) of size N x N ($N = m + n$) to be solved 2^m times. Obviously, it can only be applied for tolerance analysis of circuits of moderate size.

The efficiency of the general method for dc tolerance analysis may be improved if the circuit studied permits application of Theorem 1.2 to part of the independent input parameters. To illustrate this possibility we shall take up Example 2.3 from Chapter 2. It was shown there that using the explicit formulation for the tolerance problem therein considered the lower endpoint i_L^* of the tolerance on the input current i for the bridge circuit studied can be found as the global minimum of the function (2.10) where the variables $v = \underline{v}$, $r_1 = \overline{r}_1$, $r_4 = \overline{r}_4$ and $r_5 = \overline{r}_5$ are fixed while r_2, r_3 and r_6 are allowed to vary within their tolerances. Based on this result and the implicit formulation for the circuit at hand i_L^* can be determined as the lower endpoint of the corresponding solution component of an associated linear interval system in which only R_2, R_3 and R_6 are intervals. In a similar way, i_U^* can be found by solving another linear interval system in which the variables v, r_1, r_4 and r_5 are fixed at v, $\underline{r}_1$, $\underline{r}_4$ and $\underline{r}_5$.

Several other cases where the general method can be substantially improved will be considered in the next section.

3.2.3. Improved efficiency method

In some special cases when the interval matrix A^I has certain specific properties the general method from the previous section can be modified resulting in a substantial improvement of its numerical efficiency.

Case A

In this case the matrix A^I must first satisfy the condition

$$\rho(A^I) < 1 \tag{3.47}$$

A sufficient condition for (3.47) to hold is the validity of the following inequality [13]

$$\rho(D) < 1 \tag{3.48}$$

where

$$D = |A_c^{-1}|\Delta$$

Let

$$C = D(E-D)^{-1}$$

Now the following matrices are computed

$$\tilde{B} = A_c^{-1} - C|A_c^{-1}|$$

$$\tilde{B} = A_c^{-1} + C|A_c^{-1}|$$

and for each $i \in N$ the set of vectors W_i is introduced

$$W_i = \left\{ w_j: \begin{array}{lll} w_j = 1 & \text{if} & \tilde{B}_{ij} > 0 \quad (3.49a) \\ w_j = -1 & \text{if} & \tilde{B}_{ij} > 0 \quad (3.49b) \\ |w_j| = 1 & & \text{otherwise} \quad (3.49c) \end{array} \right.$$

It is proven in [14] that in this instance

$$\overline{y}_i = \max \{(y_w)_i : w \in W_i\} \tag{3.50a}$$

$$\underline{y}_i = \min \{(y_w)_i : w \in (-W_i)\} \tag{3.50b}$$

where

$$-W_i = \{-w : w \in W_i\}, \quad i = \overline{1,N}$$

As seen from (3.49) the set

$$\tilde{W} = (\bigcup_{i=1}^{N} W_i) \cup (\bigcup_{i=1}^{N} -W_i)$$

may be smaller that the set W' from (3.45) if (3.49a) and (3.49b) are valid for several indices i (only the components w_j corresponding to (3.49c) are not fixed and can be either +1 or −1). In the extreme case if (3.49a) and (3.49b) hold for each i then W_i consists of one single vector. Therefore, the exact solution of the d.c. tolerance problem considered can be obtained rather efficiently since now $\tilde{W}$ consists of $2N$ vectors. Thus, on account of(3.50), we need to solve problem (3.42) only $2N$ times.

Case B

In this case the matrix A^I is inverse-stable [13], i.e.

$$|A^{-1}| > 0 \quad \text{for} \quad \forall A \in A^I$$

An interval matrix A^I is inverse-stable if (3.47) holds and

$$C\,|A_c^{-1}| < |A_c^{-1}| \tag{3.51}$$

Here again the optimal solution is determined by (3.50) but now the set $\tilde{W}$ is guaranteed to consist of only $2N$ vectors. Indeed, it is proven that if A^I is inverse-stable then

$$W_i = \operatorname{sgn}(A_c^{-1})_i \tag{3.52}$$

where $(A_c^{-1})_i$ is the ith row of A_c^{-1}. Thus, the exact solution Y of the d.c. tolerance problem is, in this case, guaranteed to be obtained by solving problem (3.42) only $2N$ times.

If additionally A^I is narrow enough (which is generally the case in practice) the sign-accord algorithm converges in only one iteration. (A sufficient condition for this is the condition $D\,|y_w| < |y_w|$ [15]). Thus, the exact tolerances on all N variables (branch currents and node or branch voltages) are found in this case by solving only $2N$ ordinary linear systems (3.46).

Example **3.2.** The circuit studied is given in Fig. 3.2. It has $m = 11$ branches and $n + 1 = 6$ nodes. Every resistor has a nominal resistance $r_k^c = 100\ \Omega$, $k = \overline{1, m}$, and an equal tolerance radius $\Delta_k = w(R_k)/2 = 2\ \Omega$. The source voltages are $e_1^c = e_2^c = 100$ V, $e_5^c = e_7^c = 10$V and are assumed to have zero tolerances. The problem is to find the intervals of all branch currents i_k, $k = \overline{1, m}$ and the intervals of all node voltages V_k, $k = \overline{m + 1, N}$ (the last $(n+1)$th node is grounded, i.e. $V_{17} = 0$).

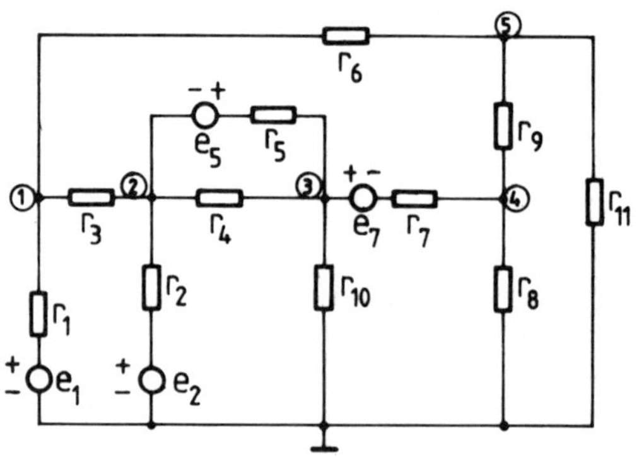

Fig. 3.2. Illustrative example.

The problem considered was solved using the simplified method based on formulae (3.48) to (3.52). The condition (3.48) was tested in a simple manner (which circumvents the need for determining the eigenvalues of D) to be presented later in Chapter 4, section 4.2.1. A numerical program implementing the present method has been developed. For the example considered, the following results have been obtained.

$i_1 \in [\ 0.379369,\ \ 0.400715]\,A,\qquad i_2 \in [\ 0.421617,\ \ 0.440620]\,A$

$i_3 \in [\ 0.034411,\ \ 0.052416]\,A,\qquad i_4 \in [\ 0.179552,\ \ 0.197247]\,A$

$i_5 \in [\ 0.278983,\ \ 0.297855]\,A,\qquad i_6 \in [\ 0.337637,\ \ 0.355695]\,A$

$i_7 \in [-0.105624, -0.090322]\,A,\qquad i_8 \in [-0.389692, -0.367897]\,A$

$i_9 \in [-0.089821, -0.076092]\,A,\qquad i_{10} \in [\ 0.172867,\ \ 0.188959]\,A$

$i_{11} \in [-0.272800, -0.254743]\,A,\qquad v_{12} \in [60.023969, 61.999363]\,V$

$v_{13} \in [55.671529, 57.690911]\,V,\qquad v_{14} \in [36.918763, 38.814119]\,V$

$v_{15} \in [17.316638, 18.851887]\,V,\qquad v_{16} \in [25.517193, 27.226597]\,V$

Since the sign-accord algorithm converged every time in one iteration the above results were found by solving $2N = 32$ linear system of type (3.46).

Several other examples have been solved using the numerical program developed. It has turned out that in each example the associated interval matrix A^I was inverse-stable and the sign-accord algorithm terminated in just one iteration. Thus, for all the set of circuits studied the d.c. tolerance analysis problems were solved in a rather efficient way by solving $2N$ systems of N ordinary linear equations for each circuit.

Case C

In this case A^I must be inverse-positive, that is,

$$A^{-1} > 0 \quad \text{for} \quad \forall A \in A^I \tag{3.53}$$

Then [13]

$$\underline{y} = y_f, \quad \overline{y} = y_e \tag{3.54}$$

where y_f and y_e are the solutions of (3.42) for $w = f$ and $w = e$ respectively, where e and f are given by (2.29). In this case the solution Y is found by solving (3.46) only twice. However, the requirement A^I to be inverse-positive seems to be rather restrictive for arbitrary circuits of the class studied. Moreover, nowadays there exists no simple condition for verifying (3.53)

Case D

Now the set S of solutions to (3.1) lies in one single orthant $R_z^N = \{y \in R^N\colon T_z\, y \geq 0\}$, that is,

$$S \subset R_z^N \tag{3.55}$$

Let

$$D = |A_c^{-1}|\Delta, \quad d = |A_c^{-1}|\delta, \quad y_c = A_c^{-1} b_c$$

The following sufficient condition for (3.55) to hold is given in [15]. If

$$\rho(D) < 1 \tag{3.56}$$

and

$$(E - D)^{-1}(|y_c| + d) < 2\,|y_c| \tag{3.57}$$

then $S \subset R_z^N$ for

$$z = \operatorname{sgn} y_c \tag{3.58}$$

If the condition (3.55) is fulfilled, then by Property 3.2 the set S is a convex bounded polyhedron. Therefore, $\underline{y}_i$ (and $\overline{y}_i$, respectively) can be found as the solution of an associated problem from linear programming. Indeed, since in this case the signature (sgn y) of y is known the set S can be represented in the form [14]

$$S = \{y\colon A_{ez} y \le \overline{b},\ A_{fz} y \ge \underline{b},\ T_z y \ge 0\} \tag{3.59}$$

Based on (3.59) it is readily seen that $\underline{y}_i$ can be found by solving the following linear programming (LP) problem

$$\underline{y}_i = \min\ (z_i y_i) \tag{3.60a}$$

$$\begin{aligned} &A_{ez} y \le \overline{b}, \quad -A_{fz} \le \underline{b}, \\ &-z_j y_j \le 0, \quad j = \overline{1,N} \end{aligned} \tag{3.60b}$$

Obviously, $\overline{y}_i$ is the solution of the LP problem

$$\overline{y}_i = -\min\ (-z_i y_i) \tag{3.61}$$

with constraints (3.60b).

To find the intervals of all branch currents and node or branch voltages it is necessary to solve a total of $2N$ LP problems of the type (3.60) and (3.61).

The LP formulation of the tolerance problem is attractive since it permits one to use directly available software packages designed to solve general LP problems. Several d.c. tolerance analysis problem (associated with circuits comparable in complexity with that of Fig. 3.2) were easily solved using the linear programming formulation of the problems.

3.2.4. Exact solution for circuits equations with dependent coefficients

The methods considered so far are only capable of providing the exact solution to such d.c. tolerance problems that can be formulated in the form of a corresponding system of linear equations with independent coefficients. However, as was shown in section 3.1.3, there exists a large class of resistive circuits (in fact, all circuits that are not covered by the conditions related to Problems 3.1 to 3.4) for which the implicit form formulation of various tolerance problems leads to a system of linear equations with dependent coefficients. In this section a method for exact solution of such d.c. tolerance problems will be presented. The method (based on a paper of J. Rohn [35]) is applicable if certain verifiable (and seemingly not very restricting) conditions are fulfilled.

To account for the interdependence between some circuit parameters we will treat part of them as independent and the rest as dependent parameters.

Example **3.3.** Consider Problem 3.3.(ii). The system (3.26) will be written in real (noninterval) form as

$$ri + Ev_b = u \tag{3.62a}$$

$$\alpha i = i_s \tag{3.62b}$$

$$\beta v_b = 0 \tag{3.62c}$$

Now assume that the particular circuit considered has two independent current sources and $n = 3$. Then (3.62b) may be

$$\begin{aligned} \alpha_{11}i_1 + \ldots + \alpha_{1m}i_m &= i_{s_1} \\ \alpha_{21}i_i + \ldots + \alpha_{2m}i_m &= -i_{S_1} \\ \alpha_{31}i_1 + \ldots + \alpha_{3m}i_m &= i_{s_2} \end{aligned} \tag{3.63}$$

(if the second source i_{s2} is in a branch incident on the grounded $(n + 1)$th node). Assume additionally that there is one dependent VCVS, e.g. $v_{b_1} = kv_{b_n}$. Then the first equation of (3.62a) and (3.62c) will be

$$\begin{aligned} r_1 i_1 + kv_{b_n} &= u_1 \\ \beta_{11}kv_{b_n} + \sum_{j=2}^{m} \beta_{1j}v_{b_j} &= 0 \end{aligned} \tag{3.64}$$

It is seen from (3.64) that the circuit considered leads to a system of equations with dependent coefficients. Indeed, if the RHS of (3.62) is the column vector b, then from (3.63)

$$b_{m+1} = -b_{m+2} \tag{3.65}$$

Similarly, if the coefficients in the LHS of (3.62) form the matrix A, then from (3.64)

$$a_{m+1,m+n} = \beta_{11}a_{1,m+n} \tag{3.66}$$

Now we may consider b_{m+2} and $a_{1,m+n}$ as independent coefficients and b_{m+1} and $a_{m+1,\,m+n}$ as dependent coefficients (or vice versa).

At this stage we will introduce the concept of feasibility of a vector $b \in R^N$ and a matrix $A \in R^{N \times N}$. Starting with b we will distinguish two sets of indices:

J_1 - set of indices of the independent components b_j of b,

J_2 - set of indices of the dependent components b_i of b such that $J_1 \cap J_2 = \emptyset$, $J_1 \cup J_2 = \overline{1, N}$. We call b feasible if

$$b_i = \sum_{j\in J_1} c_j^i b_j, \quad i \in J_2 \tag{3.67}$$

where c_j^i are fixed prescribed real numbers. For Example 3.3 $J_1 = \overline{1, m} \cup \overline{m+2, 2m}$, $J_2 = m + 1$ and from (3.65) all $c_j^i = 0$ except for $c_{m+2}^{m+1} = -1$.

Let the independent components b_j, $j \in J_1$, lie in some prescribed intervals B_j. We now introduce the set of feasible vectors

$$b^F = \{b\colon b_j \in B_j, j \in J_1, b_i = \sum_{j\in J_1} c_j^i b_j, i \in J_2\}$$

We also form an associated interval vector b^I in the following way

$$b^I = [b^c - \delta, b^c + \delta]$$

where

$$b_i^c = \sum_{j\in J_1} c_j^i b_j^i, \quad \delta_i = \sum_{j\in J_1} |c_j^i|\,\delta_j, \quad i \in J_2$$

Obviously (because of feasibility), b^F is not an interval vector (a box) and $b^F \subset b^I$.

In a similar way, for a matrix A we distinguish two sets of indices:

I_1 - set of indices (i, j) for the independent coefficients a_{ij},

I_2 - set of indices (i, j) for the dependent coefficients a_{ij} such that $I_1 \cap I_2 = \varnothing$, $I_1 \cup I_2 = \{(i, j) : 1 \le i, j \le N\}$ We call A^1 feasible if

$$a_{lh} = \sum_{(i,j)\in I_1} c_{ij}^{lh} a_{ij}, \quad (l,h) \in I_2 \tag{3.68}$$

with c_{ij}^{lh} fixed numbers. For Example 3.3 $I_2 = m+1, m+n$, I_1 comprises all the other pairs of indices and from (3.66) $c_{1,m+n,}^{m+1,\,m+n} = ß_{11}$, while all the remaining $c_{ij}^{lh} = 0$.

We now introduce the set of feasible matrices

$$A^F = \{A\colon a_{ij} \in A_{ij},\ (i,j) \in I_1,\ a_{lh} = \sum_{(i,j)\in I_1} c_{ij}^{lh} a_{ij},\ (l,h) \in I_2\}$$

where A_{ij} are prescribed intervals and the associated interval matrix

$$A^I = [A^c - \Delta,\ A^c + \Delta]$$

with

$$a_{ij}^c = m(A_{ij}), \quad \Delta_{ij} = w(A_{ij})/2, \quad (i,j) \in I_1,$$

$$a_{lh}^c = \sum_{(i,j)\in I_1} c_{ij}^{lh} a_{ij}^c, \quad \Delta_{lh} = \sum_{(i,j)\in I_1} |c_{ij}^{lh}|\Delta_{ij}, \quad (l,h) \in I_2$$

Obviously the set A^F is not an interval matrix and $A^F \subset A^I$.

Now we are in a position to formulate the following d.c. tolerance analysis problem for circuits with dependent parameters.

Problem 3.5. Given the tolerances on the independent input parameters and the feasibility conditions (3.67) and (3.68) find the tolerances on the output variables.

In a rigorous formulation we have the following problem.

Problem 3.5′. Given the set of feasible vectors b^F and the set of feasible matrices A^F as defined above, find

$$\underline{y}_i = \min \{y_i : y \in S_F\}$$

$$\overline{y}_i = \max \{y_i : y \in S_F\}$$

where S_F is the set

$$S_F = \{y : y = A^{-1}b,\ A \in A^F,\ b \in b^F\}$$

Now consider the solution set

$$S = \{y : y = A^{-1}b,\ A \in A^I,\ b \in b^I\}$$

of the interval system

$$A^I y = b^I$$

where A^I and b^I are defined as above. Since $b^F \subset b^I$ and $A^F \subset A^I$, clearly $S^F \subset S$.

We now proceed to presenting a method for solving Problem 3.5.

As $y = A^{-1}b$ is a function of a_{ij}, $(i,j) \in I_1$ and b_i, $i \in J_1$, taking partial derivatives we obtain

$$\frac{\partial y_k}{\partial a_{ij}} = -(A^{-1}C_{ij}y)_k, \quad (i,j) \in I_1, \quad k = \overline{1,N}$$

$$\frac{\partial y_k}{\partial b_j} = \sum_{i=1}^{N} (A^{-1})_{ki} b_j^i, \quad j \in J_1, \quad k = \overline{1,N}$$

where the matrices C_{ij} are given by

$$(C_{ij}) = \begin{cases} 1 & \text{if } (l,h) = (i,j) \\ c_{ij}^{lh} & \text{if } (l,h) \in I_2 \\ 0 & \text{otherwise} \end{cases}$$

and

$$b_j^i = \begin{cases} 1 & \text{if } i = j \\ c_j^i & \text{if } i \in J_2 \\ 0 & \text{otherwise} \end{cases}$$

We now assume that:

$$\text{(A1)} \qquad \sum_{i=1}^{N} (A^{-1})_{ki} b_j^i \neq 0, \quad A \in A^F, \quad k = \overline{1,N}, \quad j \in J_1$$

$$\text{(A2)} \qquad (A^{-1}C_{ij}y)_k = (A^{-1}C_{ij}A^{-1}b)_k \neq 0$$

$$A \in A^F, \quad b \in b^F, \; k = \overline{1,N}, \quad (i,j) \in I_1$$

These assumptions ensure that the corresponding partial derivatives are either positive or negative for feasible variations of A and B. Their verification may be performed by using "crude" enclosures of A^{-1} and y (without feasibility). Thus, y can be enclosed by $\tilde{Y}$ where $\tilde{Y}$ is the optimal interval solution of

$$A^I y = b^I$$

Similarly, each column of A^{-1} can be enclosed be the vector $\tilde{Y}_k$ which is the optimal interval solution of

$$A^I y = e^k$$

where $e^k = \{e_{ij}\}$ is a diagonal matrix with $e_{ki} = 1$ if $i = k$ and $e_{ki} = 0$ if $i \neq k$.

On the basis of (A1) and (A2) we define

$$s_{ij}^k = \operatorname{sgn}\,(A^{-1}C_{ij}y)_k = \operatorname{sgn}\,(A_c^{-1}C_{ij}y_c)_k, \quad (i,j) \in I_1$$

$$s_j^k = \operatorname{sgn}\left(\sum_{i=1}^{N}(A_c^{-1})_{ki}\,b_j^{\,i}\right), \quad j \in J_1,\ k = \overline{1,N}$$

and introduce matrices D^k and vectors d^k

$$(D^k)_{lh} = \begin{cases} s_{lh}^k \Delta_{lh}, & (l,h) \in I_1 \\ \sum\limits_{(i,j)\in I_1} c_{ij}^{lh}(D^k)_{ij}, & (l,h) \in I_2 \end{cases} \qquad k = \overline{1,N}$$

$$(d^k) = \begin{cases} s_i^k \delta_i, & i \in J_1 \\ \sum\limits_{j\in J_1} c_j^{\,i}(d^k)_j, & i \in J_2 \end{cases} \qquad k = \overline{1,N}$$

The exact solution of the tolerance problem considered can be found using the following theorem [35].

Theorem 3.1. Let (A1) and (A2) hold. Then for $k = \overline{1, n}$ we have:

(1) $\underline{y}_k$ is equal to the kth component of the solution of

$$(A_c + D^k)y = b_c - d^k \tag{3.69a}$$

(2) $\overline{y}_k$ is equal to the kth component of the solution of

$$(A_c - D^k)y = b_c + d^k \tag{3.69b}$$

On the basis of the foregoing we have the following method for tolerance analysis of d.c. circuits with dependent parameters. First, assumptions (AI) and (A2) are verified using crude enclosures of A^{-1} and y (neglecting the feasibility conditions and treating all elements of A and b as independent). If they hold, then the corresponding systems (3.69) are set up and solved.

Thus the output tolerances for Problem 3.5 can be determined exactly by solving $2N$ systems of real linear equations (3.69).

It should be stressed that the formulation of this section does not cover all possible d.c. tolerance problems related to circuits with dependent sources. It is easy to verify that the following problem cannot be formulated as Problem 3.5′.

Problem 3.6. The circuit to be analyzed contains dependent current sources with interval coefficients (along with all the other possible types of independent and dependent sources). Given the tolerance on all the input parameters find the tolerances on the branch currents and node or branch voltages.

Indeed, consider first the case of voltage-controlled current sources. From (3.62) it follows that at least one equation from (3.62a) and one equation from (3.62b) will have the form

$$r_\nu k_{\nu p} v_{b_p} + \ldots = u_\nu \tag{3.70a}$$

$$k_{\nu p} v_{b_p} + \ldots = i_{s\mu} \tag{3.70b}$$

It is seen from (3.70) that the corresponding coefficients $a_{lh} = r_\nu k_{\nu p}$ and $a_{ij} = k_{\nu p}$ are dependent since

$$a_{lh} = r_\nu a_{ij} \tag{3.71}$$

Obviously, similar coefficient dependence will occur if the circuit studied contains current-controlled current sources. Unlike the feasibility condition (3.68) where each coefficient c_{ij}^{lh} is a fixed number now the coefficient r_ν from (3.71) is not a constant since $r_\nu \in R_\nu$. Therefore, the above method cannot be applied to Problem 3.6. It is, however, readily seen that Problem 3.6 is the only d.c. tolerance problem which leads to dependence between the equations coefficients other than the feasibility conditions (3.68) considered above.

3.3. TOLERANCE ANALYSIS OF A.C. CIRCUITS

3.3.1. Problems statement

Similarly to the tolerance analysis of d.c. circuits, various tolerance problems (analogous to problems 3.1 to 3.5) can be formulated in the case of a.c. circuits depending on the structure of the circuit studied and the output variables specified. However, for simplicity of presentation we shall confine ourselves to a single, relatively most simple class of a.c. circuits.

Throughout this section **N** will denote a linear circuit made of ideal resistors, inductors, capacitors and independent voltage sources. We shall consider sinusoidal steady states in this class of linear circuits.

Let the parameters R_k, L_k, C_k of the branch elements of **N** lie within some prescribed tolerances R_k^I, L_k^I and C_k^I, respectively. Moreover, if the complex branch source voltage $\dot{U}_k$, $k = \overline{1, m}$, is

$$\dot{U}_k = U_{ka} + jU_{kr}$$

let the active part U_{ka} and the reactive part U_{kr} belong to some preset intervals $U_{ka}{}^I$ and $U_{kr}{}^I$. For simplicity of exposition it is assumed that we are interested in one single output complex variable: a specified branch current $\dot{I}_\nu$ or, alternatively, a specified node voltage $\dot{V}_\mu$. To be more specific, let the output variable chosen be the branch current $\dot{I}_\nu$. Given the intervals R_k^I, L_k^I, $C_{ka}{}^I$, U_{ka} and U_{kr}, $k = \overline{1, m}$, we will formulate the following tolerance analysis problems.

Problem 3.7. Find the interval of the active part $I_{\nu a}$ of $\dot{I}_\nu$.

Problem 3.8. Find the interval of the reactive part $I_{\nu r}$ of $\dot{I}_\nu$.

Problem 3.9. Find the interval of the modulus I_ν of the current $\dot{I}_\nu$.

Problem 3.10. Find the interval of the initial phase ψ_ν of $\dot{I}_\nu$.

It should be stressed right away that the above problems are independent of each other (in the sense that the solution of a problem cannot be obtained by the solutions of the other problems). To illustrate this assertion we shall consider the following situation. Assume we have solved Problems 3.7 and 3.8. Let $\overline{I}_{\nu a}$ and $\overline{I}_{\nu r}$ denote the right endpoints of the corresponding interval solutions. Now, let $\overline{I}_\nu$ be the right endpoint of the interval solution to Problem 3.9. It is not difficult to see that, in general, the relation

$$\overline{I}_\nu = \sqrt{\overline{I}^2_{\nu a} + \overline{I}^2_{\nu r}} \tag{3.72}$$

does not hold. In fact, (3.72) is valid only in the case where all the three maxima $\overline{I}_{\nu a}$, $\overline{I}_{\nu r}$ and $\overline{I}_\nu$ occur for one and the same values of R_k, L_k, C_k and U_{ka} and U_{kr}.

It should be noted that in a special case Problems 3.7 and 3.8 may be related to finding the ranges of the active power or reactive power respectively, flowing into a one-port network. Indeed, let the input voltage $\dot{U}_{in} = U_{in}$ have zero initial phase ψ_u. Then

$$P_{in} = U_{in} I_{\nu a}$$

$$Q_{in} = U_{in} I_{\nu r}$$

If, additionally, U_{in} is kept constant, the solutions of Problems 3.7 and 3.8 are solutions of the following two problems, respectively (with given intervals R_k^I, L_k^I, C_k^I, $U_{ka}{}^I$ and $U_{kr}{}^I$).

Problem 3.11. Find the range of values for the active power P_{in}.

Problem 3.12. Find the range of values for the reactive power Q_{in}.

To solve the basic Problems 3.7 to 3.10 first we have to write the circuit equations as a system of interval linear equations. Proceeding in just the same way as in section 3.2.1. we can bring the circuit equations to the following form (similar to (3.10)).

$$\begin{bmatrix} Z^I & \alpha^T \\ \alpha & 0 \end{bmatrix} \begin{bmatrix} \dot{I} \\ \dot{V} \end{bmatrix} = \begin{bmatrix} \dot{U}_I \\ 0 \end{bmatrix} \tag{3.73}$$

Here Z^I is a diagonal matrix $Z^I = \text{diag}(Z_1^I, \ldots, I_m^I)$ whose kth element is given by the complex interval

$$Z_k^I = R_k^I + j\left(\omega L_k^I - \frac{1}{\omega C_k^I}\right) = R_k^I + jX_k^I$$

with

$$X_k^I = \left(\omega \underline{L}_k - \frac{1}{\omega \overline{C}_k},\ \omega \overline{L}_k - \frac{1}{\omega \underline{C}_k}\right)$$

(clearly, a complex interval can be viewed geometrically as a rectangular region in the complex plane) while $\dot{U}^I$ is column vector

$$\dot{U}^I = (\dot{U}_1^I, \ldots, \dot{U}_m^I)^T$$

whose kth element is given by

$$\dot{U}_k^I = U_{ka}^I + jU_{kr}^I$$

where U_{ka}^I and U_{kr}^I are intervals within which the active and reactive part of $\dot{U}_k$ vary; finally

$$Y = \begin{bmatrix} \dot{I} \\ \dot{V} \end{bmatrix} \tag{3.74}$$

is a noninterval complex vector whose first m components are

$$Y_k = \dot{I}_k = I_{ka} + jI_{kr}, \quad k = \overline{1, m}$$

and its last n components are

$$Y_k = \dot{V}_{k-m} = V_{k-m,a} + jV_{k-m,r}, \quad k = \overline{m+1, N}$$

The linear complex interval system (3.73) will serve as a basis for solving the above formulated tolerance analysis problems 3.7 to 3.10.

3.3.2. Method for approximate solution

In this section, a unified method for approximate solution of all the four a.c. tolerance problems formulated in section 3.3.1 will be suggested. The present method is based on a linearization of the original nonlinear tolerance problem considered.

First, the system (3.73) will be written as

$$Z_e Y = U_e, \quad Z_e \in Z_e^I, \quad U_e \in U_e^I \tag{3.75}$$

where

$$Z_e^I = \begin{bmatrix} Z^I & \alpha^I \\ \alpha & 0 \end{bmatrix}, \quad U_e^I = \begin{bmatrix} \dot{U}^T \\ 0 \end{bmatrix}$$

and Y is given by (3.74). Let

$$Z_e = Z_e^c + \Delta Z_e \tag{3.76a}$$

$$U_e = U_e^c + \Delta U_e \tag{3.76b}$$

$$Y = Y^c + \Delta Y \tag{3.76c}$$

and

$$Z_e^I = Z_e^c + \Delta Z_e^I, \quad U_e^I = U_e^c + \Delta U_e^I$$

(where the superscript c denotes as usual the center of the corresponding interval variables). Substituting (3.76) into (3.75) and neglecting the second-order product $Z_e \, \Delta Y$ we get

$$Z_e^c \Delta Y = \Delta U_c - Y^c \Delta Z_e \tag{3.77a}$$

with

$$\Delta Z_e \in \Delta Z_e^I, \quad \Delta U_e \in \Delta U_e^I \tag{3.77b}$$

$$Y^c = (Z_e^c)^{-1} U_e^c \tag{3.77c}$$

From (3.77a)

$$\Delta Y = (Z_e^c)^{-1} \Delta U_e - (Z_e^c)^{-1} Y^c \Delta Z_e \tag{3.78}$$

Now we will introduce the notation

$$C = [(Z_e)^{-1} | -(Z_e^c)^{-1} Y^c]$$

$$\Delta P = [\Delta U_e | \Delta Z_e]^T$$

$$\Delta P^I = [\Delta U_e^I | \Delta Z_e^I]^T$$

Thus, (3.78) can be put in the equivalent form

$$\Delta Y = C \Delta P \tag{3.79}$$

The approximate solution of the a.c. tolerance problems can be obtained if we construct the set S of solutions ΔY of (3.79) when $\Delta P \in \Delta P^I$.

The complex interval vector ΔP^I has $2m$ nonzero complex interval components ΔP_i^I. Clearly, ΔP^I can be represented geometrically as a hyper-rectangle in a complex space C^{2m} with $2m$ complex coordinates. Since C is a constant complex (N x N) matrix with $N = m + n$, the solution set S is obviously the image of ΔP^I under C. Thus, we have shown that S is, in fact, a hyper-parallelopiped in the complex space C^N.

Now, let the output variable in which we are interested be the νth component Y_ν of Y. From (3,79) it is seen that Y_ν can be written in the form

$$\Delta Y_\nu = \sum_{j=1}^{2m} c_{\nu j} \Delta P_j \tag{3.80}$$

Let S_ν denote the set of all ΔY_ν obtained by (3.80) when $\Delta P \in \Delta P_j^I$, $j = \overline{1, 2m}$. From geometrical considerations it is clear that S_ν is the projection of S onto the complex plane spanned over the νth pair of (real and imaginary) coordinates. Therefore, S_ν is a convex polygon with $4m$ vertices. An illustrative example is given in Fig. 3.3 for $m = 2$.

Clearly, to construct the set S_ν it is sufficient to determine its vertices. The following procedure is designed to find all these vertices in as simple a way as possible.

Procedure 3.1.

1. Let $\Delta P_j = \Delta P_{ja} + j\Delta P_{jr}$, $j = \overline{1, 2m}$. Let $r(\Delta P_{ji}^I) = r_{ja}$ and $r(\Delta P_{jr}^{I\ I}) = r_{jr}$ denote the radii of the corresponding intervals. First we normalize the variables as follows:

$$\Delta P_j = r_{ja} \frac{\Delta P_{ja}}{r_{ja}} + j r_{jr} \cdot \frac{\Delta P_{jr}}{r_{jr}} = r_{ja} \Delta P_j' + j r_{jr} \Delta P_{j+2m}'$$

Thus

$$\Delta P_j' \in [-1, 1], \quad \Delta P_{j+2m}' \in [-1, 1]$$

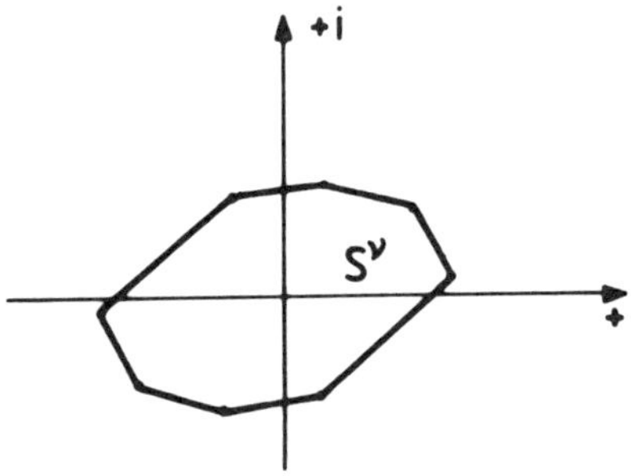

Fig. 3.3. Convex polygon S_ν with 4 m vertices for $m = 2$.

2. Calculate the complex coefficients

$$\begin{aligned} c_{\nu j}' &= c_{\nu j} r_{ja} \\ c_{\nu, j+2m}' &= j c_{\nu j} r_{jr} \end{aligned} \qquad j = \overline{1, 2m}$$

and order them in decreasing moduli. Rename (for simplicity of notation) the corresponding coefficients and normalized variables again as $c_{\nu j}$ and ΔP_j, $j = \overline{1, 4m}$. Then (3.80) takes the equivalent form

$$\Delta Y_\nu = \sum_{j=1}^{4m} c_{\nu j} \Delta P_j$$

3. At this stage, set $\Delta P_j = 0$ for $j = \overline{2, 4m}$.

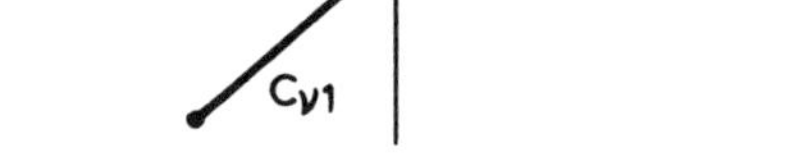

Fig. 3.4. Geometrical representation of the set $\Delta Y_\nu^{(1)}$.

Now let

$$\Delta Y_v^{(1)} = c_{v1}\Delta P_1$$

with

$$\Delta P_1 \in [-1, 1]$$

The corresponding set $\Delta Y_v^{(1)}$ is given in Fig. 3.4.

4. We now consider the set of points

$$\Delta Y_v^{(2)} = c_{v1}\Delta P_1 + c_{v2}\Delta P_2$$

with

$$\Delta P_1 \in [-1, 1], \quad \Delta P_2 \in [-1, 1]$$

which is given in Fig. 3.5. It is seen that $\Delta Y_v^{(2)}$ has 4 vertices: $(c_{v1} + c_{v2})$, $(c_{v1} - c_{v2})$, $(-c_{v1} + c_{v2})$ and $(-c_{v1} - c_{v2})$.

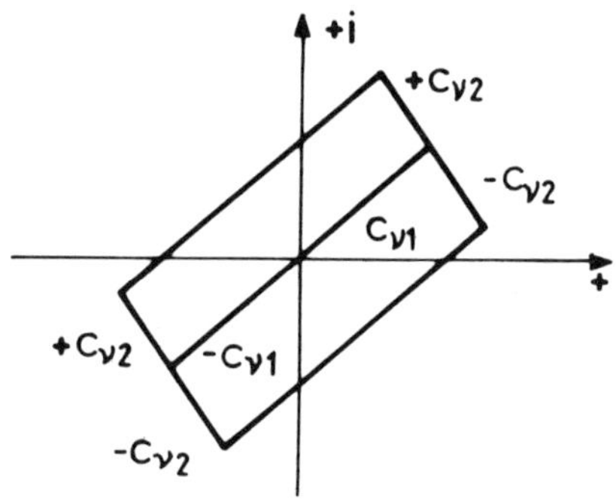

Fig. 3.5. Geometrical representation of the set $\Delta Y_v^{(2)}$.

5. Next we consider the set of points

$$\Delta Y_v^{(3)} = c_{v1}\Delta P_1 + c_{v2}\Delta P_2 + c_{v3}\Delta P_3$$

with

$$\Delta P_j \in [-1, 1], \quad j = \overline{1, 3}$$

which is given in Fig. 3.6. It is seen that now $\Delta Y_v^{(3)}$ has 6 vertices: $(c_{v1} + c_{v2} + c_{v3})$, $(c_{v1} + c_{v2} - c_{v3})$, $(c_{v1} - c_{v2} + c_{v3})$ etc.

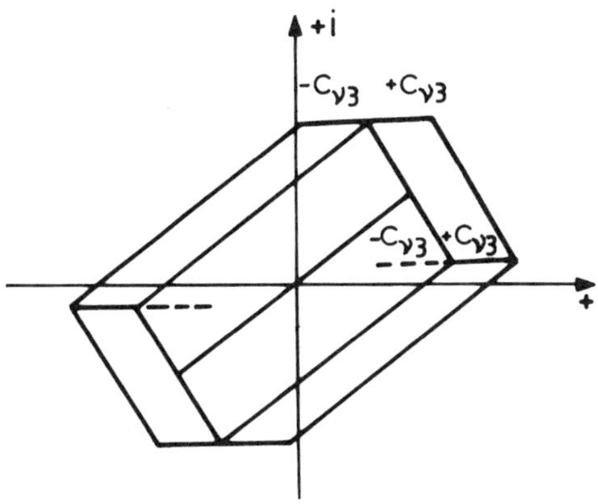

Fig. 3.6. Geometrical representation of the set $\Delta Y_v^{(3)}$.

It should be stressed that some of the possible combinations, namely $(c_{v1} - c_{v2} - c_{v3})$ and $(-c_{v1} + c_{v2} + c_{v3})$ in this instance, do not represent vertices since they correspond to points imbedded in the interior of $\Delta Y_v^{(3)}$.

6. We continue constructing $\Delta Y_v^{(4)}$, $\Delta Y_v^{(5)}$ and so on in the same manner as above. Assume we have constructed the set

$$Y\Delta_v^{(i)} = \sum_{j=1}^{i} c_{vj}\Delta P_j, \quad \Delta P_j \in [-1, 1]$$

It is clear from Fig. 3.4 to Fig. 3.6 that $\Delta Y_v^{(i)}$ has $2i$ vertices. If $i < 4m$, we go on to the set $\Delta Y_v^{(i+1)}$ which has two more vertices as compared to $\Delta Y_v^{(i)}$. (In determining the new vertices corresponding to $\Delta Y_v^{(i+1)}$ we have to avoid combinations of the vectors $\pm c_{v1}$, $\pm c_{v2}$, $\pm c_{v,i+1}$ which lead to points imbedded in $\Delta Y_v^{(i+1)}$. The technical implementation of this problem is (though tedious to present) not difficult and therefore will not be discussed here.) Finally, we will reach the set $\Delta Y_v^{(4m)}$ which is, obviously, the set S_v we set out to determine.

Up to now (using Procedure 3.1) we have found the set S^v defined by means of (3.80). Based on (3.76c), what we actually seek is the set

$$\tilde{S}_v = y_v^c + S_v \tag{3.81}$$

where y_v^c is the νth component of Y^c given by (3.77c). From (3.81) it follows that $\tilde{S}_v$ is, in fact, the set S_v translated by the vector y_v^c (see Fig. 3.7).

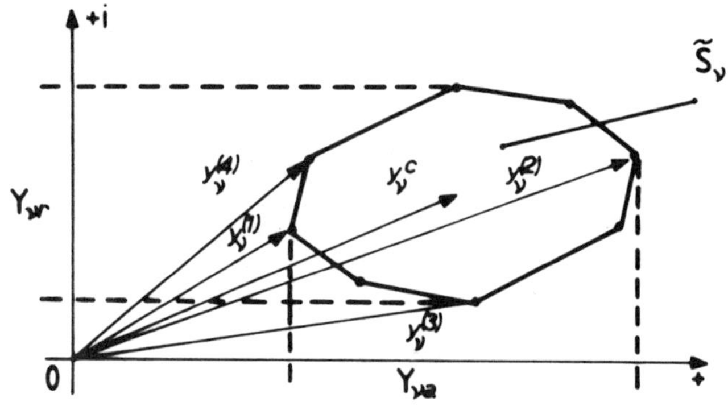

Fig. 3.7. Geometrical representation of the set $\tilde{S}_v$.

Now using the set $\tilde{S}_v$ we are in a position to approximately solve all the four problems 3.7 to 3.10. Indeed, from Fig. 3.7 it is seen that the solution of Problem 3.7 is the interval Y_{va} while the interval Y_{vr} is the solution of Problem 3.8. Similarly, the solution of Problem 3.9 is given by the interval $[\,|y_v^{(1)}|,\ |y_v^{(2)}|\,]$. Finally the solution to Problem 3.10 is found on the basis of the vectors $y_v^{(3)}$ and $y_v^{(4)}$ as the interval

$$\left[\tan^{-1}\frac{y_{vr}^{(3)}}{y_{va}^{(3)}},\ \tan^{-1}\frac{y_{vr}^{(4)}}{y_{va}^{(4)}}\right]$$

In actual computation, we have chosen to proceed as follows. We pass successively through all vertices of $\tilde{S}_v$. At each vertex $v^{(s)}$ we compute all the data related to the solution of the problems considered, namely the real part, the imaginary part, the length and the argument of the vector $y_v^{(s)}$ connecting the origin of the complex plane with the vertex $v^{(s)}$. We store the minimum and the maximum for the corresponding value, thus obtaining the solution of the respective problem.

With the exception of the lower endpoint of the solution interval of Problem 3.9, the endpoints of the solution intervals of all the problems always occur at some vertices of $\tilde{S}_v$ (for a geometrical justification see Fig. 3.7). Indeed, in the general case, the lower endpoint of $|y_v|$, $y_v \in \partial\tilde{S}_v$, may happen to take place at a point on an edge between two adjacent vertices (Fig.3.8). In this case the vector y_v' is normal to the vector $(y_v^{(1)} - y_v^{(2)})$ and can be easily found.

Based on Procedure 3.1 a computer program has been elaborated for an approximate solution of the a.c. tolerance problems considered. The program does not account for the situation in Fig.. 3.8 and the lower endpoint of the solution interval of problem 3.9 is found using only the vertices of the polygon $\tilde{S}_v$.

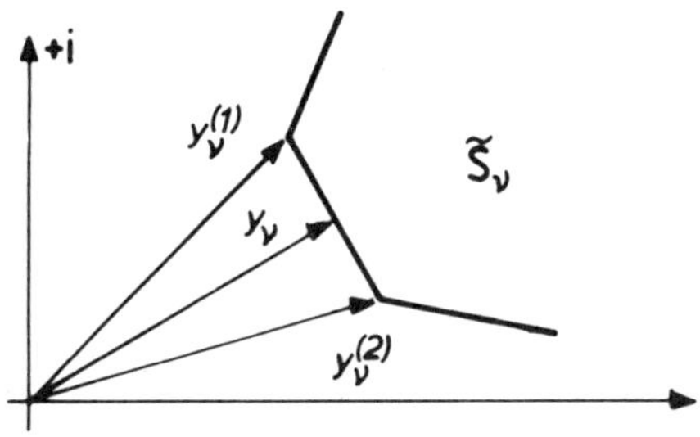

Fig. 3.8. General case of location for the lower endpoint of Y_ν.

The applicability of the program developed has been tested using several examples.

It should be pointed out that if we are interested only in problems 3.7 and 3.8, their solutions can be found in a more efficient way without appealing to Procedure 3.1. To show this possibility, we first rewrite (3.80) in the form

$$\Delta Y_\nu = \sum_{j=1}^{2m} (c_{ja} + jc_{jr})(\Delta P_{ja} + j\Delta P_{jr})$$

Then we separate the real and imaginary terms to get

$$\Delta Y_{\nu a}^I = \sum_{j=1}^{2m} (c_{ja}\Delta P_{ja}^I - \Delta P_{jr}^I) \tag{3.82a}$$

and

$$\Delta Y_{\nu r}^I = \sum_{j=1}^{2m} (c_{jr}\Delta P_{ja}^I + c_{ja}\Delta P_{jr}^I) \tag{3.82b}$$

It is seen that $\Delta Y_{\nu a}^I$ and $\Delta Y_{\nu r}^I$ being linear expressions of the intervals ΔP_{ja}^I and ΔP_{jr}^I can be computed directly by (3.82). It can be easily seen that we finally have

$$r(\Delta Y_{\nu a}^I) = \sum_{j=1}^{2m} [\,|c_{ja}|r(\Delta P_{ja}^I) + |c_{jr}|r(\Delta P_{jr}^I)]$$

$$r(\Delta Y_{\nu r}^I) + \sum_{j=1}^{2m} [\,|c_{jr}|r(\Delta P_{ja}^I) + |c_{ja}|r(\Delta P_{jr}^I)]$$

where $r(X^I) = w(X^I)/2$ is the radius of the interval X^I. Having found the radii $r(\Delta Y^I_{\nu a})$ and $r(\Delta Y^I_{\nu r})$ the tolerances sought are determined approximately as

$$Y^I_{va} = Y^c_{va} + [-r(\Delta Y^I_{va}),\ r(\Delta Y^I_{va})]$$
$$Y^I_{vr} = Y^c_{vr} + r(\Delta Y^I_{vr})[-1, 1]$$

where Y^c_{va} and Y^c_{vr} are the respective center (nominal) values of the output variables.

The approach adopted in this section is based on a linearization of the original (nonlinear) tolerance problem. It is, therefore, applicable for relatively small widths of the input variable intervals. A better approximate solution to the a.c. tolerance problems considered, accurate enough (under certain conditions) for much wider input variable intervals will be proposed in subsection 3.3.4. It is based on an appropriate real variable representation of the circuit equations to be presented in the next subsection.

3.3.3. Equivalent real variable representation

Consider again system (3.73) which can be written as a family of noninterval systems

$$\begin{bmatrix} Z & \alpha^T \\ \alpha & 0 \end{bmatrix} \begin{bmatrix} \dot{I} \\ \dot{V} \end{bmatrix} = \begin{bmatrix} \dot{U} \\ 0 \end{bmatrix}, \quad Z \in Z^I, \quad \dot{U} \in \dot{U}_I \tag{3.83}$$

Without any loss of generality we assume that each branch contains either R_k or jX_k. (If the above assumption is not fulfilled for the original circuit, we can always introduce artificial quasinodes between R_k and L_k or C_k of the kth branch to satisfy it). We suppose that the first m_1 branches ($m_1 \le m$) include only R elements while the remaining $m - m_1$ branches contain only X elements. With this in mind, we rewrite (3.83) in the form

$$\left[\begin{array}{c:c} \begin{matrix} R_1 & & & & & & \\ & \cdot & & & & & \\ & & \cdot & & 0 & & \\ & & & R_{m1} & & & \\ & & & & jX_{m+1} & & \\ & 0 & & & & \cdot & \\ & & & & & & jX_m \end{matrix} & \alpha^T \\ \hdashline \alpha & 0 \end{array}\right] \bullet \left[\begin{array}{c} I_{1a} + jI_{1r} \\ \cdot \\ \cdot \\ I_{m1a} + jI_{m1r} \\ \cdot \\ \cdot \\ \cdot \\ I_{ma} + jI_{mr} \\ \hdashline V_{1a} + jV_{ar} \\ \cdot \\ \cdot \\ v_{na} + jV_{nr} \end{array}\right] = \left[\begin{array}{c} U_{1a} + jU_{1r} \\ \cdot \\ \cdot \\ U_{m1a} + jU_{m1r} \\ \cdot \\ \cdot \\ \cdot \\ U_{ma} + jU_{mr} \\ \hdashline \\ 0 \\ \\ \end{array}\right] \tag{3.84}$$

Separating real and imaginary parts in (3.84) and regrouping the resulting real equations we can bring the (N x N) complex system (3.84) in the form of the following system of $2N$ equations in $2N$ real variables

$$
\begin{array}{r} \\ m_1 \\ m \\ N \\ N+m_1 \\ N+m \\ 2N \end{array}
\begin{array}{c}
\begin{array}{cccccc} m_1 & m & N & N+m_1 & N+m & 2N \end{array} \\
\left[\begin{array}{cc:c:c:c:c}
R & 0 & \alpha^T & & & \\
0 & 0 & & & -X & \\ \hdashline
\multicolumn{2}{c:}{\alpha} & 0 & & & \\ \hdashline
 & & & R & & \alpha^T \\
 & X & & & & \\ \hdashline
 & & & \multicolumn{2}{c:}{\alpha} & 0
\end{array}\right]
\end{array}
\bullet
\left[\begin{array}{c} I^a \\ \hdashline V^a \\ \hdashline I^r \\ \hdashline V^r \end{array}\right]
=
\left[\begin{array}{c} U^a \\ \hdashline 0 \\ \hdashline U^r \\ \hdashline 0 \end{array}\right]
\tag{3.85}
$$

where R is a diagonal matrix diag$[R_1, \ldots, R_{m1}]$, X is a diagonal matrix diag$[X_1, \ldots, X_{m-n}]$, $I^a = [I_1^a, \ldots, I_m^a]$, $I^r = [I_1^r, \ldots, I_m^r]^T$, $V^a = [V_1^a, \ldots, V_n^a]^T$, $V^r = [V_1^r, \ldots, V_n^r]^T$, $U_a = [U_1^a, \ldots, U_m^a]^T$ and $U^r = [U_1^r, \ldots, U_m^r]^T$. Each vector I^a, I^r, U^a and U^r can be further separated into two parts. We shall illustrate this with I^a. The first m_1 entries of I^a will form the vector I^{1a} and the remaining $m - m_1$ entries will form the vector I^{2a}, i.e.

$$
I^a = \begin{bmatrix} I^{1a} \\ --- \\ I^{2a} \end{bmatrix} \tag{3.86}
$$

Similarly

$$
I^r = \begin{bmatrix} I^{1r} \\ --- \\ I^{2r} \end{bmatrix}, \quad U^a = \begin{bmatrix} U^{1a} \\ --- \\ U^{2a} \end{bmatrix}, \quad U^r = \begin{bmatrix} U^{1r} \\ --- \\ U^{2r} \end{bmatrix} \tag{3.87}
$$

Finally, we will split the matrix α into two submatrices α^1 and α^2 regrouping the first m_1 columns and the last $m - m_1$ columns of α, respectively

$$
\alpha = [\alpha^1 \vdots \alpha^2] \tag{3.88}
$$

Based on (3.84) to (3.88) we can rearrange the equations of (3.85) to get the following equivalent system

$$\left[\begin{array}{cc|c|cc|c} R & & \alpha^T & & & \\ & -X & \alpha^T & & & \\ \hline \alpha^1 & & & & \alpha^2 & \\ \hline & & & R & & \alpha^T \\ & & & & X & \alpha^T \\ \hline & \alpha^2 & & \alpha^1 & & \end{array}\right] \cdot \begin{bmatrix} I^{1a} \\ I^{2r} \\ V^a \\ I^{1r} \\ I^{2a} \\ V^r \end{bmatrix} = \begin{bmatrix} U^{1a} \\ U^{2r} \\ 0 \\ U^{1r} \\ U^{2a} \\ 0 \end{bmatrix} \tag{3.89}$$

Now we first multiply the rows of the system corresponding to the diagonal matrix $-X$ by -1. Then we rewrite the modified system (3.89) as

$$Ay = b, \quad A \in A^I, \quad b \in B \tag{3.90}$$

Thus, we have managed to transform the original family (3.83) of systems with N complex variables into an equivalent family (3.90) of systems with $2N$ real variables.

At first glance the family of systems (3.90) is similar to the family of systems (3.10) from section 3.1.1 describing the d.c. tolerance analysis problem. However, the resemblance of (3.90) with (3.10) is purely formal. Indeed, all the interval variables from (3.10) are independent intervals; at the same time some of the variables in (3.90) are interdependent. From the matrix in (3.90) it is seen that actually

$$a_{N+i,N+i} = a_{ii}, \quad i = \overline{1,m} \tag{3.91}$$

Therefore, (3.90) cannot be treated as a linear interval system since such a treatment presupposes independence of all the interval coefficients involved. In fact, (3.90) represents a system of linear equations with dependent coefficients whose feasibility conditions are given by (3.91). Therefore, the methods from sections 3.2.1 to 3.2.3 are not directly applicable. In the next section we shall, however, show that under certain conditions problems 3.1 to 3.4 can be solved rather accurately (although theoretically not always exactly) by adopting the general method from section 3.2.2.

Remark 3.1. If we are interested only in problems 3.1 and 3.2 their exact solutions may be found using the method from section 3.2.4 (provided the corresponding assumptions A1 and A2 are fulfilled).

3.3.4. Improved accuracy method

In this section a method [36] will be presented for handling problems 3.7 to 3.10 which constitutes an improvement over the approximate method of section 3.3.2. The new method is based on the equivalent real variable representation (3.90), (3.91) introduced in the previous section, on one hand, and on the theoretical considerations of section 3.2.2

on the other. It is applicable in the case where the set S of solutions to the unconstrained problem (3.90) (when the relations (3.91) are ignored) belongs to an orthant of the parameter space.

In order to account for the constraints (3.91) we first reconsider formula (3.39)

$$A_{wz} = A_c - T_w \Delta T_z$$

which can be written as

$$(A_{wz})_{ij} = (A_c)_{ij} - w_i \Delta_{ij} z_j \ , \quad i,j = \overline{1,N} \tag{3.92}$$

It follows from (3.89) and (3.90) that $\Delta_{ij} \neq 0$ only for $i = j$, $i \in \overline{1,m} \cup \overline{N+m}$. Now write (3.92) for the nonzero elements of Δ_{ii}

$$(A_{wz})_{ii} = (A_c)_{ii} - w_i \Delta_{ii} z_i$$

$$(A_{wz})_{N+i,N+i} = (A_c)_{N+i,N+i} - w_{N+i} \Delta_{N+i,N+i} z_{N+i}$$

Subtracting one equation from the other we get

$$w_i \Delta_{ii} z_i = w_{N+i} \Delta_{N+i,N+i} z_{N+i} \tag{3.93}$$

(due to the feasibility condition (3.91) $(A_c)_{ii} = (A_c)_{N+i,\,N+i}$, and $(A_{wz})_{ii} = (A_{wz})_{N+i,\,N+i}$).

From (3.91) it is clear that $\Delta_{ii} = \Delta_{N+i,\,N+i}$. Using (3.93) we thus obtain the following relationship

$$w_{N+i} = \frac{z_i}{z_{N+i}} w_i \ , \quad i = \overline{1,m} \tag{3.94}$$

Now recall the assumption that S belongs to a given orthant, that is,

$$S \subset R_z^{2N} \tag{3.95}$$

To verify (3.95) the sufficient condition (3.57) is again applicable to the matrix in (3.90) provided we ignore the feasibility relations (3.91) and let all the diagonal elements vary in an independent way. If (3.57) holds for the associated "freed" interval matrix, then the parameters z_i, $i = \overline{1, 2N}$, are all known. From (3.94) it is seen that in this case all the elements w_{N+i} are determined once we have fixed the first m elements w_i of an arbitrary vector w containing $2N$ entries. Thus, it is sufficient to solve a real element $2N$ x $2N$ linear system of the type (3.46) 2^m times to find those extreme points of S which correspond to the constrained problem when the relations (3.94) are taken into account. Having determined these points it is straightforward to solve any of the Problems 3.7 to 3.10.

The amount of computation can be reduced if the following iterative procedures are used.

Procedure 3.2 (for approximating the upper endpoint of the output variable tolerance).

Step 0. Compute the vector

$$z = \operatorname{sgn} y_c$$

where y_c is the center (nominal) solution of the problem corresponding to A_c and b_c.

Step 1. Choose the first m components w_i, e.g. set $w_i = 1$, $i = \overline{1, m}$. Using (3.94) find w_{N+i}, $i = \overline{1, m}$. Fix the remaining components w_i, $i \in \overline{m+1, N} \cup \overline{N+m+1, 2N}$ at $+1$ (these components can be arbitrary since the corresponding $\Delta_{ii} = 0$).

Step 2. Form the corresponding system

$$A_{wz}\, y = b_w \tag{3.96}$$

and denote its solution by y'. Let f' be the corresponding value of the output variable for the problem considered (e.g. if we solve problem 3.9 with respect to I_1 then $f' = \sqrt{(y_1^2 + y_{N+1}^2)}$ as is seen from (3.89)).

Step 3. Let $i_0 = 1$.

Step 4. Permute the component w_{i0} from, say, $+1$ to -1 (or vice versa) and the corresponding component w_{N+i_0}.

Step 5. Form and solve the corresponding new system (3.96) for a new solution y'' and compute the corresponding value f''.

Step 6. If $f'' < f'$ restore the previous value of w_{i0} and w_{N+i0}. Go to the next step.

Step 7. Let $i_0 = i_0 + 1$ if $i_0 < m$; otherwise set $i_0 = 1$.

Step 8. If the following *termination criterion* is not fulfilled go back to Step 4 (with $f' := f''$ if $f'' > f'$).

Termination criterion: if $f'' \leq f'$ successively m times then stop (indeed, in this case there is no better combination of the parameter lower and upper endpoints which would improve the value of f).

Procedure 3.3 (for approximating the lower endpoints of the output variable tolerance).

This procedure is, essentially, the same as Procedure 3.2. The only difference occurs in Step 6 and the termination criterion where the condition $f'' \leq f'$ is replaced by $f'' \geq f'$.

Remark 3.2. It should be noted that unlike the d.c. problems solved exactly in the previous section, the a.c. tolerance problem considered here may, in general, not be solved exactly by the present method. Indeed, the present solution has each of its components fixed at lower or upper interval endpoint while the exact solution may have a component whose value lies in-between these endpoints. Indeed, Example 2.7 (solved by the first-order (optimization) method from Chapter 2) shows that the exact solution for the tolerance problem therein considered (Problem 3.9 in terms of this section's terminology) is obtained as the image of a corresponding vertex of the input parameter box X for all frequencies but one (the resonance frequency). Since the disagreement (if it exists) between the exact and approximate solution is obviously very small, the present method can be recommended in practice for solving a.c. tolerance problems of increased size. It has been successfully applied in illustrative examples containing up to six complex variables.

Comments

Section 3.1. As was shown in this section each of the d.c. tolerance problems 3.1 to 3.4 can be formulated in implicit form as a corresponding linear interval system. This approach permits the original d.c. tolerance problem considered to be handled by solving (exactly or approximately) the resulting linear interval system.

The idea to formulate the worst-case d.c. tolerance analysis problem as a system of linear equations with independent interval coefficients was apparently proposed for the first time in [25], [26]. However, only approximate solutions were obtained along these lines at the time. In fact, each endpoint of the output variable tolerances is bracketed by an outward and inward bound.

In some cases (e.g. in circuits comprising controlled sources with interval coefficients) the implicit tolerance problem formulation may lead to a system of linear equations with dependent coefficients. Since the exact solution of such systems is considerably more difficult to obtain as compared with the case of linear systems with independent coefficients it is natural to try to reduce the former kind of systems to the latter one whenever possible. If the cause for coefficient interdependence is the presence of independent current sources, then the only dependent coefficients appear in the RHS of (3.25) or (3.26) as illustrated in (3.24) and the reduction to an equivalent system with independent coefficients is in this instance quite straightforward. Indeed, if equation (3.24a) is added to equation (3.24b) the RHS of the transformed equation (3.24b) will be zero which leads to an equivalent linear system with no coefficient dependence.

Section 3.2. The exact method for solving tolerance problems 3.1 to 3.5 are based on the works of the Czech mathematician J. Rohn. It seems that presently his results are best suited for treating the d.c. tolerance analysis problems considered. However, it should be noted that only direct methods for solving the resulted linear interval system have been presented in this section. Nowadays, there exist iterative methods for exact solution of (3.1) [37] which may, in some cases, prove more efficient for solving some of the tolerance problems 3.1 to 3.4 (for example whenever the special methods from section 3.2.3 are not applicable).

The general method from section 3.2.2 is quite universal – the only assumption that the interval matrix A^I involved in the circuit equations is regular is always satisfied in practice. Its numerical efficiency can be improved by a factor of N if the set of real linear systems (3.46) is solved in the following manner. First, the vectors w belonging to W' are ordered in such a way that any adjacent vectors w^1 and w^2 differ only in one component (this is always possible). Then it is readily seen that the corresponding two linear systems (3.46) will have matrices A_{w1_z} and A_{w2_z} differing only in one column. That is why the inverse of the second matrix can be calculated in a most economical way using the inverse of the first. Such an approach is, of course, recommendable if N is not very large since the inverted matrices are almost always full although the matrices A_{wz} may be rather sparse. If the circuit studied is of increased size (N is large) it might prove a better policy to use some sparse matrix method for solving (3.46) in implementing the general method of section 3.2.2, the special methods of section 3.2.3 or the method of section 3.2.4.

The special methods from section 3.2.3 are more efficient than the general method but are applicable only when some sufficient conditions are met. If these conditions do not hold for some particular circuits (e.g. having large tolerances on the input parameters) then the general method has to be applied. If, however, N is rather large and at the same time the accuracy on the output variable tolerances needs not be very high (as is the case at the early stages of design) it is expedient to resort to some approximate methods for solving the tolerance problems considered.

Various methods for obtaining approximate solutions to the tolerance problems 3.1 to 3.4 can be suggested. The most direct approach is to use some of the existing methods for nonoptimal solution of (3.1). For instance, an approximate solution was obtained for the tolerance problem from Example 3.2 by means of the Gauss elimination method applied to the associated linear interval system of type (3.10). As expected it yielded an approximation Y with larger intervals than those obtained by the corresponding exact method.

If the tolerance problem investigated leads to a system of equations with dependent coefficients the method from section 3.2.4 may provide the exact solution. If, however, any of the assumptions A1 or A2 adopted is not fulfilled, one is led once again to resort to some approximate solution. The simplest approach is to neglect the interdependence among the coefficients and to solve (better exactly) a corresponding interval system of the type (3.1) (with independent coefficients). A better approach would be to apply some recent results on the approximate solution of linear systems with dependent coefficients suggested in [38] which is expected to ensure narrower solution intervals than the former approach. This is, however, a matter of future research.

It should be noted that in contrast to the traditional (noninterval) approximate solutions (based, for example, on sensitivity analysis) the approximations obtained by the interval analysis techniques mentioned above are guaranteed to enclose the exact output variable tolerances.

Finally, it is worth pointing out that whenever the interval methods from section 3.2.3 and 3.2.4 are applicable they provide the exact solutions of the d.c. tolerance problem considered with infallible accuracy and require by far less computer time than the statistical methods now in use for "exact" tolerance analysis of linear electrical circuits.

Section 3.3. In this section the implicit tolerance problem formulation is generalized to tolerance analysis of a.c. circuits by writing down the corresponding linear system in complex form. It should, however, be stressed that nowadays no suitable interval methods seem to exist for exactly solving linear interval systems with complex coefficients. (The situation may, hopefully, change for the better in the near future.) Therefore, an appropriate real variable representation was devised in the form of a system of linear equations (3.90) with dependent coefficients – feasibility conditions (3.91).

Due to the special (diagonal) form of the feasibility conditions (3.91) a method for solving all the a.c. tolerance analysis problems 3.1 to 3.6 was suggested in section 3.3.4. In its present form this method is only applicable if the interval solution of the problem considered is known to lie in an orthant of R^{2N} – assumption (3.95). It is, however, hoped that it can be generalized to the case where (3.95) is not needed (by using the sign-accord algorithm of section 3.2.2). Such a generalization would be useful since the method of section 3.3.4 is more accurate than the method of 3.3.2 – in fact, most often it provides the optimal solution of the a.c. tolerance problems considered.

General Remark. All the methods presented in this chapter are designed to solve worst-case tolerance analysis problems. However, the exact (or approximate) solution to the worst-case tolerance problem can be used to find an approximate solution to the basic tolerance problem in probabilistic setting (section 2.1.3). The approximate solution can then be improved using some (possibly statistical) method for local optimization. Such a combined approach to solving the probabilistic tolerance problem is believed to be more efficient than the traditionally used methods [27]–[29], [39]–[41] since the latter ones are applied to the whole box $X^{(0)}$ while the search by the local optimization method for the exact solution would be confined to a small subregion of $X^{(0)}$.

CHAPTER 4

STABILITY OF LINEAR CIRCUITS WITH INTERVAL PARAMETERS

In this chapter some aspects of the problem of stability of linear electric circuits with interval parameters are considered for both the continuous time and discrete time case. Sufficient or necessary and sufficient conditions for checking the circuit stability are presented. Most of them are obtained by generalizing certain known stability criteria to interval form. These results can be useful in stability analysis of robust circuits and systems.

4.1. PROBLEM STATEMENT

In this chapter we shall be studying the stability of linear lumped electric circuits when their elements values are not known exactly. More precisely, let $N(p)$ denote such a circuit whose constitutive element parameters (passive element values R, C, L, M or controlled source coefficients k) form the vector $p = (p_1, \ldots, p_n)$. Furthermore let $p \in P$ where P is an interval vector ($P \in I(R^n)$). The basic problem to be investigated here can be formulated as follows.

Problem 4.1. Check that each individual circuit $N(p)$ is stable when $p \in P$.

Alternate stability problem formulations (assessing certain margins of stability) will be considered in the sequel.

4.1.1. Interval polynomial stability

In this section the stability of the electric circuit studied $N(p)$ with $p \in P$ will be assessed by checking the stability of an associated interval polynomial (polynomial with interval coefficients). To highlight the basic features of such an approach we shall consider the following example.

***Example* 4.1.** The circuit whose stability we are interested in is shown in Fig. 4.1. Assuming that the resistances R_i, i = 1, 2, 3, may be negative let $R_i \in R_i^I$, $L \in L^I$ and $C \in C^I$ where R^I, L^I and C^I are specified intervals. In this example $p = (R_1, R_2, R_3, L, C)$ and $P = (R_1^I, \ldots, C^I)$.

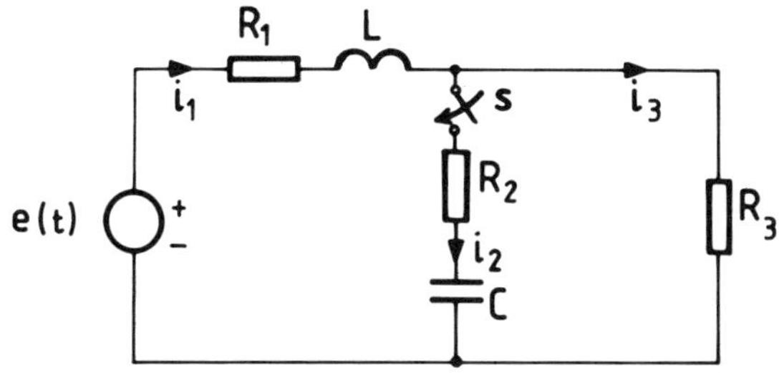

Fig. 4.1. Stability analysis of an electric circuit with interval parameters.

It is readily seen that the characteristic equation for any of the circuit branch currents i_1, i_2 or i_3 is

$$L(R_2 + R_3)s^2 + \left(R_1R_2 + R_2R_3 + R_1R_3 + \frac{L}{C}\right)s + \frac{R_1 + R_3}{C} = 0 \tag{4.1}$$

Consider the polynomial

$$q(s) = a_0s^2 + a_1s + a_2 \tag{4.2}$$

where

$$a_0 = L(R_2 + R_3) \tag{4.3a}$$

$$a_1 = R_1R_2 + R_2R_3 + R_1R_3 + L/C \tag{4.3b}$$

$$a_2 = (R_1 + R_3)/C \tag{4.3c}$$

Each individual circuit is stable iff the zeros of the corresponding polynomial (4.2) lie in the open left-hand half of the complex plane. For the sake of brevity such a polynomial will also be called stable. (To simplify the terminology the terms "stable" or "stability" will be used in the sequel rather than the exact terms "asymptotically stable" or "asymptotical stability" when referring to the circuit property or "strictly Hurwitz", "strict Hurwitzness" when referring to the polynomial property.) Note, however, that the polynomial coefficients a_0, a_1 and a_2 are according to (4.3) nonlinear functions of the circuit parameters p_i :

$$a_0 = f_0(p_2,p_3,p_4) \tag{4.4a}$$

$$a_1 = f_1(p_1,p_2,p_3,p_4,p_5) \tag{4.4b}$$

$$a_2 = f_2(p_1,p_3,p_5) \tag{4.4c}$$

Now we shall introduce the vector $a = (a_0, a_1, a_2)$ and the vector function $f = (f_0, f_1, f_2)$ so that (4.4) may be written in vector form as

$$a = f(p) \tag{4.5}$$

where f: $R^5 \to R^3$. We need two more notations. Let

$$S = \{a : a = f(p),\ p \in P\} \tag{4.6}$$

Obviously, S is the set of the images of p (when $p \in P$) under f. Moreover, let

$$\tilde{A}_i = f_i(P), \quad i = 0,1,2 \tag{4.7}$$

($\tilde{A}_i$ is the range of f_i over the box P). Now form the interval vector

$$\tilde{A} = (\tilde{A}_0, \tilde{A}_1, \tilde{A}_2) \tag{4.8}$$

It should be stressed that due to the nonlinearity of the functions (4.4) S is properly contained in A, i.e.

$$S \subset \tilde{A} \tag{4.9}$$

In fact, $\tilde{A}$ is the interval hull of S.

Now we shall consider the stability of the circuit at hand. Let (4.2) be written as

$$q(s,a) = a_0 s^2 + a_1 s + a_2 \tag{4.10}$$

to express explicitly the dependence of q on the vector a. Obviously, each individual circuit (for a fixed $p \in P$) is stable iff each real polynomial (4.10) is stable when the corresponding $a \in S$.

Since the set S is practically impossible to determine we shall resort to a simplified approach which is based on the notion of an interval polynomial (e.g. [1]). An interval polynomial is a family of real polynomials whose coefficients a_i may take on values independently one of another from some intervals A_i. Thus, (4.10) defines an interval polynomial if $a_0 \in A_0$, $a_1 \in A_1$ and $a_2 \in A_2$. Let these intervals be determined as follows

$$A_i = F_i(P) \tag{4.11}$$

i.e. as some interval extension of f_i in P, $i = 0, 1, 2$. Thus, if we form the interval vector

$$A = (A_0, A_1, A_2) \tag{4.12}$$

the interval polynomial associated with (4.10) will be denoted as

$$q(s,A) = \{q(s,a) : a \in A\} \tag{4.13}$$

We shall also introduce the set of polynomials

$$q(s,S) = \{q(s,a) : a \in S\} \tag{4.14}$$

Any of the above two sets (4.13) and (4.14) will be called stable if all its elements are stable polynomials. It should be realized that because of (4.9) and the inclusion $\tilde{A} \subseteq A$ we have

$$q(s,S) \subset q(s,A) \tag{4.15}$$

We are now in a position to state an important result: if $q(s, A)$ is a stable interval polynomial then the circuit studied is stable for all $p \in P$. (The proof of this assertion follows directly from the inclusion (4.15).)

Thus, the stability of all circuits $N(p)$ from Fig. 4.1 (with $p \in P$) can be guaranteed if the corresponding interval polynomial $q(s, A)$ is proven to be stable.

Based on the example considered we can state the following general result. Let $N(P)$ denote the set of circuits when the parameter vector $p = (p_1, \ldots, p_n) \in P \in I(R^n)$. We assume that the characteristic polynomial

$$q(s,a) = \sum_{i=0}^{m} a_i s^{m-i} \tag{4.16}$$

has been determined in explicit form with

$$a_i = f_i(p), \; i = \overline{0,m} \tag{4.17}$$

Similarly to the above example introduce the vector function

$$a = f(p), \quad f: R^n \to R^m$$

and the corresponding set S using (4.6). Then the set of polynomials $q(s, S)$ is defined by (4.16) when $a \in S$.

Let $A = (A_0, A_1, \ldots, A_m)$ be an interval vector whose components are some interval extensions of (4.17). Furthermore, let $q(s, A)$ be the set of polynomials (4.16) when $a \in A$, that is, $q(s, A)$ is an interval polynomial. We have the following sufficient condition for the stability of $N(P)$.

Theorem 4.1. If $q(s, A)$ is a stable interval polynomial, then $N(P)$ is also stable.

Clearly, the inverse assertion is not true: the stability of $N(p)$ does not entail the stability of q(s, A). Indeed, if $N(P)$ is stable then $q(s, S)$ is stable and vice versa. However, since

$$q(s,S) \subset q(s,A) \tag{4.18}$$

$q(s, A)$ may happen to contain unstable polynomials.

A set of circuits (or polynomials) will be called unstable if at least one element of the set is unstable. It should be pointed out straight away that the instability of $q(s, A)$ does not entail the instability of $N(P)$. This assertion follows directly from the inclusion (4.18).

The following result due to Kharitonov [42], [43] permits the assessment of the stability (or instability) of the interval polynomial to be carried out in a most efficient way.

Define the four polynomials

$$\begin{aligned}
q_1(s) &= \underline{a}_0 s^m + \underline{a}_1 s^{m-1} + \overline{a}_2 s^{m-2} + \overline{a}_3 s^{m-3} + \underline{a}_4 s^{m-4} + \ldots \\
q_2(s) &= \overline{a}_0 s^m + \underline{a}_1 s^{m-1} + \underline{a}_2 s^{m-2} + \overline{a}_3 s^{m-3} + \overline{a}_4 s^{m-4} + \ldots \\
q_3(s) &= \overline{a}_0 s^m + \overline{a}_1 s^{m-1} + \underline{a}_2 s^{m-2} + \underline{a}_3 s^{m-3} + \overline{a}_4 s^{m-4} + \ldots \\
q_4(s) &= \underline{a}_0 s^m + \overline{a}_1 s^{m-1} + \overline{a}_2 s^{m-2} + \underline{a}_3 s^{m-3} + \underline{a}_4 s^{m-4} + \ldots
\end{aligned} \tag{4.19}$$

where $\underline{a}_i$ and $\overline{a}_i$, $i = \overline{0, m}$, are the lower and upper endpoints, respectively, of the interval A_i with A_i being some interval extension of the corresponding function (4.17).

Theorem 4.2. The polynomial (4.16) is stable for all $a_i \in A_i = [\underline{a}_i, \overline{a}_i]$, $i = \overline{0, m}$, iff the polynomials $q_j(s)$, $j = \overline{1, 4}$ are stable.

Thus, the stability of the whole set $N(P)$ of infinitely many linear circuits $N(p)$ can be guaranteed if only the four particular polynomials defined by (4.18) are proved to be stable.

The stability of each of the above four polynomials can be assessed in a very effective way. Let

$$q(s) = a_0 s^m + a_1 s^{m-1} + \ldots + a_{m-1} s + a_m$$

be one of these polynomials (with $a_0 > 0$). Now form the polynomials $\alpha(s)$ and $\beta(s)$ by taking alternate terms from $q(s)$, starting with $a_0 s^m$ and $a_1 s^{m-1}$, respectively. Thus,

$$\begin{aligned}
\alpha(s) &= a_0 s^m + a_2 s^{m-2} + a_4 s^{m-4} + \ldots \\
\beta(s) &= a_1 s^{m-1} + a_3 s^{m-3} + a_5 s^{m-5} + \ldots
\end{aligned}$$

Next form the ratio $\alpha(s)/\beta(s)$ and express it as a continued fraction as follows:

$$\frac{\alpha(s)}{\beta(s)} = \gamma_1 s + \cfrac{1}{\gamma_2 s + \cfrac{1}{\gamma_3 s + \cfrac{1}{\ddots \, \cfrac{1}{\gamma_m s}}}}$$

Then the following theorem (presented e.g. in [44]) is valid.

Theorem 4.3. The polynomial $q(s)$ is stable iff $\gamma_i > 0$ for all $i = \overline{1, m}$.

The above approach to assessing the stability of the set of circuits $N(P)$ is very attractive since it reduces the original problem to that of checking the stability of four real polynomials. Its application may, however, be difficult in some cases: the derivation of the functions (4.17) usually presents serious difficulties for large size circuits. Another shortcoming of this approach is that the stability criterion based on Theorem 4.1 is usually rather conservative: due to the inclusion (4.18) the circuit may be stable although the associated interval polynomial turns out to be unstable.

An alternate approach to assessing the stability of linear electric circuits with interval parameters will be proposed in the next subsection. This approach being based directly on the matrix formulation of the circuit equations in transient analysis circumvents the need for the derivation of the characteristic polynomial in explicit form.

4.1.2. Interval matrix stability

In this subsection the stability of the set of circuits $N(P)$ will be assessed through the stability of a related interval matrix. The approach herein adopted will be initially introduced by means of the circuit from Fig. 4.1.

Using the inductor current i_L and the capacitor voltage v_C as state variables it is readily seen that the state variable equations for the circuit considered are given in matrix form as

$$\frac{dx}{dt} = Ax + Bf(t) \tag{4.20a}$$

where

$$A = \begin{bmatrix} -\left(\frac{R_2}{L_k} + \frac{R_1}{L}\right) & \frac{R_2}{R_3 L_k} - \frac{1}{L} \\ \frac{1}{C_k} & \frac{1}{R_3 C_k} \end{bmatrix}, \quad k = 1 + R_2/R_3 \tag{4.20b}$$

and

$$x = (i_L, v_C)^T$$

As is well known the particular circuit studied is stable iff all the eigenvalues of A have negative real parts.

Based on the example considered it is clear that in the general case of an arbitrary circuit with l state variables the matrix $A = \{a_{ij}\}$ will have elements a_{ij} that are nonlinear functions of the parameter vector p, i.e.

$$a_{ij} = a_{ij}(p)\ , \quad i,j = \overline{1,l} \tag{4.21}$$

Thus, we shall be interested in studying the stability of the following family of normal differential equations

$$\frac{dx_i}{dt} = \sum_{j=1}^{l} a_{ij}(p)x_j, \quad i = \overline{1,l} \tag{4.22a}$$

$$p \in P \tag{4.22b}$$

Since the elements $a_{ij}(p)$ are all dependent on the parameter vector p the stability analysis of (4.22) is a very difficult problem. That is why the original stability problem (4.22) will be imbedded (similarly to the development from the previous section) in the following simpler stability problem:

$$\frac{dx_i}{dt} = \sum_{j=1}^{l} a_{ij}x_j, \quad i = \overline{1,l} \tag{4.23a}$$

$$a_{ij} \in A_{ij} \tag{4.23b}$$

Here A_{ij} are independent intervals defined as some interval extensions $A_{ij}(P)$ of $a_{ij}(p)$ over P. Now form the interval matrix A^I with elements A_{ij}. The matrix A^I will be called stable if each real matrix $A \in A^I$ is stable, that is, if Re $[\lambda_i(A)] < 0$, $i = \overline{1, l}$, where $\lambda_i(A)$ is the ith eigenvalue of A. Using exactly the same argument as for the interval polynomial formulation from the previous section the following theorem is easily seen to be valid.

Theorem 4.4. If the interval matrix A^I is stable, then the set of circuits $N(P)$ is also stable.

We shall postpone the assessment of the stability of the interval matrix A^I for paragraph 4.2.

Theorem 4.4 provides only a sufficient condition for the stability of $N(P)$: if matrix A^I happens to be unstable that does not necessarily mean that $N(P)$ is unstable. It is clear that the assessment of the circuit stability will be less conservative if the elements A_{ij} of A^I are defined from (4.21) as the ranges $a_{ij}(P)$ of $a_{ij}(p)$ over P since, in general, $a_{ij}(P) \subset A_{ij}(P)$. However, determination of all the ranges $a_{ij}(P)$ requires the global solution of $2l^2$ optimization problems which for larger l may prove too difficult a task.

Here we shall discuss an alternative way of reducing the stability problem of the circuit studied to the stability of an associated matrix which might, in some cases, be preferable to the approach based on the state variable description of the circuit.

For notational simplicity we shall confine ourselves to considering circuits made of resistors, capacitors, independent voltage sources and voltage-controlled voltage sources, the latter ones being only in branches incident on the grounded node of the circuit (as in section 3.1.1 each circuit is assumed to have m branches, n + 1 nodes one of which is grounded). Then using matrix notation as in sections 3.1.1 and 3.1.3 we have

$$ri(t) + d\int_0^t i(\tau)d\tau + \alpha_1 v(t) = u \tag{4.24a}$$

$$\alpha i(t) = 0 \tag{4.24b}$$

where r is the (m x m) diagonal matrix of branch resistances, d is a (m x m) diagonal matrix whose ith diagonal element is $1/C_i$ (C_i being the ith branch capacitance),α is the (reduced) incidence matrix, α_1 is some modification of α due to the presence of controlled sources, $i(t)$ is the branch current vector and $v(t)$ is the node voltage vector. Assuming the vector u of independent sources to be constant and differentiating (4.24) we get

$$r\frac{di}{dt} + di + \alpha_1\frac{dv}{dt} = 0 \tag{4.25a}$$

$$\alpha\frac{di}{dt} = 0 \tag{4.25b}$$

Equation (4.25) can be put in vector form

$$\begin{bmatrix} r & \alpha_1 \\ \alpha & 0 \end{bmatrix}\begin{bmatrix} di/dt \\ dv/dt \end{bmatrix} = -\begin{bmatrix} d & 0 \\ 0 & 0 \end{bmatrix}\begin{bmatrix} i \\ v \end{bmatrix} \tag{4.26}$$

Now introduce the notation

$$B = \begin{bmatrix} r & \alpha_1 \\ \alpha & 0 \end{bmatrix} \tag{4.27a}$$

$$\tilde{d} = -\begin{bmatrix} d & 0 \\ 0 & 0 \end{bmatrix} \tag{4.27b}$$

$$F = B^{-1} = \begin{bmatrix} F_{11} & F_{12} \\ F_{21} & F_{22} \end{bmatrix} \tag{4.27c}$$

$$x = \begin{bmatrix} i \\ v \end{bmatrix} \tag{4.27d}$$

(where F_{11} is a (m x m) matrix). Then (4.26) can be rewritten as

$$\begin{bmatrix} di/dt \\ dv/dt \end{bmatrix} = \frac{dx}{dt} = F\tilde{d}x = \begin{bmatrix} F_{11} & F_{12} \\ F_{21} & F_{22} \end{bmatrix} \begin{bmatrix} -d & 0 \\ 0 & 0 \end{bmatrix} x = \begin{bmatrix} -F_{11}d & 0 \\ -F_{21}d & 0 \end{bmatrix} \begin{bmatrix} i \\ v \end{bmatrix}$$

Hence

$$\frac{di}{dt} = -F_{11}di \tag{4.28a}$$

$$\frac{dv}{dt} = -F_{21}di \tag{4.28b}$$

We shall now assume that the elements of matrices r and d and some of the elements of matrix α_1 (corresponding to coefficients of the controlled sources) lie in some prescribed intervals. Then, obviously B, d and F from (4.27a), (4.27b) and (4.27c) become interval matrices B^I, d^I, F^I. We shall be interested in the stability of system (4.28a) when $B \in B^I$ and $d \in d^I$. Obviously, the partition F_{11} of F is an interval matrix $F_{11}{}^I$ when $B \in B^I$. The columns of $F_{11}{}^I$ can be determined by means of the first m columns of $(B^I)^{-1}$. Hence from (4.28a)

$$\frac{di}{dt} = Ai \tag{4.29a}$$

with

$$A \in A^I, \ \ A^I = -F_{11}^I \alpha^I \tag{4.29b}$$

Thus, the problem of the stability of the given class of circuits has been reduced to determining the stability of the interval matrix A^I defined by (4.29b).

As compared with the approach to deriving A^I by means of the state variable description of the circuit studied formula (4.29b) has the merit that the elements of A^I are defined in an automatic way by solving m systems of linear interval equations (in order to determine the first m columns of the matrix $(B^I)^{-1}$), thus circumventing the need for explicitly deriving and evaluating the ranges $a_{ij}(P)$. Moreover, the matrix $F_{11}{}^I$ does not depend on the elements of d^I which may prove advantageous in assessing the stability of A^I. It should however be borne in mind that the dimension of A^I obtained by the state variable description is smaller than that of A^I obtained by (4.29b).

Regardless of the way A^I is derived it is seen from the foregoing that the stability of the linear continuous circuits with interval parameters can be determined by considering the normal differential system

$$\frac{dx}{dt} = Ax \tag{4.30a}$$

with

$$A \in A^I \tag{4.30b}$$

where A^I is some associated interval matrix.

Finally, we shall consider the case where the circuit studied is a discrete linear circuit. We shall confine ourselves to discrete circuits obtained from the continuous circuits (4.25) by means of the implicit Euler integration method. Thus

$$r(i^{\nu+1}-i^{\nu}) + h\alpha i^{\nu+1} + \alpha_1(v^{\nu+1} - v^{\nu}) = 0$$
$$\alpha(i^{\nu+1} - i^{\nu}) = 0$$

(ν is the number of the integration step and h is the step size) or equivalently

$$(r + hd)i^{\nu+1} + \alpha_1 v^{\nu+1} = ri^{\nu} + \alpha_1 v^{\nu} \tag{4.31a}$$

$$\alpha i^{\nu+1} = \alpha i^{\nu}, \quad \nu \geq 0 \tag{4.31b}$$

Using matrix notation (4.31) becomes

$$B_1 x^{\nu+1} = Bx^{\nu} \tag{4.32}$$

where B is given by (4.27a),

$$B_1 = \begin{bmatrix} r + hd & \alpha_1 \\ \alpha & 0 \end{bmatrix} = B - h\tilde{d} \tag{4.33}$$

with d defined by (4.27b) and

$$x^{\nu} = [i^{\nu} \vdots v^{\nu}]^T$$

From (4.32) and (4.33)

$$x^{\nu+1} = B_1^{-1}Bx^{\nu} = (B - h\tilde{d})^{-1}Bx^{\nu}, \quad \nu \geq 0 \tag{4.34a}$$

If $B \in B^I$ and $d \in d^I$, the stability of the discrete system (4.34a) can be assessed by studying the properties of the associated interval matrix

$$A^I = (B^I - h\tilde{d}^I)^{-1}B^I \tag{4.34b}$$

The matrix obtained has dimensions N x N where $N = m + n$.

Another discrete circuit of reduced dimensions may be obtained if the implicit Euler method is applied directly to (4.28a). It is readily seen that the discrete circuit is then

$$i^{\nu+1} = (E + hF_{11}d)^{-1}i^{\nu}, \quad \nu \geq 0 \tag{4.35a}$$

If the circuit parameters are intervals, i.e. $B \in B^I$ and $d \in d^I$ then the matrix from (4.35a) becomes an interval matrix

$$A^I = (E + hF_{11}^I d^I)^{-1} \tag{4.35b}$$

of dimensions m x m.

Based on the concrete examples (4.34) and (4.35) it is clear that the description of any discrete circuit with interval parameters can be put in the form

$$x^{\nu+1} = A\,x^{\nu} \tag{4.36a}$$

with

$$A \in A^I \tag{4.36b}$$

where A^I is some (l x l) interval matrix associated with the stability problem. The interval matrix A^I will be called stable if each real matrix $A \in A^I$ is stable, that is, if $|\lambda_i(A)| < 1$, $i = \overline{1, l}$, where $\lambda_i(A)$ is the ith eigenvalue of A.

Similarly to the continuous-time case (Theorem 4.4) the following result is valid.

Theorem 4.5. If the interval matrix A^I from (4.36b) is stable, then the set of discrete circuits with interval parameters $N(P)$ is also stable.

Based on Theorems 4.4 and 4.5 the stability of $N(P)$ can be checked by way of the stability of the corresponding interval matrices.

4.2. SUFFICIENT CONDITIONS FOR INTERVAL MATRIX STABILITY

As was shown in section 4.1.2 the stability of the original set of linear continuous circuits $N(P)$ can be assessed by means of the family of the normal differential systems (4.30):

$$\dot{x}(t) = Ax(t), \quad A \in A^I \tag{4.37}$$

(where $\dot{x}(t)$ stands for $dx(t)/dt$). Recall that the original stability analysis problem (4.22) with dependent coefficients $a_{ij}(p)$ was imbedded in the simpler stability problem (4.37) where A^I is an interval matrix with independent coefficients. To distinguish the two stability problems (and by analogy with the terminology from Chapter 3) the family of systems (4.37) will be called an interval linear dynamic system with continuous time and for simplicity of notation will be written in the sequel as

$$\dot{x} = A^I x \tag{4.38}$$

If the set of circuits with interval parameters $N(P)$ is made up of discrete circuits the corresponding family of finite difference systems is given by (4.36):

$$x^{\nu+1} = Ax^{\nu}, \quad A \in A^I, \quad \nu \geq 0 \tag{4.39}$$

Similarly to the continuous-time case the family (4.39) will referred to as an interval linear dynamic system with discrete time and will be denoted by the symbolic notation

$$x^{\nu+1} = A^I x^{\nu} \tag{4.40}$$

In view of its practical applications (especially in robust stability of control systems) the problem of assessing the stability of the interval dynamic systems (4.38) and (4.40) has been intensively investigated over the last years (e.g. [45]–[49]).

In this section simple sufficient conditions for testing the stability of interval dynamic systems will be presented. By Theorems 4.4 and 4.5 these conditions guarantee the stability of the original set of linear circuits with interval parameters.

4.2.1. Stability of discrete circuits

In this subsection the stability of discrete linear circuits with interval parameters will be assessed by introducing simple sufficient conditions for checking the stability of the interval discrete system (4.40).

First, we shall consider the discrete system (4.36a):

$$x^{k+1} = Ax^k, \quad k \geq 0 \tag{4.41}$$

where A is a fixed real (l x l) matrix. As was mentioned in section 4.1.2 the system (4.41) is stable iff the spectral radius $\rho(A)$ of A satisfies the inequality

$$\rho(A) < 1 \tag{4.42}$$

Consider the sequence $A, A^2, A^4, \ldots, A^{\nu}, \ldots, \nu = 2^m, m \geq 0$ and suppose that (4.42) is valid. Then it follows that $\|A\|^{\nu} \to 0$ as $\nu \to \infty$ where $\|.\|$ is some matrix norm. It is clear that in this case there exists a finite ν such that

$$\|A^{\nu}\| < 1 \tag{4.43}$$

It is known [50] that in fact (4.43) is a necessary and sufficient condition for the system (4.41) to be stable. In practice the following norms have been used

$$\|A\|_{\infty} = \max_{i} \sum_{j=1}^{l} |a_{ij}| \tag{4.44a}$$

$$\|A\|_{1} = \max_{j} \sum_{i=1}^{l} |a_{ij}| \tag{4.44b}$$

$$\|A\|_{2} = \left(\sum_{i,j=1}^{l} a_{ij}^{2} \right)^{1/2} \tag{4.44c}$$

The criterion (4.43) has the merit that it does not necessitate the computation of $\rho(A)$. It is known [50] that if

$$|\mathrm{tr} A^{\nu}| > l \tag{4.45}$$

then $\rho(A) > 1$ and (4.41) is unstable (here $\mathrm{tr}\, A^{\nu}$ denotes the trace of the matrix A^{ν}).

Now we turn back to the linear interval system (4.40)

$$x^{k+1} = A^{I} x^{k} \tag{4.46}$$

Obviously, (4.46) is stable iff

$$\rho(A^{I}) < 1 \tag{4.47a}$$

that is, iff

$$\rho(A) < 1 \quad \text{for} \quad \forall A \in A^{I} \tag{4.47b}$$

Unfortunately, there does not exist, for the time being, efficient methods for testing (4.47). Therefore, we appeal to the equivalent condition (4.43) which, generalized to encompass the interval case, leads to the following result.

Theorem 4.6. If for some $\nu = 2^{m}$, $m \geq 0$, the condition

$$\|(A^{I})^{\nu}\| < 1 \tag{4.48}$$

holds ($\|.\|$ is some interval matrix norm), then (4.46) is stable.

On the basis of (4.44) the norms of the interval matrix which can be used are

$$\|A^I\|_\infty = \max_i \sum_{j=1}^{l} |a_{ij}^I| \tag{4.49a}$$

$$\|A^I\|_1 = \max_j \sum_{i=1}^{l} |a_{ij}^I| \tag{4.49b}$$

$$\|A^I\|_2 = \left(\sum_{i,j=1}^{l} |a_{ij}^I|^2 \right)^{1/2} \tag{4.49c}$$

where

$$|a_{ij}^I| = \max(|\underline{a}_{ij}|, |\overline{a}_{ij}|) \tag{4.49d}$$

It should be stressed that unlike (4.43) the criterion (4.48) is only a sufficient condition for the stability of (4.46). Indeed, the map $F = A^\nu$ is a nonlinear map $F: R^{l\times l} \to R^{l\times l}$ as regards the element a_{ij} of A. Therefore, its interval extension

$$F^I = (A^I)^\nu$$

contains properly the image of A^ν under F with $A \in A^I$. Thus, (4.48) is only a sufficient condition.

Let $c^I = [\underline{c}, \overline{c}]$ be an interval and define (the so-called mignitude)

$$\langle c^I \rangle = \min(|\underline{c}|, |\overline{c}|) \tag{4.50}$$

Based on condition (4.45) the following interval matrix result is straightforward.

C o r o l l a r y 4.1. If for some ν

$$\langle \operatorname{tr} F^I \rangle = \left\langle \sum_{i=1}^{l} F_{ii}^I \right\rangle > l \tag{4.51}$$

then (4.46) is unstable.

Based on the preceding results the following procedure for testing the stability or instability of (4.46) is suggested.

Procedure 4.1.

Step 0. Let $B^I = A^I$ and $m = 0$.

Step 1. Compute $\|B^I\|_\infty$ defined by (4.49a). If $\|B^I\|_\infty < 1$ go to Step 7; otherwise proceed to the next step.

Step 2. Compute $\|B^I\|_1$ defined by (4.49b). If $\|B^I\|_1 < 1$ go to Step 7; otherwise proceed to the next step.

Step 3. Compute $\|B^I\|_2$ defined by (4.49b). If $\|B^I\|_2 < 1$ go to Step 7; otherwise proceed to the next step.

Step 4. Using (4.50) calculate

$$d = \left\langle \sum_{i=1}^{n} B_{ii}^I \right\rangle$$

If $d > l$ go to Step 8; otherwise proceed to the next step.

Step 5. In $m < \bar{m}$ where $\bar{m}$ is a prefixed integer proceed to the next step; otherwise go to Step 9.

Step 6. Evaluate $C^I = (B^I)^2$ with elements

$$c_{ij}^I = \sum_{l=1}^{n} b_{il}^I \, b_{lj}^I$$

if $i \neq j$ and

$$c_{ij}^I = (b_{ii}^I)^2 + \sum_{\substack{l+1 \\ l \neq i}}^{n} b_{il}^I b_{li}^I$$

if $i = j$. Let $B^I = C^I$. Put $m = m + 1$ and go back to Step 1.

Step 7. Stop. The discrete interval system (4.46) is stable.

Step 8. Stop. The system (4.46) is unstable.

Step 9. Stop. No decision concerning the stability or the instability of (4.46) can be made.

Remark 4.1. If the procedure terminates in Step 8 no decision concerning the instability of the set of discrete linear circuits with interval parameters $N(P)$ can be made. This follows from the fact that the original stability problem concerning $N(P)$ has been imbedded in the broader stability problem (4.46). Therefore, the instability of (4.46) does not necessarily entail instability of $N(P)$.

Remark 4.2. As is seen from the above procedure, Steps 1 to 6 form a loop. In order to prevent occurrence of too many iterations m when the procedure converges rather slowly to either Step 7 or Step 8 or to the inconclusive result (Step 9) the number m must be bounded by a limit $\bar{m}$ as this is done in Step 5. Experimental evidence shows (Example 4.6 from 4.2.3) that good results (with a fairly low percentage of cases where the procedure becomes inconclusive) are obtained if $\bar{m}$ equals 10 to 15.

Remark 4.3. In Step 6, when computing the diagonal elements c_{ii}^I the square $(b^I_{ii})^2$ is used since

$$(b^I)^2 \subset b^I b^I$$

if $0 \in b$.

Now we proceed to considering the so-called M-margin stability of system (4.46). The discrete interval system is said to be stable with stability margin M if $\rho(A) < 1 - M$ for every $A \in A$ where $0 < M < 1$. Let $r = 1 - M$ and

$$B^I = A^I / r \tag{4.52}$$

It easily seen that the new matrix B^I is stable if the original matrix A^I is M-margin stable. Thus, in order to guarantee the M-margin stability of (4.46) it suffices to introduce the following condition.

Theorem 4.7. If for some $\nu = 2^m$, $m \geq 0$

$$\|(B^I)^\nu\| < 1 \tag{4.53}$$

where B^I is defined by (4.52), then the system (4.46) is stable with stability margin M.

Corollary 4.2. If for some ν

$$\langle \mathrm{tr}(B^I)^\nu \rangle > l \tag{4.54}$$

then (4.46) is not M-margin stable. (The fulfilment of (4.54) does not, of course, imply that (4.46) is unstable.)

To implement (4.53) Procedure 4.1 is again used with the following obvious modifications.

In Step 0, B^I is defined by (4.52).

In Steps 7, 8 and 9 the corresponding conclusions refer to the M-margin stability of (4.46).

Remark 4.4. If the above (modified) procedure terminates in Step 8 then (for reasons similar to those mentioned in Remark 4.1) no decision concerning the M-margin stability of the set $N(P)$ of discrete circuits with interval parameters can be made.

4.2.2. Stability of continuous circuits

In this section, the stability of the set $N(P)$ of continuous linear circuits with interval parameters will be assessed by means of simple sufficient conditions for checking the stability of the corresponding interval dynamic system (4.38).

First we shall consider the continuous system

$$\dot{x} = A x \tag{4.55}$$

where A is a fixed real (l x l) matrix. The system (4.55) will be called D-stable if

$$\lambda_i(A) \in D \ , \quad i = \overline{1,l} \tag{4.56}$$

where D is the interior of a disk of radius R centred on the point ($-R$, $j0$) in the complex plane (Fig. 4.2a). The system will be called D_η-stable if again (4.56) is fulfilled but now the above disk is displaced by η ($\eta > 0$) to the left, i.e. its centre is the point ($-R - \eta$, $j0$) (Fig.4.2b).

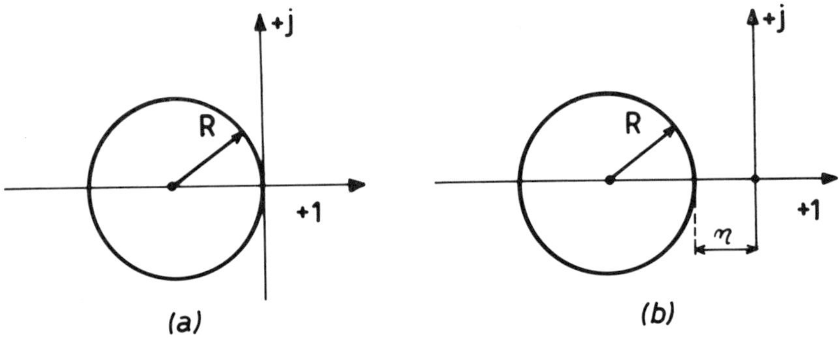

Fig. 4.2. Stability region for continuous systems:
a) D-stability b) D_η-stability

Let

$$B = E + A/R \tag{4.57}$$

(E is the identity matrix). It is known [50] that system (4.55) is D-stable iff $\rho(B) < 1$ or, equivalently, iff there exist a finite $\nu = 2^m$, $m \geq 0$, such that

$$\|B^\nu\| < 1 \tag{4.58}$$

where B is defined by (4.57) and $\|.\|$ is any of the norms (4.44). Conversely, system (4.55) is not D-stable if for some ν

$$|\mathrm{tr} B^\nu| > l \tag{4.59}$$

Similarly, let

$$B = (1 + \eta/R)E + A/R \tag{4.60}$$

Then (4.55) is D_η-stable [50] iff there exists a $\nu = 2^m$, $m \geq 0$, such that (4.58) is fulfilled with B defined by (4.60). The system (4.55) is not D_η-stable if for some ν the condition (4.59) occurs with B defined by (4.60).

Now we turn back to the continuous interval system

$$\dot{x} = A^I x \tag{4.61}$$

The interval system (4.61) will be called D-stable or D_η-stable if every real system (4.55) is D-stable or D_η-stable when $A \in A^I$.

On the basis of the above results concerning the stability of the real system (4.55) it is readily seen that the following results about the stability of the interval system (4.61) are valid. Let

$$B^I = E + A^I/R \tag{4.62}$$

Theorem 4.7. If for some ν

$$\|(B^I)^\nu\| < 1 \tag{4.63}$$

where B is defined by (4.62) and $\|.\|$ is any of the norms (4.49), then the interval system (4.61) is D-stable.

Corollary 4.3. If for some ν

$$\langle \mathrm{tr}(B^I)^\nu \rangle > l \tag{4.64}$$

then (4.61) is not D-stable.

Now let

$$B^I = (1 + \eta/R)E + A^I/R \tag{4.65}$$

Theorem 4.8. The interval system (4.61) is D_η-stable if for some ν the condition (4.63) holds with B^I defined by (4.65).

C o r o l l a r y 4.4. The system (4.61) is not D_η-stable if for some ν the condition (4.64) becomes valid with B^I defined by (4.65).

Next we shall consider the "classical" stability problem concerning the interval system (4.61) when D is the open left half of the complex plane. It is well known that for the noninterval case the system (4.55) is stable iff [50]

$$\rho(B) < 1 \tag{4.66}$$

where

$$B = E - 2(E - A)^{-1} \tag{4.67}$$

The stability of (4.55) is not affected if we consider the stability of the matrix εA rather than that of the original matrix A whenever $0 < \varepsilon \leq 1$. Indeed, if all $\lambda_i(A)$ lie in the open left half of the complex plane, so do the eigenvalues $\lambda_i(\varepsilon A)$, $i = \overline{1, l}$. Thus, instead of (4.67) we can use the equivalent relation

$$B = E - 2(E - \varepsilon A)^{-1}$$

The constant ε can be always chosen small enough so that $\|\varepsilon A\| < 1$. Then the series

$$(E - \varepsilon A)^{-1} = E + \varepsilon A + \varepsilon^2 A^2 + \ldots + \varepsilon^k A^k + \ldots$$

is convergent. Therefore, the system (4.55) is stable iff (4.66) is valid with

$$B = -E - 2\sum_{i=1}^{\infty} (\varepsilon A)^i$$

Based on these results we proceed to the stability of the interval system (4.61). First we have to compute the matrix

$$B^I = -E - 2\sum_{i=1}^{\infty} (\varepsilon A^I)^i \tag{4.68}$$

In practice we evaluate B_I by truncating the series, i.e.

$$B^I \approx \tilde{B}^I = -E - 2\sum_{i=1}^{\overline{n}} (\varepsilon A^I)^i$$

On account of the subdistributivity property (1.20) sharper bounds are in general obtained if B^I is evaluated by Horner's scheme:

$$\tilde{B}^I = -E - 2[\varepsilon A^I(E + \varepsilon A^I(E + \varepsilon A^I(\ldots)))] \tag{4.69}$$

The value of the constant ε can be determined in the following manner. The series (4.68) is convergent if

$$\|\varepsilon\tilde{A}\| = \varepsilon\|\tilde{A}\| < 1$$

where

$$\tilde{A} = |A^I| = |a^I_{i,j}|$$

Hence

$$\varepsilon < 1/\|\tilde{A}\| \tag{4.70}$$

If (4.70) holds, the truncation error δ is bounded from above and can be evaluated as follows

$$\tilde{\delta} \leq \varepsilon^{\bar{n}+1}\|\tilde{A}\|^{\bar{n}+1} = \bar{\delta} \tag{4.71}$$

Then using (4.69) and (4.71), the elements of B^I are finally computed as

$$B^I_{ij} = \tilde{B}^I_{ij} + [-\bar{\delta}, \bar{\delta}] \tag{4.72}$$

Thus we have the following result.

Theorem 4.9. If for some ν the condition (4.63) holds with B^I defined by (4.72) then the system (4.81) is stable.

Corollary 4.5. The interval system (4.61) is unstable if for some ν the condition (4.64) is fulfilled with B^I defined by (4.72).

Finally we shall consider the M-margin stability of (4.61). The interval system is said to have an M-margin stability if

$$\mathrm{Re}[\lambda_i(A)] < -M, \quad i = \overline{1,l}, \quad A \in A^I \tag{4.73}$$

Similarly to the previous results we have the following theorem.

Theorem 4.10. If for some ν the condition (4.63) holds with

$$B^I = E - 2[(1 + M)E - A^I] \tag{4.74}$$

then (4.61) is stable with an M stability margin.

The interval matrix $C^I = (1 + M)\,E - A^I$ involved in (4.74) can be inverted in the following manner. First, C^I is written in the form

$$C^I = \alpha(E - \frac{1}{\alpha}A^I), \quad \alpha = 1 + M$$

But

$$(C^I)^{-1} = \frac{1}{\alpha}(E - \frac{1}{\alpha}A^I)^{-1} = \frac{1}{\alpha}(C_1^I)^{-1}$$

and the matrix (C_1^I) can now be computed (after multiplying, if necessary, A^I by a small $\varepsilon > 0$) by a truncated series as this was shown for Theorem 4.9.

Corollary 4.6. The interval system (4.61) is not M-margin stable if for some ν the condition (4.64) is fulfilled using (4.74).

It is evident that all the sufficient conditions of interval form suggested above can be tested by means of Procedure 4.1 from section 4.2.1 (on introduction of corresponding obvious modifications in Steps 0, 7, 8, 9).

Remark 4.5. The assertion of Remarks 4.1 and 4.4 from 4.2.1 concerning Step 8 of Procedure 4.1 remain (after obvious modifications) also valid in the case of continuous linear circuits with interval parameters.

4.2.3. Illustrative examples

Example **4.2.** [48]. We consider the interval discrete system (4.46) with

$$A^I = \begin{bmatrix} [\ 0.2\ 0.6] & [0.1\ 0.4] \\ [-0.4\ 0.2] & [0.0\ 0.5] \end{bmatrix} \tag{4.75}$$

Using formulae (4.49a) and (4.49b) we have $\|A^I\|_\infty = 1$, $\|A^I\|_1 = 1$. However, calculating (4.49b) yields $\|A^I\|_2 = 0.9644 < 1$. Thus the sufficient condition (4.48) is fulfilled already with $\nu = 1$, so the interval discrete system (4.46) with A^I given by (4.75) is stable.

Example **4.3.** [49]. The system under consideration is again an interval discrete system with

$$A^I = \begin{bmatrix} [-0.20\ 0.16] & [-0.31\ 0.02] \\ [-0.24\ 0.12] & [-0.16\ 0.20] \end{bmatrix} \tag{4.76}$$

Now $\|A^I\|_\infty = \|A^I\|_1 = 0.54 < 1$ and according to (4.48) the system is stable.

Let us consider the M-margin stability of (4.76). From (4.52) and (4.53) $\|B^I\| = \|A^I\|_\infty/R < 1$; hence $R < 0.54$ for this example. On the other hand $R < 1 - M$, thus $M > 1 - R = 0.46$. It is worth nothing that using the sufficient condition from [49] a smaller stability margin $M = 0.11$ has been obtained.

Example **4.4.** [49]. In this example we consider the continuous time system (4.61) with the following interval matrix

$$A^I = \begin{bmatrix} [-4.1 \;\; -3.5] & [1.3 \;\; 1.91] \\ [0.3 \;\; 0.9] & [-4.5 \;\; -3.9] \end{bmatrix} \tag{4.77}$$

The centre matrix A_c of (4.77) has two real eigenvalues $\lambda_1 = -3$ and $\lambda_2 = -5$. Using the corresponding (rather complicated) sufficient condition it has been shown in [49] that the dynamic system with A^I given by (4.68) is stable with stability margin $M = 1.78$ (defined by (4.73)).

Now we shall apply the simple criterion (4.52), (4.53) to (4.77) to verify the D-stability of the associated dynamic system. We have chosen $R = 5$ for the radius of the disc D with centre $(-R, j0)$. The modified matrix B^I is

$$B^I = E + A^I/R = \begin{bmatrix} [0.13 \;\; 0.30] & [0.26 \;\; 0.38] \\ [0.06 \;\; 0.18] & [0.10 \;\; 0.22] \end{bmatrix}$$

Computation yields $\|B^I\|_\infty = 0.68$ and $\|B^I\|_1 = 0.6$ and by the criterion (4.63) the dynamic interval system considered is D-stable. Since the region D covers (for this example) all possible displacement of $\lambda_i(A)$ when $A \in A^I$, the dynamic system is, in fact, stable.

Finally, we shall consider the D_η-stability of (4.68). We have chosen $\eta = 2$ and $R = 5$, so the corresponding disc D has its centre at the point $(-7, j0)$. With these data the modified matrix B^I evaluated by (4.65) is

$$B^I = \begin{bmatrix} [0.58 \;\; 0.70] & [0.26 \;\; 0.30] \\ [0.06 \;\; 0.18] & [0.50 \;\; 0.62] \end{bmatrix}$$

We have $\|B^I\|_\infty = 1.08$, $\|B^I\|_1 = 1$ and $\|B^I\|_2 = 1.025$, so we have to calculate $\|(B^I)^2\|$. To do this, first B^I is put in the form

$$B^I = B_C + \Delta^I =$$

$$\begin{bmatrix} 0.64 & 0.32 \\ 0.12 & 0.56 \end{bmatrix} + \begin{bmatrix} [-0.06 \;\; 0.06] & [-0.06 \;\; 0.06] \\ [-0.06 \;\; 0.06] & [-0.06 \;\; 0.06] \end{bmatrix}$$

Using the formula $(B_c + \Delta^I)^2 = B_c^2 + \Delta^I B_c + B_c\Delta^I + (\Delta^I)2$ we find

$$(B^I)^2 = \begin{bmatrix} [0.3832 \;\; 0.52001] & [0.2736 \;\; 0.5016] \\ [0.0960 \;\; 0.1992] & [0.2584 \;\; 0.4528] \end{bmatrix}$$

Now $\|(B^I)^2\|_\infty = 1.2016$ but $\|(B^I)^2\|_1 = 0.9544$ and $\|(B^I)^2\|_2 = 0.8756 < 1$ and according to Theorem 4.8 the system considered is D_η-stable with $\eta = 2$. Since $\|(B^I)^2\|_2 = 0.8756$ it is obvious that in fact η can be taken greater then 2. Moreover, for this example the D_η -stability is actually M-margin stability (since $\lambda_l(A)$ remain real for all $A \in A^I$). Thus, we have shown that $M = 2$ by the new test while $M = 1.78$ by the criterion from [49] and at the same time the less conservative estimate $M = 2$ is obtained in a very simple way.

***Example* 4.5.** In this example we are interested in testing the D-stability of a continuous-time system with the following 5 x 5 interval matrix

$$A^I = \begin{bmatrix} a_{11}^I & 1 & 0 & 0 & 0 \\ 0 & a_{22}^I & 1 & 0 & 0 \\ 0 & 0 & -1.5 & 1 & 0 \\ 0 & 0 & 0 & -0.5 & 1 \\ a_{51}^I & a_{52}^I & a_{53} & a_{54} & a_{55} \end{bmatrix} \tag{4.78}$$

where

$$a_{11}^I = [-2, 1.8], \; a_{22}^I = [-1.5, -1]$$
$$a_{51}^I = [-0.085924, -0.002524], \; a_{52}^I = [-0.057426, 0.736574]$$
$$a_{53} = -0.112133, \; a_{54} = -0.633343, \; a_{55} = -1.505552$$

First, we took $R = 1$ and applied the (modified) Procedure 4.1. It turned out that the matrix studied is not D-stable in the disc of radius $R = 1$ (the computation process ended up in Step 8). However, for $R = 2$ the procedure stopped in Step 7, thus guaranteeing that A^I given by (4.78) is D-stable. To study the relationship between the radius R of the disk D and the number of iterations m needed to establish the corresponding D-stability of A^I, the present example has been solved several times with increasing value for R. The results are given in the following table.

Table 4.1.

R	2	5	10	100	1000
m	2	4	5	8	11

These results are in accordance with the theoretical prediction: as the radius R grows the number of iterations m increases. Indeed, it is seen from (4.62) that B^I approaches the identity matrix E as R grows which explains the decreasing rate of convergence of the method used.

Example **4.6.** This example has been designed to show that the inconclusive results possible in applying Procedure 4.1 or its modifications occur in practice very seldom. To this end, a large set of tests has been carried out with randomly generated interval matrices A^I.

Each individual matrix A^I was formed in the following manner. First, a real (n x n) matrix A^c (centre matrix) is generated by assigning different random numbers from a preset range $[\underline{r}, \overline{r}]$ of values to the elements a^c_{ij} of A^c. Then, for a chosen value of a parameter τ (which determines the tolerance of each a_{ij}^I in percentage with respect to the centre value a_{ij}^c) the elements a_{ij}^I of the corresponding interval matrix A^I are formed

$$a_{ij}^I = a_{ij}^c + [-\tau a_{ij}^c, \tau a_{ij}^c]$$

The tests were performed by means of Procedure 4.1 (for checking the stability of discrete interval systems) or its first modification from Section 4.2.2 (for checking the D-stability of continuous interval systems). The bulk of the tests was, however, carried out using Procedure 4.1 since the results concerning its modification were practically the same as for the basic Procedure 4.1.

Each test consisted in applying the chosen procedure to a large number N of randomly generated interval matrices for fixed values of the following input parameters:

- n – size of the matrix;
- $[\underline{r}, \overline{r}]$ – range of random value for generating the elements of the real matrix A^c ;
- τ – tolerance parameter determining the width of the interval matrix A^I ;
- m – maximum number of iterations in the inner loop formed by Steps 1 to 6 of Procedure 4.1 (or its modification).

The output parameters for a test are the numbers μ_1 : number of occurrences of exit in Step 7 (stable or D-stable), μ_2 : number of occurrences of exit in Step 8 (unstable or not D-stable) and μ_3 : number of occurrences of exit in Step 9 (inconclusive results).

A great deal of different tests has been performed. Typically N (number of different interval matrices generated in a test) was 100 (although at the beginning of the numerical experiment 200 or even greater values for N were used). The size n of A^I was 3, 4 and 5. The smallest range $[\underline{r}, \overline{r}]$ was [0.5, 0.5] and the largest one [−2, 2]. The parameter τ varied from 0.05 to 0.8 through 0.1, 0.2, 0.4 and 0.6. Most often $\overline{m}$ = 10 or $\overline{m}$ = 15 (higher values were used at the beginning of the experiment).

The results from the numerical experiment can be summarized as follows. The number μ_3 of inconclusive outcomes in each test never exceeded 5. The average of μ_3 over the whole set of tests is about 2% of the sum of $\mu_1 + \mu_2 + \mu_3$. Quite often, especially when the set of the randomly generated matrices A^I in a test is such that most of the matrices are stable (or unstable) $\mu_3 = 0$. Therefore, changing the range $[\underline{r}, \overline{r}]$ and the parameter τ more balanced tests were achieved for which $\mu_1 \approx \mu_2$ (number of stable matrices is approximately equal to the number of unstable ones on a test). The average of μ_3 over this subset of tests is around 3%.

In accordance with the theoretical consideration it has been observed that keeping all the remaining parameters unchanged while increasing $\bar{m}$ the number of inconclusive cases μ_3 decreases. For instance, in a test using Procedure 4.1 and the parameters: $n = 3$, $N = 100$, $[\underline{r}, \bar{r}] = [-1, 1]$, $\tau = 0.05$ and $\bar{m} = 10$ the number μ_3 in this test was $\mu_3 = 2$ (with $\mu_1 = 45$ and $\mu_2 = 55$). However, increasing $\bar{m}$ to 15 the next test yielded $\mu_3 = 1$ (with $\mu_1 = 48$ and $\mu_2 = 51$). Likewise changing N to $N = 200$, a test with $\bar{m} = 10$ (and the same remaining parameters as before) led to $\mu_3 = 2$ (with $\mu_1 = 77$ and $\mu_2 = 21$) while μ_3 became $\mu_3 = 1$ when $\bar{m}$ was increased to 15.

The experimental evidence of this example seems to show that Procedure 4.1 and its modifications can effectively be applied for checking all types of stability considered in sections 4.2.1 and 4.2.2, the number of inconclusive results being relatively rather low (on the order of 3%).

4.2.4. A less conservative criterion

In this section a less conservative (but computationally much more involved) M-margin stability criterion for continuous interval systems will be presented. It is based on a theorem due to A. Neumeier [51].

Consider the interval matrix A^I from (4.61). The centre matrix of A_c is represented in the form

$$A_c = Q\Lambda Q^{-1} \tag{4.79}$$

Here Λ is a block-diagonal matrix whose (2 x 2) blocks (if any) are of the form

$$\begin{pmatrix} \mathrm{Re}\lambda_i & \mathrm{Im}\lambda_i \\ -\mathrm{Im}\lambda_i & \mathrm{Re}\lambda_i \end{pmatrix}$$

where λ_i is the ith complex eigenvalue of A_c; Q is a matrix whose columns are made of the real and imaginary parts of the eigenvectors v of A_c. Now the following matrices are formed:

$$B^I = Q^{-1}A^IQ \tag{4.80}$$

$$B^I_{sym} = \frac{1}{2}[B^I + (B^I)^T] \tag{4.81}$$

and the so-called companion matrix of the interval matrix B^I_{sym}

$$C = \langle B^I_{sym} \rangle = \{c_{ij}\} \tag{4.82}$$

The elements c_{ij} of the companion matrix are defined as follows:

$$c_{ij} = -|(B^I_{sym})_{ij}|, \quad i \neq j \tag{4.83a}$$

$$c_{ii} = \langle (B^I_{sym})_{ii} \rangle \tag{4.83b}$$

here the magnitude $|.|$ is calculated by (4.49d) and the mignitude is evaluated by (4.50). Thus, C is a real symmetrical matrix.

It is seen from (4.79) and (4.80) that $B_c = \Lambda$. Therefore, it follows from (4.81), that if the diagonal entries of $B^c{}_{sym}$ are negative, then A_c is stable and vice versa. Based on these considerations and the properties of the companion matrix (4.82) the following theorem has been proven in [51].

T h e o r e m 4.11. If

$$(B^I)_{ii} < 0, \quad i = \overline{1,l} \tag{4.84a}$$

and there exists a vector $u > 0$ ($u_i > 0$, $i = \overline{1, l}$) such that

$$\langle B^I_{sym} \rangle u \geq Mu \tag{4.84b}$$

then

$$\mathrm{Re}[\rho(A^I)] \leq -M.$$

The theorem provides a sufficient and "almost necessary" condition [51] for M-margin stability of continuous interval systems (provided the width of the matrix A^I is not very large). Therefore, the estimate for M seems to be less conservative than that obtained by the method from section 4.2.2.

We shall first consider the case where M is given. When (4.84a) is fulfilled the condition (4.84b) is verified in the simplest way if u is chosen to be

$$u = (1,1,\ldots,1)^T \tag{4.85}$$

If the condition (4.84b) is fulfilled, then the interval system (4.61) is stable. Very often the choice (4.85) will work. If not we have to verify that the set of linear inequalities

$$\sum_{j=1}^{n} (c_{ij} - M) u_j \geq 0 \tag{4.86a}$$

$$u_j > 0, \quad j = \overline{1,n} \tag{4.86b}$$

is compatible. This can be done quite easily by any linear programming method if (4.86b) is approximated by

$$u_j - \varepsilon \geq 0, \quad j = \overline{1,n} \tag{4.86c}$$

where $\varepsilon > 0$ is a small constant.

Next we shall consider the case where the original problem is to determine the maximum possible value $\bar{M}$ of the stability margin M for a given interval system. Using Theorem 4.11 a very good approximation $\tilde{M}$ of $\bar{M}$ can be found in the following manner. First (since we are interested in $\bar{M}$), the inequality (4.84b) is replaced by the equality

$$Cu = Mu \tag{4.87}$$

Problem (4.87) is obviously an eigenvalue problem with a symmetrical matrix. We next solve (4.87) and the maximum eigenvalue M for which the corresponding eigenvector $\tilde{u}$ satisfies the condition $\tilde{u} > 0$ yields the sought approximation of $\bar{M}$. It should be noted that $\tilde{M} < \bar{M}$ because of the fact that Theorem 4.11 provides only a sufficient condition for stability. Thus, the approximation introduced can never lead to wrong conclusions about the M-margin stability of (4.61) (which would be the case if $\tilde{M}$ could be greater than $\bar{M}$).

As is seen from the foregoing the present criterion for checking the M-margin stability requires a much greater computational effort than the corresponding condition from section 4.2.2. Indeed, the representation (4.79) is obtained by solving the complete eigenvalue problem

$$A_c v = \lambda v \tag{4.88}$$

where A_c is, in general, not symmetrical. If M is given the compatibility of the linear inequalities (4.86a), (4.86c) must be checked. One more eigenvalue problem (4.87) (with symmetrical matrix) must be solved if $\bar{M}$ is sought. These problems can be, however, handled by standard subroutines. Thus, the present M-margin stability criterion can be easily implemented and should be applied in cases where the criterion from section 4.2.2 fails to assess the stability of the interval system studied.

4.3. NECESSARY AND SUFFICIENT CONDITIONS FOR ROBUST STABILITY

In the previous section various sufficient conditions for assessing the stability (D-stability, D_η-stability or M-margin stability) of electric circuits with interval parameters were considered. With the exception of the condition from section 4.2.4 the stability criteria presented so far are very simple to implement numerically. However, they all share the common drawback that they provide a conservative estimate of the stability properties of the circuit studied whether the original stability problem be formulated equivalently as an interval polynomial stability problem or as an interval matrix stability problem. This is due to the fact that the coefficients of the corresponding interval polynomial or interval matrix are assumed to be independent intervals whereas the examples considered (formulae (4.3) and (4.20b)) clearly indicate that they are interdependent through the parameter vector p. Ignoring this interdependence leads to conservativeness of the stability tests. Thus, no assertion concerning the stability or instability of the circuit studied can be made whenever the sufficient stability condition is not fulfilled.

In this section necessary and sufficient conditions for checking the stability (or instability) of linear circuits with interval parameters will be considered for the case where the original stability Problem 4.1 is equated to that of assessing the stability of an associated set of polynomials. In Section 4.3.1 a stability test is presented which checks the positiveness of only three functions over the parameter box. It requires the characteristic polynomial of the circuit studied (open- or closed-loop configuration) in explicit form. Based on an interval generalization of the Nyquist criterion a stability test for the case of closed-loop circuit configuration is proposed in section 4.3.2. Several sufficient stability conditions are also derived in this manner.

4.3.1. Frazer – Duncan criterion

We take up the stability problem for linear circuits with interval parameters in its polynomial formulation. More specifically the class of linear circuits to be considered is defined as follows:

$$\sum_{j=0}^{m} a_j(p) \frac{d^{(m-j)}x}{dt^{(m-j)}} = 0 \tag{4.89a}$$

$$p \in P \tag{4.89b}$$

where $p = (p_1, \ldots, p_n)$ is the parameter vector, P is a given interval vector and $a_j(p)$ are arbitrary nonlinear functions $a_j : P \subset R^n \rightarrow R$ which are at least continuous (i.e. $aj \in C^{(h)}$, $h \geq 0$). The set of all circuits (4.89) when $p \in P$ will be denoted (as in section 4.1.1) by the symbol $N(P)$. Recall that $N(P)$ is stable iff (if and only if) each circuit (4.89a) is stable for $p \in P$.

According to the approach adopted in section 4.1.1 the stability analysis of (4.89) will be effected by studying the associated family of polynomials

$$q(s,p) = \sum_{j=0}^{m} a_j(p) s^{(m-j)} \tag{4.90a}$$

$$p \in P \tag{4.90b}$$

where $q(s, p)$ is stable iff all its roots lie in the open left half-plane of the complex plane. Let $q(s, P)$ denote the set of all polynomials (4.90). Then $q(s, P)$ is stable iff each $q(s,p)$ is stable with $p \in P$.

In this section a necessary and sufficient condition for determining the stability of (4.89) is presented. Consider the Hurwitz matrix associated with (4.90)

$$H(p) = \begin{bmatrix} a_1 & a_3 & a_5 & a_7 & . & . & . & 0 & 0 \\ a_0 & a_2 & a_4 & a_6 & . & . & . & 0 & 0 \\ 0 & a_1 & a_3 & a_5 & . & . & . & 0 & 0 \\ 0 & a_0 & a_2 & a_4 & . & . & . & 0 & 0 \\ . & . & . & . & . & . & . & . & . \\ 0 & 0 & . & . & . & . & . & a_{m-1} & 0 \\ 0 & 0 & . & . & . & . & . & a_{m-2} & a_m \end{bmatrix} \tag{4.91}$$

Let $\Delta_m(p)$ denote the determinant of $H(p)$; furthermore, let $\Delta_{m-1}(p)$ denote the determinant derived from $H(p)$ by deleting the last row and column of $H(p)$ (these determinants are called Hurwitz determinants of order m and $m - 1$, respectively). We have the following theorem [57].

Theorem 4.12. The set $q(s, P)$ given by (4.90) is stable iff:

i) there exists a $p = p^0 \in P$ such that $q(s, p^0)$ is stable and

ii) the coefficients a_0, a_m and the Hurwitz determinant of order $m - 1$ are different from zero over the parameter box , that is,

$$\begin{aligned} a_0(p) &\neq 0 \\ a_m(p) &\neq 0 \\ \Delta_{m-1}(p) &\neq 0 \end{aligned}$$

for all $p \in P$.

Based on the above theorem and some well-known facts related to the stability of polynomials – positiveness of the polynomial coefficients (necessary condition, e.g. [44]), positiveness of all Hurwitz determinants (necessary and sufficient condition, e.g. [58]) – the following result is straightforward.

Theorem 4.13. (Necessary and sufficient condition). The set $N(P)$ of circuits (4.89) is stable iff:

i) the nominal circuit $N(p^c)$ (with p^c being the centre of P) is stable and

ii) the coefficients a_0, a_m and the Hurwitz determinant of order $m - 1$ are all positive in P, i.e.

$$\begin{aligned} a_0(p) &> 0 \\ a_m(p) &> 0 \\ \Delta_{m-1}(p) &> 0 \end{aligned}$$

for all $p \in P$.

Since the verification of Condition i) of Theorem 4.13 presents no difficulties we shall henceforth assume that it is fulfilled and we shall concentrate on checking Condition ii).

Let $f(p)$ denote any of the functions involved in Condition ii) of the theorem. Thus, we are led to solve three times the following problem.

Problem 4.2. Check that

$$f(p) > 0, \quad p \in P \tag{4.92}$$

There are various ways to verify (4.92). The simplest approach is to use some interval extension $F(P) = [\underline{F}, \overline{F}]$ of $f(p)$ in P. By Theorem 1.1 from Chapter 1

$$F(P) \supseteq f(P) = [\underline{f}^*, \overline{f}^*] \tag{4.93}$$

where $f(P)$ is range of p over P. Hence, (4.92) is satisfied if $\underline{F} > 0$.

Now let $F_i(P) = [\underline{F}_i, \overline{F}_i]$ be some interval extensions of $f_i(p)$ in P, with $f_1(p) = a_0(p)$, $f_2(p) = a_m(p)$ and $f_3(p) = \Delta_{m-1}(p)$. Based on Theorem 4.13 and inclusion (4.93) the following results are obvious.

Corollary 4.7. (Sufficient condition for stability). Let the nominal circuit $N(p^c)$ be stable. If

$$\underline{F}_i > 0, \quad i = 1, 2, 3,$$

then the set $N(P)$ of circuits (4.89) is stable.

Corollary 4.8. (Sufficient condition for instability). If at least one of the endpoints

$$\overline{F}_i \leq 0$$

$i = 1, 2, 3$, then the set $N(P)$ is not stable.

Various interval extensions can be used in implementing Corollaries 4.7 and 4.8: natural extensions, mean-value form extensions etc. (section 1.2). For the case where $f(p) \in C^{(1)}$ $F(P)$ can be evaluated in a most efficient way by means of the modified *MT*-form and *MV*-form of the mean-value extension (section 2.2) since they provide very narrow intervals $F(P)$ and thus lead to sharp bounds on the range $f(P)$.

If the condition of Corollary 4.7 is not satisfied then one must resort to solving Problem 2.4 three times as needed in Theorem 4.13.

One way to do this is to find the lower endpoint $\underline{f}^*$ of the range of f in P. If

$$\underline{f}^* \geq 0 \tag{4.94}$$

then obviously Problem 4.2 has a solution.

The lower endpoint $\underline{f}^*$ of $f(P)$ can be determined as the global solution of the minimization problem:

$$\underline{f}^* = \min_{p \in P} f(p) \tag{4.95}$$

Thus, the problem of determining the stability of the set (4.89) has been transformed to globally solving 3 constraint minimization problems of the type (4.95). It will be noted that the minimization problem of type (4.95) (with constraints in the form of an interval vector) arises in tolerance analysis of linear circuits. Loosely speaking, using the approach herein adopted the stability problem considered has been equated to 3/2 tolerance problems which can be tackled by the tolerance analysis methods presented in Chapter 2, section 2.3.

There exists a better way to solve Problem 4.2 which circumvents the need to find $\underline{f}$. It is based on the following considerations (section 2.4). Recall that the interval tolerance methods are iterative and have the appealing feature that at each iteration they provide infallible bounds on $\underline{f}^*$; more precisely

$$\underline{f}^* \in [\underline{f}^\nu, \overline{f}^\nu] \tag{4.96}$$

where $\underline{f}^\nu$ is obtained as the lower endpoint of $F(P^\nu)$ where P^ν is the current subregion of P while $\overline{f}^\nu$ is the current upper bound on the global minimum $\underline{f}^*$. Moreover, $\underline{f}^\nu$ are nondecreasing and $\overline{f}^\nu$ are nonincreasing, i.e.

$$\underline{f}^{\nu+1} \geq \underline{f}^\nu, \quad \nu \geq 0 \tag{4.97a}$$

$$\overline{f}^{\nu+1} \leq \overline{f}^\nu, \quad \nu \geq 0 \tag{4.97b}$$

with $\underline{f}^\nu$ tending to $\underline{f}^*$ from below and $\overline{f}^\nu$ approaching $\overline{f}^*$ from above when $\nu \to \infty$. Thus, there is no point in finding $\underline{f}^*$ when solving Problem 4.2. Instead, on account of (4.97a) and condition (4.94) the iterative process can be terminated whenever the inequality

$$\underline{f}^\nu > 0$$

is fulfilled for the first time, since by virtue of (4.96) $\underline{f}^*$ is guaranteed to satisfy the inequality (4.94). Therefore, the condition of Problem 4.2 is satisfied.

If at some iteration ν

$$\overline{f}^\nu \leq 0$$

then obviously $\underline{f}^* \leq 0$ and hence $f(p)$ is not positive in P.

Let $\overline{f}_i^{\nu}$ be the current upper bound on the global minimum $\underline{f}_i^*$ of $f_i(p)$ in P at some iteration ν. In conjunction with Theorm 4.13 we have the following sufficient condition for instability.

C o r o l l a r y 4.9. If at least one of the elements $\overline{f}_i^{\nu}$, $i = 1, 2, 3$ is nonpositive, then the set $N(P)$ is not stable.

The above sufficient condition for instability is, of course, less conservative than that of Corollary 4.8.

The bounds $\underline{f}^{\nu}$ and $\overline{f}^{\nu}$ on $\underline{f}^*$ for each of the arising problems (4.92) can be determined by using some interval method for worst-case tolerance analysis. If the coefficients $a_j(p)$ are only continuous functions ($a_j(p) \in C^{(0)}$ in P or the evaluation of the derivatives of $a_j(p)$ is rather costly), then the interval $[\underline{f}^{\nu}, \overline{f}^{\nu}]$ can be found by the method from section 2.3.1. If $a_j(p) \in C^{(1)}$ then $[\underline{f}^{\nu}, \overline{f}^{\nu}]$ is recommended to be determined by the methods from sections 2.3.2, 2.3.3 or 2.4.2; if $a(p) \in C^{(2)}$ the methods from sections 2.3.4 or 2.4.4 may be the best choice.

Based on the foregoing the following procedure for assessment of the stability or instability of (4.89) is suggested.

P r o c e d u r e 4.2.

S t e p 1. Choose a suitable interval method for global optimization and set $\nu = 0$ (ν is the number of the current iteration) and $P^{\nu} = P$.

S t e p 2. Apply the method (its νth iteration) sequentially (or parallelly) to all functions $f_i(p)$, $i = \overline{1, 3}$, with $f_i(p)$ being either $a_0(p)$, $a_m(p)$ or $\Delta_{m-1}(p)$ and $p \in P^{\nu}$. At the end of this step the corresponding values $\underline{f}_i^{\nu}$ and $\overline{f}_i^{\nu}$ are obtained.

S t e p 3. Check the condition of Corollary 4.7 substituting $\underline{f}_i^{\nu}$ for $\underline{F}_i$. If it is satisfied, go to Step 5; otherwise go to the next step.

S t e p 4. Check the condition of Corollary 4.9. If it is satisfied go to Step 6; otherwise put $\nu = \nu+1$ and go back to Step 2 with a new (smaller) subregion P^{ν} (each subsequent subregion P^{ν} is automatically generated by the interval minimization method used).

S t e p 5. Termination 1: the set $N(P)$ of circuits (4.89) is stable.

S t e p 6. Termination 2: the set $N(P)$ of circuits (4.89) is not stable.

It should be stressed that in most cases Procedure 4.2 will terminate either in Step 5 or Step 6 long before the global minimum $\underline{f}^*$ of the corresponding function $f(p)$ in P is reached. Indeed, the only case where $\underline{f}^*$ needs be determined with good accuracy by rather a narrow interval $[\underline{f}^{\nu}, \overline{f}^{\nu}]$ (and hence for larger ν) occurs when a circuit (4.89a) for some $p \in P$ has a very small margin of stability. But even in this instance the interval global minimization methods from sections 2.3 and 2.4 using first- or second-order derivatives may provide $\underline{f}^*$ in a fairly small number of iterations.

R e m a r k 4.6. When Procedure 4.2 is implemented by means of a method from section 2.4.2 or section 2.4.4 use must be made of the upper bound $\overline{f}^{\nu}$ set to zero right from the beginning, i.e. $\overline{f}^0 = 0$ (formula (2.116)) as explained in section 2.4.2.

We shall illustrate the application of Procedure 4.2 through the following example.

***Example* 4.7.** The characteristic polynomial of a closed-loop control system is

$$q(s,p) = s^4 + p_1^3 p_2 s^3 + p_1^2 p_2^2 p_3 s^2 + p_1 p_2^3 p_3^2 s + p_3^3$$

The parameter interval vector P has components

$$P_1 = [1.15,\ 1.65],\ P_2 = [1.3,\ 1.7],\ P_3 = [0.6,\ 1.0]$$

The problem is to check whether the system remains stable when $p = (p_1, p_2, p_3) \in P$.

It can be verified that the nominal system with $p = p^c$ is stable. Thus, by Theorem 4.13, Condition ii) we have to check the positiveness in P of the following functions

$$a_4 = p_3^3$$

$$\Delta_2 = p_1^2 p_2^2 p_3^6 [p_1^4 p_2^4 - p_1^4 - p_2^4 p_3]$$

It is easily seen by inspection that the first function as well as the factor before the bracket in Δ_2 are positive when $p \in P$. Thus, it suffices to only check the positiveness of

$$\Delta_2' = p_1^4 (p_2^4 - 1) - p_2^4 p_3$$

by Procedure 4.2. It was implemented by means of the improved Algorithm 2.6. The following results were obtained on a PC IBM AT (6 MHz clock frequency).

Table 4.2

N_i	l_m	$t(s)$
1	0	0.38

The approach herein suggested can be also applied for assessing whether the set $N(P)$ of circuits (4.89) has a certain prescribed margin of stability M. The following example illustrates this possibility.

***Example* 4.8.** Consider the electric circuit shown in Fig.4.1. The problem is to determine whether the circuit has a desired stability margin M when $R_i \in R_i^I$, $i = 1, 2, 3$, $L \in L^I$ and $C \in C^I$ with R_i^I, L^I and C^I being specified intervals.

From (4.10) the characteristic equation for any branch current i_1, i_2 or i_3 is

$$L(R_2 + R_3)\lambda^2 + (R_1 R_2 + R_2 R_3 + R_1 R_3 + L/C) + (R_1 + R_3/C) = 0$$

The corresponding polynomial is

$$q(\lambda,p) = a_0'\lambda^2 + a_1'\lambda + a_2' \tag{4.98}$$

where

$$\begin{aligned} a_0' &= L(R_2+R_3) \\ a_1' &= R_1R_2+R_2R_3+R_1R_3+L/C \\ a_2' &= (R_1+R_3)/C \end{aligned} \tag{4.99}$$

To introduce the stability margin M, the variable λ is put in the form

$$\lambda = s + M \tag{4.100}$$

Substituting (4.100) into (4.98) the following polynomial is obtained

$$q(s,p) = a_0's^2 + (2a_0'M + a_1')s + a_0'M^2 + a_1'M + a_2'$$

Thus

$$\begin{aligned} a_0 &= a_0' \\ a_1 &= 2a_0'M + a_1' \\ a_2 &= a_0'M^2 + a_1'M + a_2' \end{aligned}$$

or taking (4.99) into account

$$a_0 = L(R_2 + R_3) \tag{4.101a}$$

$$a_1 = 2L(R_2 + R_3)M + R_1R_2 + R_2R_3 + R_1R_3 + LD \tag{4.101b}$$

$$a_2 = L(R_2 + R_3)M^2 + (R_1R_2 + R_2R_3 + R_1R_3 + LD)M + (R_1 + R_3)D \tag{4.101c}$$

where $D = 1/C$.

Assume that $M = -10$ and the circuit parameters have the following tolerances

$$\begin{aligned} R_i &\in [\underline{R}_i, \overline{R}_i] = [90, 110]\,\Omega, \quad i = 1,2,3 \\ L &\in [\underline{L}, \overline{L}] = [0.9, 1.1]\,\text{mH} \\ D &\in [\underline{D}, \overline{D}] = [0.2, 0.22]\,10^6/\mu\text{F} \end{aligned} \tag{4.102}$$

The problem is to check whether the set of circuits has the desired stability margin within the box defined by (4.102).

The exact solution of this problem can be obtained in just the same way as in the previous examples (using in fact the necessary and sufficient condition for positiveness of the functions involved) by applying Procedure 4.2 to the corresponding functions associated with (4.101) and (4.102). Here we shall present a much simpler approach for an approximate solution which is based on the sufficient condition for stability given by

Corollary 4.7. For the example considered the circuit set will have the given margin of stability M if

$$\underline{A}_1 > 0 \; ; \; \underline{A}_2 > 0 \tag{4.103}$$

where $\underline{A}_1$ and $\underline{A}_2$ denote the lower endpoint of some interval extension of (4.101b) and (4.101c), respectively.

To evaluate $\underline{A}_1$ and $\underline{A}_2$ natural interval extensions of (4.101b) and (4.101c) have been used. Thus, the following expressions for $\underline{A}_1$ and $\underline{A}_2$ are easily obtained:

$$\underline{A}_1 = -20\overline{L}(\overline{R}_2 + \overline{R}_3) + \underline{R}_1\underline{R}_2 + \underline{R}_2\underline{R}_3 + \underline{R}_1\underline{R}_3 + \underline{L}\,\underline{D} \tag{4.104a}$$

$$\underline{A}_2 = 100\underline{L}(\underline{R}_2 + \underline{R}_3) - 10(\overline{R}_1\overline{R}_2 + \overline{R}_2\overline{R}_3 + \overline{R}_1\overline{R}_3 + \overline{L}\,\overline{D}) + (\underline{R}_1 + \underline{R}_3)\underline{D} \tag{4.104b}$$

Computation shows that the conditions (4.103) are fulfilled for the given tolerances on the circuit parameters. Thus, according to Corollary 4.7 the set $N(P)$ of circuits considered is guaranteed to have the desired stability margin $M = -10$.

In the case where Corollary 4.7 is valid for some initial interval vector P^0 we may wish to determining the largest possible region (around P^0) in the parameter space within which the set $N(P)$ is still stable. More precisely, let

$$P^0 = p^0 + [-\Delta, \Delta]$$

where p^0 and Δ are the centre and radius of P^0, respectively. Now we can formulate the following problem.

Problem 4.3. Given an interval box P^0 for which the set $N(P^0)$ of circuits (systems) studied is stable, find the largest box

$$P^* = p^0 + k^*[-\Delta, \Delta]$$

for which the set $N(P)$ becomes critically stable for the first time.

The parameter k^* is an alternate measure of the stability margin (alongside the margin M) of the set $N(P^0)$.

The solution P^* of Problem 4.3 can be found (within a desired accuracy) solving the basis Problem 4.1 for different boxes $P^v \supset P^0$. The simplest approach is to use the following dichotomy process (e.g. [77]).

Procedure 4.3. We start with an initial parameter k, form the box

$$P = p^0 + k[-\Delta, \Delta]$$

and solve the corresponding Problem 4.1 (associated with the current box P). If $N(P)$ is stable the parameter k is given an increment Δk, a new box P' with $k' = k + \Delta k$ is formed and the stability of the set $N(P')$ is again checked. This continues until the current box P^ν leads for the first time to an unstable set $N(P^\nu)$. Now the last interval $[k^{\nu-1}, k^\nu]$ is halved to give a new value $k^{\nu+1}$ for the parameter k^* and a corresponding box $P^{\nu+1}$. The stability of the set $N(P^{\nu+1})$ is again checked and depending on whether it is stable or not the corresponding interval $[k^\nu, k^{\nu+1}]$ or $[k^{\nu+1}, k^{\nu-1}]$ is again halved. This process of halving (dichotomy) continues until the solution k^* (and hence P^*) of Problem 4.3 is obtained within a preset accuracy.

Problem 4.3 can be easily generalized to include the case when we wish to find the largest box P^* within which the set $N(P)$ has a desired margin of stability M. To illustrate this more general stability problem the following example will be considered.

Example **4.9.** We take up the circuit from Example 4.7. Now we wish to determine approximately the largest possible interval region P^* (around the centre of the initial interval vector P^0 given by (4.102)) in the parameter space within which $N(P)$ has still the desired stability margin $M = -10$.

Let each component of the parameter vector P^0 be represented in the equivalent form

$$P_i^0 = p_i^c + [-\Delta p_i, \Delta p_i]$$

Then, the corresponding component of a new enlarged parameter vector P' can be written as

$$P_i' = p_i^c + t[-\Delta p_i, \Delta p_i] \tag{4.105a}$$

where $t > 1$ and

$$p_i^c - t\Delta p_i > 0, \quad i = \overline{1,5} \tag{4.105b}$$

(the latter condition ensures that the interval $[\underline{P}_i', \overline{P}_i']$ is positive which is required by the positiveness of the circuit parameters). The maximum possible value t_m of t ensuring the largest vector P_i' within which the circuit is guaranteed to have the desired stability margin can be determined approximately by means of (4.104) and (4.105) in the following manner. Starting from (4.105a) the endpoints of P_i' are obviously given by

$$\underline{P}_i' = p_i^c - t\Delta p_i \tag{4.106a}$$

$$\overline{P}_i' = p_i^c + t\Delta p_i \tag{4.106b}$$

Substituting (4.106) for the corresponding circuit parameters into (4.104a) yields

$$\begin{aligned}\underline{A}_1 &= -20(L^C + t\Delta L)(R_2^C + R_3^C + t(\Delta R_2 + \Delta R_3)) \\ &+ (R_1^C - t\Delta R_1)(R_2^C - t\Delta R_2) + (R_2^C - t\Delta R_2)(R_3^C - t\Delta R_3) \\ &+ (R_1^C - t\Delta R_1)(R_3^C - t\Delta R_3) + (L^C - t\Delta L)(D^C - t\Delta D)\end{aligned} \tag{4.107}$$

Setting $\underline{A}_1$ as defined by (4.107) to zero the following quadratic is obtained

$$\alpha_0 t^2 + \alpha_1 t + \alpha_2 = 0 \tag{4.108}$$

Proceeding in exactly the same way with $\underline{A}_2$ and letting $\underline{A}_2 = 0$ a second quadratic will be derived

$$\beta_0 t^2 + \beta_1 t + \beta_2 = 0 \tag{4.109}$$

Now let t_1 and t_2 be the smallest (positive) solution of (4.108) and (4.109), respectively. The approximate value $\tilde{t}$ of t_m can obviously be estimated as the smallest of the solutions t_1 and t_2. For the example considered $t_1 = 9.79$ and $t_2 = 6.66$; hence

$$\tilde{t} = 6.66 \tag{4.110}$$

It can be verified that $\tilde{t}$ satisfies condition (4.105b) for each circuit component. Finally, the new parameter vector P' can be determined by formula (4.105a) where t is replaced by $\tilde{t}$ from (4.110). It should be noted that $\tilde{t}$ is always less or equal to t_m. Thus, the approximation P' is never larger than the exact largest vector P^* ensuring the desired stability margin.

4.3.2. Nyquist criterion

In this section we shall consider circuits whose flow-graph representation can be reduced to an equivalent graph of the form depicted in Fig. 4.3.

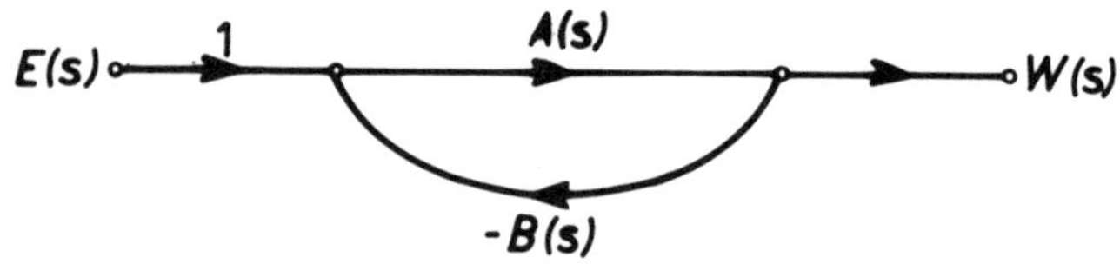

Fig. 4.3. Equivalent feedback flow-graph

The transfer function associated with this feedback flow-graph is

$$H(s,p) = \frac{A(s,p)}{1 + A(s,p)B(s,p)} \tag{4.111a}$$

where p is the parameter vector. The problem herein studied is to establish whether the set of circuits $N(p)$ having (4.111a) as their transfer function is stable when

$$p \in P \tag{4.111b}$$

where P is prespecified interval vector.

Let

$$T(s,p) = A(s,p)B(s,p) \tag{4.112}$$

For simplicity and ease of explanation it will be assumed that:

A1) $T(s, p)$ has no poles in the right half-plane for any $p \in P$

A2) $A(s, p)$ has left-half-plane zeros only for any $p \in P$.

Assumptions A1 and A2 can be verified by the method of the previous section. Throughout this section it will be assumed that they are fulfilled. Then the following theorem is valid.

Theorem 4.14. The set $N(P)$ of circuits whose transfer function is given by (4.111) is stable iff the Nyquist diagram of $T(s, p)$ defined by (4.112) does not encircle or intersect the point $(-1, j0)$ for all $p \in P$.

This theorem is obvious interval generalization of the well-known Nyquist stability criterion.

Recall that the Nyquist diagram (for fixed p) is the locus of $T(j\omega, p)$ in the complex plane when ω increases from $\omega = 0$ to $\omega = \infty$. Therefore, Theorem 4.14 is impractical since it deals with all possible Nyquist diagrams generated by all $p \in P$. However, it serves as the basis for derivation of the following result which is computationally amenable.

Let

$$T_1(\omega,p) = \mathrm{Re}[T(j\omega,p)], \tag{4.113a}$$

$$T_2(\omega,p) = \mathrm{Im}[T(j\omega,p)]. \tag{4.113b}$$

Consider the following global minimization problem

$$T_1^* = \min T_1(\omega,p) \tag{4.114a}$$

$$T_2(\omega,p) = 0 \tag{4.114b}$$

$$p \in P \tag{4.114c}$$

$$\omega > 0 \tag{4.114d}$$

We have the following theorem (which, of course, presupposes that assumptions A1 and A2 about $T(s, p)$ and $A(s, p)$ are fulfilled).

Theorem 4.15. The set $N(P)$ of circuits whose transfer function is given by (4.111) is stable iff $T_1^* > -1$.

The proof of this theorem (which is in fact a corollary of the basic Theorem 4.14) is straightforward and is therefore omitted.

By analogy with classical (point) stability analysis (when p is fixed) the so-called gain margin (GM) can be introduced for the whole set $N(P)$ of circuits with interval parameters. Clearly, GM is defined by the formula

$$GM = 1 + T_1^* \tag{4.115}$$

In some circuits the gain margin alone does not adequately describe the margin of stability. Therefore, a second stability characteristic, namely the so-called phase margin (PM) is also used. In the case considered of circuits with interval parameters PM can be defined and evaluated (once again by analogy with the case of circuits with constant parameters) in the following way. Consider the global minimization problem

$$T_2^* = \min T_1(\omega,p) \tag{4.116a}$$

$$|T(j\omega,p)| = 1 \tag{4.116b}$$

$$p \in P \tag{4.116c}$$

$$\omega > 0 \tag{4.116d}$$

Now the angle θ ($\theta < 0$) is introduced as follows

$$\theta = \tan^{-1} \frac{\sqrt{1 - (T_2^*)^2}}{T_2^*} \tag{4.117a}$$

Finally, PM is defined by the formula

$$PM = 180^0 + \theta \tag{4.117b}$$

It is evident from the above that the following corollary is valid.

C o r o l l a r y 4.10. The set $N(P)$ considered is stable iff PM defined by (4.117) is greater than zero.

Similarly with the case of circuits (systems) with exact (noninterval) parameters the true margin of stability of the set of circuits $N(P)$ can be best characterized by using both GM and PM or, equivalently, T_1^* and T_2^*.

Now we shall consider some computational aspects associated with the resolution of problems (4.114) and (4.116). These are complex global minimization problems in $n + 1$ variables with constraints of the general type including equality and inequality restrictions. Moreover, the interval for the variable ω is not explicity known except that it contains the point ω_Φ where ω_Φ is the so-called phase crossover frequency for $p = p^c$ (when p is fixed the phase crossover frequency is the frequency at which $\text{Im}\,[T(j\omega, p)] = 0$).

Therefore, the solution of (4.114) or (4.116) seems at first approach to be rather a difficult task. We shall, however, show that these problems can be solved in a most efficient manner by fixing ω at several appropriately chosen discrete values. Thus, each minimization problem is reduced to a small number of a.c. tolerance problems of type P1 (cf. section 3.3.1). We shall demonstrate this possibility by way of problem (4.114).

P r o c e d u r e 4.4.

To begin with, we set $p = p^c$ (p^c is the centre of the interval vector P) and find the phase crossover frequency ω_Φ. We let $\omega^{(1)} = \omega_\Phi$ and solve the following a.c. tolerance problem

$$t_1^{(1)} = \min_{p \in P} T_1(\omega^{(1)}, p) \tag{4.118}$$

Let $p^{(1)}$ be the corresponding parameter vector providing the global minimum $t_1^{(1)}$. At this stage the imaginary part $T_2(\omega, p)$ is calculated for $\omega = \omega^{(1)}$ and $p = p^{(1)}$. Generally, $T_2(\omega^{(1)}, p^{(1)}) \neq 0$. Now keeping $p = p^{(1)}$ we seek a new frequency $\omega^{(2)}$ at which

$$T_2(\omega, p^{(1)}) = 0 \tag{4.119}$$

Obviously, the solution $\omega^{(2)}$ of (4.119) is the phase crossover frequency when $p = p^{(1)}$. Next, a new a.c. tolerance problem is solved, namely

$$t_1^{(2)} = \min_{p \in P} T_1(\omega^{(2)}, p) \tag{4.120}$$

and the corresponding solution vector $p^{(2)}$ is found. If $p^{(2)} = p^{(1)}$ (which will be most often the case) the pair $(\omega^{(2)}, p^{(1)})$ is the solution of the original problem (4.114). Indeed, by (4.120) $(\omega^{(2)}, p^{(1)})$ minimizes the real part of $T(j\omega, p)$ and at the same time by (4.119) it equates the imaginary part of $T(j\omega, p)$ to zero.

If $p^{(2)} \neq p^{(1)}$ we set $\omega^{(1)} = \omega^{(2)}$ and start the computation process over again from problem (4.118) until the newly labelled vector $p^{(2)}$ obtained by (4.120) becomes equal to the preceding vector $p^{(1)}$ (in practice, of course, until some norm of the difference $p^{(2)} - p^{(1)}$ becomes smaller than a specified accuracy ε).

Problem (4.116) needed to determine the phase margin PM can be solved in much the same way as the above problem (4.114) by slightly modifying Procedure 4.4. The only difference occurs after the determination of $p^{(1)}$ from (4.118) when $\omega^{(2)}$ is found as the solution of

$$|T(j\omega, p^{(1)})| = T_3(\omega, p^{(1)}) = 1 \tag{4.121}$$

Equation (4.119) and (4.121) are nonlinear equations in one single variable and can be solved by any classical or interval method (e.g. by the interval version of the Newton method from section 1.4.1 which guarantees global convergence).

Each of the arising a.c. tolerance problems (4.118) can be solved by some of the interval methods from sections 2.3.2 to 2.3.4 (if the derivation of the function $T_1(\omega, p)$ and its derivatives with respect to p_i in explicit form is not too strenuous a task).

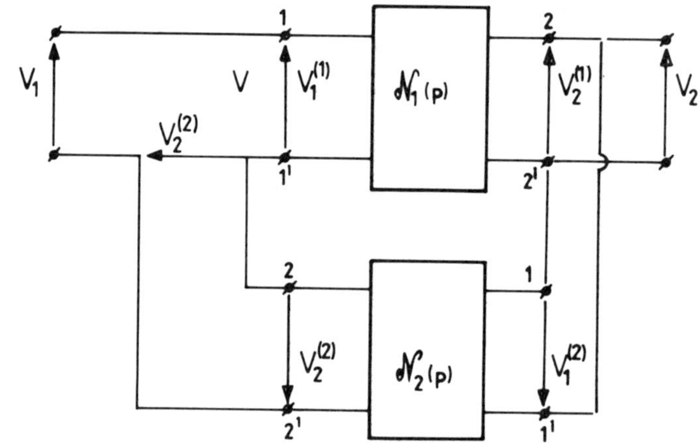

Fig. 4.4(a). Negative feedback configuration circuit.

An alternative possibility is to reformulate problem (4.118) as a corresponding a.c. tolerance problem P1 (cf. section 3.1.1) in implicit form. We shall illustrate this approach by way of the following example. Consider the circuit shown in Fig. 4.4(a). We shall first assume that p is fixed. The voltage transfer function of this negative feedback configuration circuit is

$$\frac{V_2(s,p)}{V_1(s,p)} = H(s,p) = \frac{A(s,p)}{1 + A(s,p)B(s,p)} \tag{4.122}$$

where $A(s, p)$ and $B(s, p)$ are the corresponding voltage transfer functions of networks $N_1(p)$ and $N_2(p)$, respectively. So $T(s, p) = A(s, p)\, B(s, p)$ and obviously $T(j\omega, p)$ is the output voltage V_2 of the circuit shown in Fig. 4.4(b) when V_1 is a unit voltage (i.e. when V_1 is sinusoidal with $v_1 = \sqrt{2}\sin\omega t$ and hence $V_1 = 1$). Thus, (for any fixed ω) $T_1(\omega, p)$ can be found as the real part of the output voltage V_2. Now let $p \in P$; then the initial problem (4.118) is clearly equivalent to the following a.c. tolerance problem of the type P1 from section 3.1.1:

(**TP1**) Find the lower endpoint of the real part of the output voltage V_2 from Fig. 4.4(b) when $p \in P$.

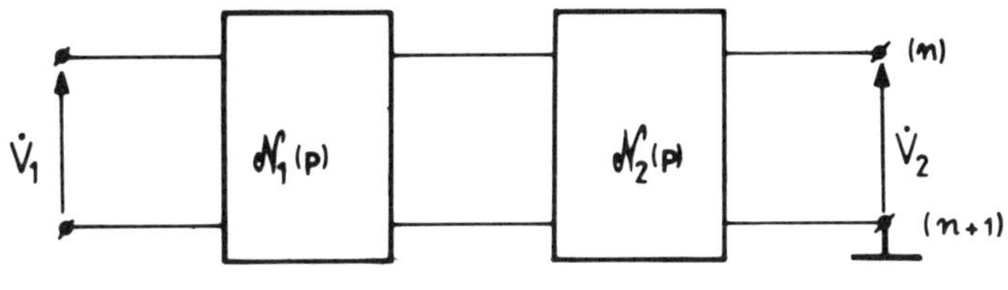

Fig. 4.4(b) Resultant open-loop congiguration circuit.

Problem (TP1) can be formulated in the form (3.10) if the grounded node in the circuit from Fig.4.4(b) is the node $(n + 1)$. The resultant systems of linear equations with dependent coefficients (3.90) may be solved exactly (if the corresponding assumptions A1 and A2 are fulfilled) by the method from section 3.2.4. These systems can always be solved rather accurately and most often exactly by the method from section 3.3.4 (Procedure 3.3). However, it should be borne in mind that theoretically Procedure 3.3 provides only an upper bound $\overline{f}_U$ on the lower endpoint $\underline{f}^*$ of the output interval variable f where f stands for the real part of the output voltage V_2. A lower bound f_L on $\underline{f}^*$ is therefore needed at the last iteration of the solution of problem (4.114) (when $\|p^{(2)} - p^{(1)}\| \le \varepsilon$). Such a bound f_L can be easily found by solving the (associated with the last iteration) complex linear interval system (3.73) using some method for solution of interval linear systems with complex coefficients (e.g. [2], [10]). For instance, system (3.73) can be solved (after obvious modifications to account for the complex quantities involved) by the iterative procedure defined by formula (1.59). By Theorem 1.9 the complex solution X^* obtained by (1.59) contains the exact solution X of (3.73). Therefore, the lower endpoint of the real part of the corresponding component X_n^* of X^* provides a lower bound f_L on $\underline{f}^*$ guaranteeing that $f_L \le \underline{f}^*$. Thus, at the last iteration problem (TR1) has been solved twice: once by Procedure 3.3 to obtain f_U and secondly by Procedure (1.59) to evaluate f_L. In general $f_L < f_U$ and in practice $\varepsilon = f_U - f_L$ has rather a small value. Most often $f_L = f_U$ thus guaranteeing that the interval global minimization problem (4.114) has been solved exactly.

The application of Theorem 4.15 will be illustrated by the following example.

***E x a m p l e* 4.10.** To show the applicability of the present approach to control engineering problems we shall consider the stability robustness of the feedback system shown in Fig. 4.5.

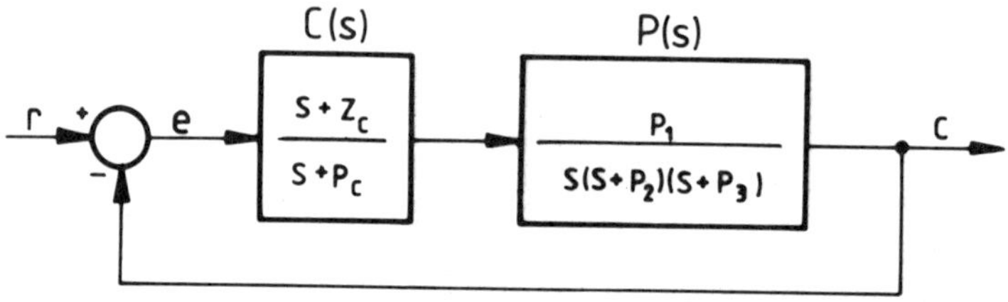

Fig. 4.5. Feedback control system with uncertain plant.

The parameters z_c and p_c of the controller are assumed to have exactly known values. The value of the gain p_1 and the location of the poles p_2 and p_3 of the plant are not known exactly, i.e. p_i are allowed to lie in some intervals P_i with midpoints (nominal values) p_i^0, $i = 1, 2, 3$. The problem considered is Problem 4.2: find the largest possible box P^* (centred at the point $p^0 = (p_1^0, p_2^0, p_3^0)$ in the parameter space) within which the closed-loop system is still stable (it is assumed that the closed-loop system with the nominal parameter plant is stable) This problem will be solved using an approach based on problem (4.114).

From Fig. 4.5 the open-loop transfer function $T(s, p)$ is

$$T(s,p) = \frac{(s + z_c)p_1}{(s + p_c)s(s + p_2)(s + p_3)} \tag{4.123a}$$

so

$$|T(j\omega,p)| = \frac{p_1\sqrt{\omega_2 + z_c^2}}{\omega\sqrt{\omega^2 + p_c^2}\sqrt{\omega^2 + p_2^2}\sqrt{\omega^2 + p_3^2}} \tag{4.123b}$$

Let $P_i = [\underline{p}_i, \overline{p}_i] > 0$, $i = 1, 2, 3$ be some initial parameter intervals centred at p_i^0. It is obvious from (4.123b) that for any ω the modulus $|T(j\omega, p)|$ is maximum in the region $P = (P_1, P_2, P_3)$ if $p_i = \tilde{p}_i$ with $\tilde{p}_1 = \overline{p}_1$, $\tilde{p}_2 = \underline{p}_2$ and $\tilde{p}_3 = \underline{p}_3$.

Let

$$P_i = p_i^0 + [-\Delta_i, \Delta_i], \quad i = 1,2,3.$$

Now, we shall form a new interval vector P' whose components are defined as follows

$$P_i' = p_i^0 + k[-\Delta_i, \Delta_i] > 0, \quad I = 1,2,3. \tag{4.124a}$$

so that

$$\tilde{p}_i' = p_i^0 + k\tilde{\Delta}_i, \quad \tilde{\Delta}_1 = \Delta_1, \quad \tilde{\Delta}_2 = -\Delta_2, \quad \tilde{\Delta}_3 = -\Delta_3 \tag{4.124b}$$

The solution P^* can be found in the following manner. First, (4.114) will be modified into the equivalent form

$$T^* = \max |T(j\omega, p)|$$
$$T_2(\omega, p) = 0 \tag{4.125}$$
$$p \in P, \ \omega > 0$$

We next observe that the maximum of (4.123b) when $p \in P'$ is again obtained (at any ω) for $p_i = \tilde{p}_i'$ defined by (4.124b). Based on formulae (4.125) and (4.124b) it is easily seen that P^* is determined uniquely by the values of k and ω for which the following two equalities hold:

$$|T(j\omega, \ p_i^0 + k\tilde{\Delta}_i)| = 1 \tag{4.126a}$$

$$T_2(\omega, \ p_i^0 + k\tilde{\Delta}_i) = 0 \tag{4.126b}$$

where the LHS of (4.126a) is defined by means of (4.123b) and (4.124) and the LHS of (4.126b) is the corresponding imaginary part of (4.123a) (when $s = j\omega$). Let k^* and ω^* be the solutions of the system (4.126). Finally the largest parameter box P^* within which the closed-loop system remains stable is determined by the formula (1.124a) with $k = k^*$.

Comments

Section 4.1. The problem treated in the present chapter is of considerable practical significance. Indeed, the circuit considered may be an amplifier or a control system (or part of it) and it is of paramount importance to know that the stability of the circuit (whatever its function) is guaranteed even in the presence of some uncertainties about the values of various component parameters. This problem (usually referred to as robust stability) has been intensively studied over the last years in the control literature (e.g. [42], [43], [45]-[49], [52]-[59] both in the framework of subsection 4.1.1 (as stability of interval polynomials) and in that of subsection 4.1.2 (as stability of interval matrices).

Theorem 4.1 provides only a sufficient condition for the set $N(P)$ of circuits with interval parameters to be stable. However, due to the computational simplicity of Kharitonov's result (Theorem 4.2 on which it is based) this approach to testing the stability of $N(P)$ should be tried out first before the more complex necessary and sufficient condition from subsection 4.3.1.

Several attempts to extend Kharitonov's approach to more general stability problems have been made in the recent years. In [52] Kharitonov's theorem is generalized to polynomials which have all its zeros only in a sector in the complex plane. A second extension guaranteeing that the corresponding dynamic system has only aperiodic behaviour is obtained in the same paper. An attempt to extend Kharitonov's stability test to polynomials with dependent coefficients has been made in [53]; the applicability of the latter generalization seems, however, to be limited to rather simple cases.

If the derivation of the characteristic polynomial is impractical (as in the case of circuits of larger size) it may prove computationally more advantageous to assess the stability of $N(P)$ by testing the stability of an associated interval matrix. Two approaches to deriving such a matrix have been considered in subsection 4.1.2.

Section 4.2. The problem of determining the stability of interval matrices has received much attention over the last few years (e.g. [45]–[49]). In [45] and [46] rather simple conditions were proposed, limited however, to the special case where $a_{ii} < 0$, $i = \overline{1, n}$ for every $A \in A^I$. For the general case of arbitrary matrices a complex criterion requiring the solution of two Lyapunov matrix equations was suggested in [47]. By means of similarity transformation and Gershgorin's theorem, new criteria applicable to the M-margin stability of both continuous and discrete dynamic systems were developed in [49]; however, these require the computation of all the eigenvalues of a real matrix.

A necessary and sufficient condition for the stability of two special subclasses of discrete systems (the so-called D^+ and D^- type systems imposing rather restrictive requirements on the signs of the system coefficients) were derived in [48].

The stability criteria (including D-stability, D_η-stability and M-margin stability cases) suggested in subsections 4.2.1 and 4.2.2 are extremely easy to implement on a computer. The only precaution concerns the realization of the mignitude operation defined by formula (1.50). Indeed, while the magnitude operation (defined by (4.49c)) requires the usual outward rounding of the quantities involved the mignitude operation must be implemented using inward rounding to ensure correct results. The portion of inconclusive outcomes (termination in Step 9 of Procedure 4.1 and its modifications) can be reduced if corresponding matrix measures [62] are used instead of the norms (4.49a) to (4.49c) in implementing the above criteria.

The most crucial moment in applying this section's approach for assessment of the stability of the set $N(P)$ of continuous-time or discrete-time circuits is the obtainment of the associated interval matrix A^I. In subsection 4.1.2 two possible ways of deriving A^I are exposed. The first one is based on the state-variable description for transient analysis of the circuit studied. While the real matrix A with elements $a_{ij}(p)$ thus obtained is of relatively small size l this method presents the drawback that each element A_{ij} of A^I should be evaluated (in order to reduce the conservativeness of the stability test) as the range $a_{ij}(P)$ of $a_{ij}(p)$ over P. Therefore, $2l^2$ global optimization problems are to be solved to obtain the narrowest possible interval matrix A^I. The second method for computing A^I is based on a dynamic generalization of the implicit form formulation of the d.c. tolerance problem from section 3.1.1. It is presented for the case where the circuit studied includes only resistors and capacitors but can be also applied for analyzing arbitrary R, L, C circuits; indeed, it is known [28] that a R, L, C circuit can be modelled equivalently by a R, C circuit. Although this method for obtaining A^I results in larger matrices (as compared to the state-variable method) it has the advantage that no optimization problems are to be solved. It would be of interest to assess the computational efficiency of both methods for certain classes of circuits (for instance, for filters of a particular type).

The criterion of subsection 4.2.4 is apparently less conservative than the criteria considered in the previous sections. This is achieved at the cost of greater computational effort. However, using this criterion one is able to determine the maximum possible value M of the stability margin M for the interval system studied.

Section 4.3. The idea to solve the robust stability problem by finding the global minima of certain (approximately chosen) multivariable functions has been pursued by several authors (see, e.g. [55], [58], [77] and references cited therein). The basic approach used is to transform the original problem to an equivalent problem which is then solved globally by a special method applicable only to the transformed problem introduced. Thus, in [55] all the uncertain parameters (gain, phase, pole and zero locations, etc.) must be rearranged into a diagonal feedback structure (and this preliminary stage may be rather time-consuming). Then the closed-loop characteristic polynomial is found and the corresponding Hurwitz matrix (4.91) is formed. By the Routh–Hurwitz criterion the closed-loop system is stable iff all the principle minors of (4.91) are positive. Most of the corresponding m global minimization problems are, however, much more complex than the minimization problems associated with Theorem 4.13 where only the coefficients a_0, a_m and the determinant Δ_{m-1} are to be evaluated and minimized. Similarly, in [58] the original problem is transformed to a generalized geometrical programming problem which is then solved by a signomial algorithm. Moreover, some restrictions on the nonlinear functions are imposed: e.g. they are multilinear in [55] or at most multivariable polynomials in [58] .

The interval approach adopted in section 4.3.1 addresses directly the original problem of checking the positiveness of the functions involved. It can be solved by the highly efficient methods from section 2.4 for any nonlinear functional relationships $a_i = a_i(p)$.

It is important to stress that Theorem 4.13 can be extended by appropriate modification of the characteristic polynomial to encompass more complex cases such as M-margin stability (cf. example 4.8) and the two problem statements from [52] mentioned in the comments to section 4.1.

Whenever Corollary 4.7 is valid the problem of determining the largest possible interval region P^* within which the set of circuits $N(P)$ is still stable – Problem 4.3 – or still has the requisite stability margin can be solved in a most simple manner as a finite series of the usual stability check problems by Procedure 4.3. A significant reduction of the amount of computation is possible if Problem 4.3 is to be only solved approximately (e.g. at an early stage of the design). As can be easily seen from Example 4.8 this problem is reduced to finding the smallest zeros $t_1 > 1$ and $t_2 > 1$ of two real polynomials of degree m (polynomials (4.108) and (4.109) for $m = 2$).

The interval extension of the Nyquist criterion from section 4.3.2 seems to be a most effective tool for robust stability analysis of feedback circuits or systems. Indeed, according to Procedure 4.4 – formulae (4.118) to (4.120) or the modified Procedure

(4.118), (4.119) and (4.121) – the assessment of the gain *GM* or phase *PM* margin of the closed-loop circuit (system) is reduced to several a.c. tolerance analysis problems related to the open-loop transfer function. The latter problems can be solved by the methods from section 2.3 or section 3.2.4.

Procedure 4.4. has been tested on the open-loop transfer functions from Example 4.10 and Example 5 of [77]. Each problem (4.118), (4.120) was solved by means of the global minimization algorithm A4 from section 2.3.3. The results obtained so far are quite encouraging.

General remark. In the present chapter the problem of stability robustness of circuits (systems) with interval parameters has been considered. A closely related problem is that of performance robustness. More precisely, loop performance robustness requires that some specified values of the closed-loop systems response are achieved regardless of given plant and sensor performance variations. According to the performance robustness theorem of Doyle, Wall and Stain [59] robust performance is assured iff stability robustness is achieved with a fictitious complex-disc-bounded uncertainty. The latter can be modified by two bilinear transformations into a two-dimensional real vector [54]. Thus, it seems that the robust performance problem can be also handled by the methods of this chapter.

CHAPTER 5

TRANSIENT ANALYSIS OF LINEAR CIRCUITS WITH INTERVAL DATA

In Chapters 2 and 3, tolerance analysis problems dealing with d.c. or sinusoidal steady-states in linear lumped electric circuits have been considered. In this chapter, tolerance analysis is extended to cover problems related to transient analysis of circuits of the same class. More specifically, several dynamic worst-case tolerance problems will be formulated in which some or all of the input data are given as intervals. Methods for obtaining approximate and, in some cases, exact solutions to the dynamic tolerance problems formulated will be presented.

5.1. PROBLEM STATEMENT

Similarly to the static tolerance analysis from Chapters 2 and 3 we shall distinguish between input parameters and output variables in formulating dynamic tolerance problems.

Transient tolerance analysis of linear electric circuits gives rise to a great variety of problems depending on the mathematical description of the transients, on one hand, and the number and nature of the interval input parameters as well as the number of output variables, on the other. In this section three basic approaches to formulating (and solving) transient tolerance analysis problems will be presented.

5.1.1. Explicit form formulation

In the case of transient analysis problems in explicit form formulation there is only one output variable which may be some transient current or voltage in the circuit studied. The input parameters may be component values, amplitudes of d.c. or sinusoidal excitations and initial conditions values. The relationship between the input parameters and the output variable must be given in a closed explicit form. Obviously, this is possible only for circuits whose order of complexity l (as defined, e.g. in [44]) does not exceed the number 2 or 3. To introduce this simplest form of dynamic tolerance analysis we shall consider the following example.

Example **5.1.** The circuit studied is shown in Fig 5.1. (the supply v is constant and $v_C(0) = 0$). The dynamic tolerance analysis problem herein considered is to find for a fixed (but arbitrary) time t the interval $I_3(t)$ of all possible values of the branch current $i_3(t)$ when L, C, R and v belong to some prescribed intervals L^I, C^I, R^I and v^I.

We shall assume that the quantity

$$\delta = \frac{1}{R^2C^2} - \frac{1}{LC} \tag{5.1}$$

is positive for all $R \in R^I$, $L \in L^I$ and $C \in C^I$. Then the point solution $i_3(t)$ (for fixed R, L, C and v) is given by the formula

$$i_3(t) = \frac{v}{R}\left[1 + \frac{1}{2CR\sqrt{\delta}}\left(e^{k_1 t} - e^{k_2 t}\right)\right] \tag{5.2}$$

where δ is defined by (5.1) and k_1, k_2 are

$$k_{1,2} = -\frac{1}{2RC} \pm \sqrt{\delta} \tag{5.3}$$

It should be stressed that the assumption about the positiveness of δ is essential since only then is formula (5.2) valid for all $R \in R^I, L \in L^I, C \in C^I$ and $v \in v^I$.

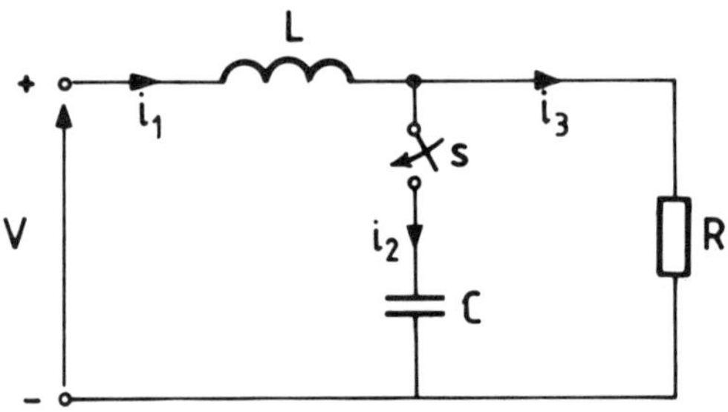

Fig. 5.1. Tolerance analysis of the transient $i_3(t)$.

Let $p = (R, L, C, v)$ and $P = (R^I, L^I, C^I, v^I)$. To underline the dependence of the point solution $i_3(t)$ on the parameter vector p the notation

$$i_3(t) = f(t,p) \tag{5.4}$$

will be used with $f(t, p)$ given by the RHS of (5.2), (5.1) and (5.3). Thus, the sought interval $I_3(t)$ can be determined by the range $f(t, P)$ of $f(t, p)$ over P, i.e

$$I_3(t) = \{(f(t,p): p \in P)\} \tag{5.5}$$

For fixed t problem (5.5) is a static tolerance problem in explicit form and can be solved exactly using one of the methods from section 2.3.

Based on this example it is now straightforward to present the explicit formulation of the transient analysis of circuits with interval data. Let N(p) denote a linear circuit with $p = (p_1, \ldots, p_n)$ being a real parameter vector which influences the (scalar) transient $x(t)$. In the general case p may include component values, initial conditions values and amplitudes of d.c. or sinusoidal excitations. Let $p \in P = (P_1, \ldots, P_n)$ where P is an interval vector. Each individual transient corresponding to some fixed p will be denoted as

$$x(t) = f(t,p) \tag{5.6}$$

We shall assume that the set N(P) of circuits N(p) is stable when $p \in P$. Then the set of time functions

$$X(t) = \{f(t,p): p \in P,\ t \in [0, \infty)\} \tag{5.7}$$

will be referred to as the interval transient since for each fixed t, $X(t)$ is an interval. Thus the dynamic tolerance problem considered can be formulated in explicit form as follows.

Problem 5.1. Given the linear circuit N(p), $p \in P$ and the function $f(t,p)$ in explicit form, find the interval transient $X(t)$.

In practice Problem 5.1 is solved for a series of discrete time moments.

It should be noted that approximate enclosing solutions are, of course, readily obtained if for fixed t one uses one interval extension or another of (5.6) in P.

As is seen from the example considered above the explicit form formulation of the dynamic tolerance analysis problem is possible only when an analytical expression for the relationship between input parameters and output variable can be derived. Its application is, therefore, limited to tolerance analysis of circuits of low complexity excited by d.c. or sinusoidal sources. It should be, however, mentioned that nonzero initial conditions are no obstacle for the present section's formulation since they are naturally accounted for in deriving the expression for the output variable (note that for the example considered $i_1(0) = v/R \neq 0$).

5.1.2 Frequency-domain formulation

An alternate explicit form formulation of a special dynamic tolerance analysis problem when there is only one single step excitation and one output variable will be introduced in this section. It is based on an equivalent representation of the output signal using frequency-domain analysis and is therefore applicable only if the initial conditions are zero. Unlike the explicit formulation introduced in the previous section, the new approach (called frequency-domain formulation) may, however, be used for tolerance analysis of linear circuits having (theoretically) an arbitrarily high order of complexity.

Now $x(t)$ is the step response of the circuit studied and the vector p includes only the component values and the step excitation magnitude (all the initial conditions are zero). In this case the set of functions $X(t)$ generated by $p \in P$ will be called the interval step response of the circuit. It is well known that there exists a relationship between the step response $x(t)$ and the real part $r(\omega)$ of the frequency response $F(j\omega)$ of the circuit investigated, namely (e.g. [44])

$$x(t) = \frac{2}{\pi}\int_0^\infty \frac{r(\omega)}{\omega}\sin\omega t\,d\omega \tag{5.8}$$

The basic approach adopted here to estimate the interval step response $X(t)$ is the interval generalization of (5.8) when $r(\omega)$ and hence $x(t)$ depends on the parameter vector p defined above. To underline the dependence of $r(\omega)$ on p the notation $r(\omega, p)$ will be used. Thus, (5.8) can be rewritten as

$$x(t) = \frac{2}{\pi}\int_0^\infty \frac{r(\omega,p)}{\omega}\sin\omega t\,d\omega \tag{5.9}$$

The dynamic tolerance analysis problem considered can be formulated as follows.

Problem 5.2. Given the linear circuit N(p) with zero initial conditions, $p \in P$ and the real part $r(\omega, p)$ of the frequency circuit response $F(j\omega)$ (as an analytical expression), find the interval step response $X(t)$.

Comparison of (5.6) and (5.9) reveals that now

$$f(t,p) = \frac{2}{\pi}\int_0^\infty \frac{r(\omega,p)}{\omega}\sin\omega t\,d\omega \tag{5.10}$$

Thus, it is seen from (5.10) that the solution $X(t)$ of Problem 5.2 for fixed t can be found as the range of $f(t, p)$ defined by (5.10) over P. The discussion of this problem requiring an interval analysis technique for evaluating integrals will be postponed for section 5.2.

The two transient analysis formulations introduced so far are explicit form formulations since both functions $f(t, p)$ from (5.6) and $r(\omega, p)$ from (5.10) are assumed to be known analytically. In the next subsection a third approach will be presented whereby the dynamic tolerance problem will be stated in an implicit form formulation. The latter formulation permits the study of a considerably larger class of circuits as compared with the former (explicit) formulations.

5.1.3. Time-domain formulation

In this section we shall consider linear circuits whose transients are described in the time interval $[0,\tau]$ by the system of differential equations (written in vector form)

$$\dot{x} = Ax + b(t), \quad t \in [0,\tau], \quad \tau \leq \infty \tag{5.11}$$

with initial conditions

$$x(0) = c \tag{5.12}$$

The excitation vector $b(t)$ is most often of the form

$$b(t) = \varphi(t)b \tag{5.13}$$

where $\varphi(t)$ is a diagonal matrix and b is a column. Each component $\varphi_i(t)$ of $\varphi(t)$ is bounded (also in the case where $\tau \to \infty$).

In the most general case the elements a_{ij} of A, b_i of b and c_i of c may all depend on the input parameter vector p. Indeed, for the example from section 4.1.2, Fig. 4.1 where $x = (i_L, v_C)^T$ we have

$$A = \begin{bmatrix} -\frac{1}{L}\left(\frac{R_2}{k} + R_1\right) & \frac{1}{L}\left(\frac{R_2}{R_3 k} - 1\right) \\ \frac{1}{C_k} & \frac{1}{R_3 C_k} \end{bmatrix} \tag{5.14a}$$

$$k = 1 + \frac{R_2}{R_3}$$

$$b(t) = \begin{bmatrix} \frac{1}{L}e(t) \\ 0 \end{bmatrix}, \quad b_1 = \frac{1}{L}, \quad b_2 = 0 \tag{5.14b}$$

If $e(t) = e =$ const., then

$$c_1 = i_L(0) = \frac{e}{R_2 + R_3} \tag{5.14c}$$

$$c_2 = V_C(0) = V_{Co} \tag{5.14d}$$

Let $p = (R_1, R_2, R_3, L, C, v_{Co}, e)$. It is seen that for the example considered

$$\begin{aligned} a_{11} &= a_{11}(p_1, p_2, p_3, p_4), \quad a_{12} = a_{12}(p_2, p_3, p_4), \\ a_{21} &= a_{21}(p_2, p_3, p_5,), \quad a_{22} = a_{22}(p_2, p_3, p_5), \\ b_1 &= b_1(p_4), \quad c_1 = c_1(p_2, p_3, p_7), \quad c_2 = p_6 \end{aligned}$$

showing that in this instance the elements a_{ij} of A, b_i of b and c_i of c all depend on one or other components of the parameter vector p and hence are not independent.

On account of (5.13) the system (5.12) may be written as

$$\dot{x} = Ax + \varphi(t)b \tag{5.15a}$$

$$x(0) = c \tag{5.15b}$$

Having the above example in mind (5.15) will be rewritten in the form

$$\dot{x} = A(p)x + \varphi(t)b(p), \quad t \in [0, \tau] \tag{5.16a}$$

$$x(0) = c(p) \tag{5.16b}$$

where $a_{ij}(p)$, $b_i(p)$ and $c_i(p)$, $i, j = \overline{1, l}$ are, in general, nonlinear functions of p.

Let

$$x(t) = f(t,p) \tag{5.17}$$

denote the solution of (5.16) for some fixed $p \in P$ where P is a given interval vector. The set

$$S(0, \tau) = \{f(t,p): t \in [0, \tau], p \in P\} \tag{5.18}$$

will be called the solution set of (5.16) for the time interval $[0, \tau]$ The set of values

$$S(t) = \{f(t,p): p \in P\} \tag{5.19}$$

of $S(0, \tau)$ for a fixed time t will be referred to as the reachability set at time t.

To encompass the case where $\tau \to \infty$ the following assumption is needed.

Assumption 5.1. The set of matrices $A(p)$ with $p \in P$ is either stable or D-stable in some large enough domain D.

It follows from Assumption 5.1 that $S(t)$ is a bounded region of R^l for each $t \in [0, \infty)$.

Let $X(t)$ denote the interval hull of $S(t)$, i.e. $X(t)$ is the smallest interval vector still containing the reachability set $S(t)$. The symbol Ih will be used for the interval hull; thus

$$X(t) = \mathrm{Ih}\, S(t) \tag{5.20}$$

The set of interval vectors (5.20) for $t \in [0, \tau]$ will be called the interval solution of problem (5.16) (and for notational simplicity will be denoted by the same symbol $X(t)$). On account of Assumption 5.1 the interval solution always exists even in the case where $t \in [0, \infty)$.

Now we are in a position to state the following (rather general) dynamic tolerance problem.

Problem 5.3. Given the linear circuit N(p) with the dynamic description (5.16), find the interval solution (5.20) when $p \in P$.

The above problem is extremely difficult to solve. Therefore (by analogy with section 4.1.2) it is reasonable to simplify it by assuming that a_{ij}, b_i and c_i are independent and lie in some intervals a_{ij}^I, b_i^I and c_i^I (these intervals being in fact some extensions or the ranges of $a_{ij}(p)$, $b_i(p)$ and $c_i(p)$, respectively, in P). Thus, we arrive at the following simpler problem.

Problem 5.4. Given the circuit N(p) whose transients are described by (5.15) with $A \in A^I$, $b \in b^I$ and $c \in c^I$, find the corresponding interval solution $X(t)$ assuming A^I to be stable (or D-stable).

Even Problem 5.4 is no easy task. In trying to (approximately) solve it we shall be led to consider the following auxiliary problems.

Problem 5.5. Only the components c_i of the initial conditions vector c are given as intervals, i.e. $c_i \in c_i^I$, $i = \overline{1, l}$ or in vector form $c \in c^I$ (A and b are known exactly).

Problem 5.6. Only the components b_i of the vector b are given as intervals b_i^I or in vector notation $b \in b^I$.

Problem 5.7. Only the elements a_{ij} of A are intervals, i.e. $a_{ij} \in a_{ij}^I$, or equivalently $A \in A^I$. Furthermore, the natural assumption is made that the interval matrix A^I is stable or D-stable.

Obviously, besides the last three problems there exist alternate formulations of the transient analysis of circuits with independent interval data which are in fact various combinations of the auxiliary problems 5.5 to 5.7. Several such "mixed" cases will be considered later on (in section 5.3).

5.2. SOLVING FREQUENCY – DOMAIN FORMULATION PROBLEMS

In this section, we proceed to solving the dynamic tolerance analysis Problem 5.2 (i.e. the worst-case tolerance analysis of the step response of linear circuits with zero initial conditions) [60]. Two approximate solutions are presented in subsection 5.2.1. These solutions (called outer solutions) have the property that they enclose the exact interval response $X(t)$ of the circuit studied. In subsection 5.2.2 two alternate approximate solutions are suggested which, not guaranteed to have the enclosing property, prove to be a good approximation to the exact interval solution of Problem 5.2. Finally, several examples illustrating the frequency-domain formulation approach are given also.

5.2.1. Outer solutions

As was mentioned in section 5.1.2 the interval step response (or simply the interval solution) $X(t)$ of the circuit investigated is defined by the set of time functions

$$X(t) = \{f(t,p): p \in P, \quad t \in [0,\infty)\} \tag{5.21}$$

where $f(t, p)$ is given by (5.10):

$$f(t,p) = \frac{2}{\pi}\int_0^\infty \frac{r(\omega,p)}{\omega} \sin\omega t \, d\omega \tag{5.22}$$

The lower endpoint $\underline{X}(t)$ and the upper endpoint $\overline{X}(t)$ of $X(t)$ can be determined (for fixed t) as the global solutions of the following optimization problems:

$$\underline{X}(t) = \min_{p \in P} f(t,p) \tag{5.23}$$

$$\overline{X}(t) = \max_{p \in P} f(t,p) \tag{5.24}$$

Since problems (5.23) and (5.24) are very difficult to solve exactly the exact interval response $X(t)$ will be replaced by some approximate interval solutions $Y(t)$. If the approximation $Y(t)$ satisfies the condition

$$X(t) \subseteq Y(t) \ , \ t \in [0,\tau] \tag{5.25}$$

then $Y(t)$ will be called an outer (or enclosing) solution to the tolerance problem considered. The enclosing property (5.25) may be useful in some applications. Thus, if the outer solution $Y(t)$ does not exceed a threshold x_{max}, then all the individual responses $x(t, p)$ for all $p \in P$ will obviously satisfy the inequality $x(t, p) \le x_{max}$, $t \in [0, \tau]$.

On account of (5.22)

$$\underline{X}(t) = \min_{p \in P}\left[\frac{2}{\pi}\int_0^\infty \frac{r(\omega,p)}{\omega} \sin\omega t \, d\omega\right] \tag{5.26}$$

A similar formula is valid for $\overline{X}(t)$:

$$\overline{X}(t) = \max_{p \in P}\left[\frac{2}{\pi}\int_0^\infty \frac{r(\omega,p)}{\omega} \sin\omega t \, d\omega\right] \tag{5.27}$$

Thus, it is seen from (5.22) to (5.27) that the exact tolerance solution $X(t) = [\underline{X}(t), \overline{X}(t)]$ is defined as the range of (5.22) over the interval vector (box) P. Note that in (5.26) and (5.27) the minimizing or maximizing vector p should be constant over the whole frequency interval $[0, \infty]$.

The determination of $\underline{X}(t)$ and $\overline{X}(t)$ by (5.26) and (5.27), respectively, is no easy task. This is why an outer approximation $Y(t)$ to the exact interval tolerance solution $X(t)$ will be introduced using an interval analysis technique for evaluating integrals [2].

Let $a(\omega)$, $\omega \in [\underline{\omega},\overline{\omega}] = \Omega_0$ be a real continuous function whose interval extension $A(\Omega)$ for any interval $\Omega \subset \Omega_0$ is Lipschitz, i.e. $w(A(\Omega)) \leq lw(\Omega)$ where $w(.)$ is the width of the corresponding interval and $l \geq 0$ is some constant. By the mean-value theorem

$$\int_{\underline{\omega}}^{\overline{\omega}} a(\omega)\,d\omega = \int_{\Omega_0} a(\omega)\,d\omega = a(\xi)(\overline{\omega}-\underline{\omega}) = a(\xi)\,w(\Omega_0)$$

where $\xi \in \Omega_0$. Therefore $a(\xi) \in A(\Omega_0)$ and

$$\int_{\Omega_0} a(\omega)\,d\omega \in A(\Omega_0)\,w(\Omega_0) \tag{5.28}$$

Now let Ω_0 be divided into n subintervals $\Omega_1, \Omega_2, \ldots, \Omega_n$, $\Omega_i = [\underline{\omega}_i, \overline{\omega}_i]$, $i = \overline{1, n}$ such that $\underline{\omega} = \underline{\omega}_1 < \overline{\omega}_1 = \underline{\omega}_2 < \overline{\omega}_2 = \ldots < \overline{\omega}_n = \overline{\omega}$. According to the additivity property of the integral

$$\int_{\underline{\omega}}^{\overline{\omega}} a(\omega)\,d\omega = \sum_{i=1}^{n} \int_{\Omega_i} a(\omega)\,d\omega \tag{5.29}$$

From (5.28), (5.29) and the Lipschitz condition on $A(\omega)$ the following theorem is valid [2].

Theorem 5.1. For any partitioning of Ω_0 the inclusion

$$\int_{\Omega_0} a(\omega)\,d\omega \in \sum_{i=1}^{n} A(\Omega_i)\,w(\Omega_i) \tag{5.30}$$

and the estimate

$$w\left(\sum_{i=1}^{n} A(\Omega_i)\,w(\Omega_i)\right) \leq l \sum_{i=1}^{n} \left(w(\Omega_i)\right)^2 \tag{5.31}$$

are valid.

If the partitioning is uniform, then $w(\Omega_i) = (\overline{\omega} - \underline{\omega})/n = h$. Let

$$S_n = \sum_{i=1}^{n} A(\Omega_i)(\overline{\omega}-\underline{\omega})/n$$

From (5.31)

$$w(S_n) \leq l(\overline{\omega}-\underline{\omega})^2/n \tag{5.32}$$

Hence

$$\int_{\Omega_0} a(\omega)\, d\omega = \lim_{n \to \infty} S_n$$

Furthermore it is seen that the series $\{ Y_i \}$, $Y_1 = S_1$, $Y_{i+1} = S_{i+1} \cap Y_i$ is a series of narrower and narrower nested intervals converging to the exact value of the integral.

Now consider again formula (5.22). In practice, most often

$$\frac{r(\omega,p)}{\omega} \to 0 \quad \text{for } \omega \to \infty \tag{5.33}$$

for any $p \in P$ so the infinite frequency interval $[0, \infty]$ can be replaced with reasonable accuracy by a finite interval $[0, \overline{\omega}]$ (in actual computation $\overline{\omega}$ can be determined by truncating $r(\omega, p^c)/\omega$ where p^c is the centre of the parameter vector P). Thus, if p is fixed, the integral

$$I(t,p,\overline{\omega}) = \frac{2}{\pi} \int_0^{\overline{\omega}} \frac{r(\omega,p)}{\omega} \sin\omega t \, d\omega \tag{5.34}$$

can be evaluated for any t by formula (5.30) from Theorem 5.1, i.e.

$$I(t,p,\overline{\omega}) \in \sum_{i=1}^{n} A(t,p,\Omega_i) w(\Omega_i) \tag{5.35}$$

where $A(t, p; \Omega)$ is some interval (most often, natural) extension of

$$a(t,p,\omega) = \frac{2}{\pi} \frac{r(\omega,p)}{\omega} \sin\omega t, \qquad \omega \in \Omega_i \tag{5.36}$$

Let the lower endpoint and upper endpoint of $A(t, p; \Omega_i)$ be denoted by $A^-(t, p; \Omega_i)$ and $A^+(t, p; \Omega_i)$, respectively. It can be easily seen that A^- and A^+ are known functions of both $\underline{\omega}_i$ and $\overline{\omega}_i$.

At this stage, consider (5.35). Obviously

$$I(t,p,\overline{\omega}) \le \sum_{i=1}^{n} A^+(t,p;\Omega_i) w(\Omega_i) = S_n^+(t,p) \tag{5.37a}$$

and

$$I(t,p,\overline{\omega}) \ge \sum_{i=1}^{n} A^-(t,p;\Omega_i) w(\Omega_i) = S_n^-(t,p) \tag{5.37b}$$

for any $p \in P$. Let $\overline{S_n^+(t, P)}$ and $\underline{S_n^-(t, P)}$ denote the upper endpoint and lower endpoint of some interval extension of $S_n^+(t, p)$ and $S_n^-(t, p)$, respectively, in P. From (5.37) it follows that

$$\underline{S_n^-(t,p)} \le \frac{2}{\pi}\int_0^{\bar{\omega}} \frac{r(\omega,p)}{\omega}\sin\omega t d\omega \le \overline{S_n^+(t,P)}, \quad p \in P$$

Thus an outer solution $Y_1(t) = [\underline{Y_1(t)}, \overline{Y_1(t)}]$ to the dynamic tolerance problem considered can be determined by letting

$$\underline{Y_1(t)} = \underline{S_n^-(t,P)} - \underline{\varepsilon(t)} \tag{5.38a}$$

$$\overline{Y_1(t)} = \overline{S_n^+(t,P)} + \overline{\varepsilon(t)} \tag{5.38b}$$

where the errors $\underline{\varepsilon(t)}$, $\overline{\varepsilon(t)}$ are due to the finite value of $\bar{\omega}$. Clearly, the errors are given by

$$\underline{\varepsilon(t)} = \min \int_{\bar{\omega}}^{\infty} a(t,\omega,p)d\omega, \quad p \in P \tag{5.39a}$$

$$\overline{\varepsilon(t)} = max \int_{\bar{\omega}}^{\infty} a(t,\omega,p)d\omega, \quad p \in P \tag{5.39b}$$

A simple way of bounding $\underline{\varepsilon(t)}$ and $\overline{\varepsilon(t)}$ will be presented in the next section.

For a given n and $\bar{\omega}$ the most accurate results will be obtained if $S_n^+(t,P)$ and $S_n^-(t,P)$ are determined by the ranges of the corresponding functions in P. In this case

$$\underline{S_n^-(t,P)} = \min S_n^-(t,p), \quad p \in P \tag{5.40a}$$

$$\overline{S_n^+(t,P)} = -\min(-S_n^+(t,p)), \quad p \in P \tag{5.40b}$$

Thus, roughly speaking (neglecting the error $\varepsilon(t)$) the problem of determining (for a fixed t) the outer solution $Y_1(t)$ defined by (5.38) has been reduced to a corresponding static tolerance problem defined by (5.40). The global minimization problems (5.40) can be solved by some of the interval methods for static tolerance analysis from Chapter 2.

It is important to stress that theoretically the outer approximation $Y_1(t)$ introduced by (5.38) and (5.40) converges for a fixed t to the exact interval solution $X(t)$ of the dynamic worst-case tolerance problem considered when $\bar{\omega}$ and n tend to infinity.

In practice, the solution $X(t)$ is sought for a finite number N of discrete moments t_i. Thus, the computation of the approximate solution $Y_1(t)$ introduced above necessitates N times the global solution of the minimization problem (5.40) with $t = t_i$, $i = \overline{1, N}$, respectively. Of course, less sharp approximations are possible if some interval extensions of $S_n^-(t, p)$ and $S_n^+(t, p)$ in P are used instead of the corresponding ranges.

An alternative approach for constructing an outer solutions to Problem 5.2 will next be presented which circumvents the need for solving optimization problems of the type (5.40).

Let $R(\omega, p)$ denote some interval extension of $r(\omega, p)$ with respect to p in P. Now an alternate outer solution $Y_2(t) = [\underline{Y}_2(t), \overline{Y}_2(t)]$ of the dynamic tolerance Problem 5.2 will be defined by the formula

$$Y_2(t) = \frac{2}{\pi}\int_0^\infty \frac{R(\omega,P)}{\omega}\sin\omega t\, d\omega \tag{5.41}$$

In other words $Y_2(t)$ is defined as the interval extension of (5.22). Therefore, the outer solution (5.41) is guaranteed to have the inclusion property (5.25) since $X(t)$ is the range of (5.22) over P.

Using formula (5.41) the endpoints $\underline{Y}_2(t)$ and $\overline{Y}_2(t)$ can be determined in the following way. Consider (for fixed t) the intervals (with respect to ω) where $\sin \omega t$ is either positive or negative. Let

$$\Delta\omega = \pi/t \tag{5.42}$$

It is readily seen that $\sin \omega t$ changes sign at the frequencies

$$\omega_\nu = \nu\Delta\omega \tag{5.43}$$

where $\nu = 0, 1, 2 \ldots.$ Thus, $\sin \omega t$ is positive within the intervals $\omega_{\nu+1} - \omega_\nu$, $\nu = 0, 2, 4, \ldots$ (ν is even); it is negative within the intervals $\omega_{\nu+1} - \omega_\nu$, $\nu = 1, 3, 5, \ldots$ (ν is odd). Let the set of even ν be denoted by N_1 and the set of odd ν by N_2. Then (5.41) can be put in the form

$$Y_2(t) = \frac{2}{\pi}\sum_{\nu\in N_1}\int_{\omega_\nu}^{\omega_{\nu+1}} \frac{R(\omega,P)}{\omega}\sin\omega t\, d\omega + \frac{2}{\pi}\sum_{\nu\in N_2}\int_{\omega_\nu}^{\omega_{\nu+1}} \frac{R(\omega,P)}{\omega}\sin\omega t\, d\omega \tag{5.44}$$

Let (for fixed ω) the interval $R(\omega, P)$ be denoted as $[\underline{R}(\omega,\underline{p}), \overline{R}(\omega, \overline{p})]$. Now it is readily seen from (5.44) that the lower endpoint $\underline{Y}_2(t)$ of $Y_2(t)$ can be determined by the formula:

$$\underline{Y}_2(t) = \frac{2}{\pi}\sum_{\nu\in N_1}\int_{\omega_\nu}^{\omega_{\nu+1}} \frac{\underline{R}}{\omega}\sin\omega t\, d\omega + \frac{2}{\pi}\sum_{\nu\in N_2}\int_{\omega_\nu}^{\omega_{\nu+1}} \frac{\overline{R}}{\omega}\sin\omega t\, d\omega \tag{5.45}$$

In a similar way

$$\overline{Y}_2(t) = \frac{2}{\pi}\sum_{\nu\in N_1}\int_{\omega_\nu}^{\omega_{\nu+1}} \frac{\overline{R}}{\omega}\sin\omega t\, d\omega + \frac{2}{\pi}\sum_{\nu\in N_2}\int_{\omega\nu}^{\omega_{\nu+1}} \frac{\underline{R}}{\omega}\sin\omega t\, d\omega \tag{5.46}$$

Writing $\overline{R}$ as $\overline{R} = \overline{R} + \underline{R} - \underline{R}$ and substituting it into (5.45) we get after some manipulation

$$\underline{Y}_2(t) = \frac{2}{\pi}\int_0^\infty \frac{\underline{R}(\omega, \underline{p})}{\omega}\sin\omega t\, d\omega + S(t) \tag{5.47}$$

where

$$S(t) = \frac{2}{\pi}\sum_{\nu\in N_2}\int_{\omega_\nu}^{\omega_{\nu+1}}\frac{[\overline{R}(\omega,\overline{p}) - \underline{R}(\omega,\underline{p})]}{\omega}\sin\omega t\, d\omega < 0 \tag{5.48}$$

In the same way (5.46) can be recast in the form

$$\overline{Y_2(t)} = \frac{2}{\pi}\int_0^\infty \frac{\overline{R}(\omega,\overline{p})}{\omega}\sin\omega t\, d\omega \;-\; S(t) \tag{5.49}$$

where $S(t)$ is defined by (5.48). Thus, the new outer solution $Y_2(t)$ to the dynamic tolerance analysis problem considered can be determined using formulae (5.47) to (5.49).

It should be stressed that unlike $Y_1(t)$ introduced previously the enclosing solution $Y_2(t)$ defined by (4.41) can never be equal to the exact solution $X(t)$. Indeed, it is seen from (5.26) and (5.27) that

$$\underline{X(t)} = \frac{2}{\pi}\int_0^\infty \frac{r(\omega,p^1)}{\omega}\sin\omega t\, d\omega$$

$$\overline{X(t)} = \frac{2}{\pi}\int_0^\infty \frac{r(\omega,p^2)}{\omega}\sin\omega t\, d\omega$$

where p^1 and p^2 are the parameter vectors providing the minimum and maximum of (5.26) and (5.27), respectively. Using (5.47) and (5.49) we can introduce the "systematic" errors:

$$\underline{\delta} = \underline{X(t)} - \underline{Y_2(t)} = \frac{2}{\pi}\int_0^\infty [r(\omega,p^1) - \underline{R}(\omega,\underline{p})]\frac{\sin\omega t}{\omega}d\omega \;-\; S(t)$$

$$\overline{\delta} = \overline{X(t)} - \overline{Y_2(t)} = \frac{2}{\pi}\int_0^\infty [\overline{R}(\omega,\overline{p}) - r(\omega,p^2)]\frac{\sin\omega t}{\omega}d\omega \;-\; S(t)$$

It is seen from the above formulae that always $\underline{\delta} > 0$ and $\overline{\delta} > 0$; the smallest value $\underline{\delta} = \overline{\delta} = S(t)$ is obtained when $r(\omega,p^1) = \underline{R}(\omega,\underline{p})$ and $r(\omega, p^2) = \overline{R}(\omega, \overline{p})$ which is only possible if the interval $[\underline{R}(\omega,\underline{p}), \overline{R}(\omega, \overline{p})]$ is in fact the range of $r(\omega, p)$ over P. Thus, the obtainment of the narrowest possible interval solution $Y_2(t)$ is associated with the calculation of the range $r(\omega,P)$ of $r(\omega,p)$ for the whole frequency interval $[0,\infty)$.

5.2.2. Alternate approximate solutions

Sometimes the requirement of $Y(t)$ to have the enclosing property (5.25) may not be essential. In such cases it is reasonable to introduce two new approximate solutions $Y_3(t)$ and $Y_4(t)$ by modifying and simplifying the approaches adopted to define the outer solutions $Y_1(t)$ and $Y_2(t)$.

The first approximate solution $Y_3(t)$ is suggested as an attempt to simplify the global optimization problems (5.40). Indeed, from a computational point of view it should be emphasized that the functions $S_n^+(t, p)$ and $S_n^-(t, p)$ in (5.37) are rather complex due to the fact that they are sums of terms, each term $A^+(t, p;\ \underline{\omega}_i, \overline{\omega}_i)$ and $A^-(t, p;\ \underline{\omega}_i, \overline{\omega}_i)$ being a function of both $\underline{\omega}_i$ and $\overline{\omega}_i$. A reduction in the amount of computation is possible if the integral defined by (5.34) is solved approximately (for fixed t and p) by some numerical method. Then

$$I(t,p,\overline{\omega}) = \sum_{i=1}^{n} \alpha_i a(t,p,\omega_i) + \varepsilon(t,p) = S_n(t,p) + \varepsilon(t,p) \tag{5.50}$$

If the integration is done using Simpson's rule, n from (5.50) must be even, the partitioning of the interval $[0, \overline{\omega}]$ is uniform and α_i have the values $h/3$, $4h/3$, $2h/3$,... $2h/3$, $4h/3$, $h/3$. Since (5.50) is valid for each $p \in P$ it follows that

$$I(t,p,\overline{\omega}) \le \overline{S_n(t,P) + \varepsilon(t,P)} \tag{5.51a}$$

and

$$I(t,p,\overline{\omega}) \ge \underline{S_n(t,P) + \varepsilon(t,P)} \tag{5.51b}$$

for each $p \in P$ where RHS of (5.51a) and (5.51b) is the upper endpoint and lower endpoint of some interval extension of $S_n(t, p) + \varepsilon(t, p)$ in P.

An approximate solution $Y_3(t) = [\underline{Y_3(t)}, \overline{Y_3(t)}]$ to the dynamic tolerance problem considered can be introduced by defining

$$\underline{Y_3(t)} = \underline{S_n(t,P)} \tag{5.52a}$$

$$\overline{Y_3(t)} = \overline{S_n(t,P)} \tag{5.52b}$$

where $\underline{S_n(t,P)}$ and $\overline{S_n(t,P)}$ denote the lower endpoint and upper endpoint of some interval extension of $S_n(t,p)$ in P. For a given n and $\overline{\omega}$ the most accurate result will be obtained if

$$\underline{S_n(t,P)} = \min S_n(t,p),\quad p \in P \tag{5.53a}$$

$$\overline{S_n(t,P)} = -\min\ (-S_n(t,p)),\ p \in P \tag{5.53b}$$

that is, if $S_n(t, P)$ is the range of $S_n(t, p)$ over P.

It should be noted right away that the approximation $Y_3(t)$ defined as above is not an outer solution to the tolerance problem considered since the numerical integration $S_n(t,p)$ does not guarantee the inclusion property (5.25). However, the approximate solution $Y_3(t)$ is easier to find than $Y_1(t)$. Indeed, comparison of (5.53) and (5.40) reveals that problems (5.53) require less computation than problems (5.40) since $S_n(t, p)$ is a simpler function than $S_n^-(t, p)$ and $S_n^+(t, p)$.

Similarly to the outer solution $Y_1(t)$ the approximation $Y_3(t)$ has the property to converge to the exact solution $X(t)$ with n and ω tending to infinity.

The second approximate solution $Y_4(t) = [\underline{Y_4(t)}, \overline{Y_4(t)}]$ which does not guarantee the enclosing property (5.25) is introduced in a similar way as $Y_2(t)$. On account of (5.47) and (5.49) its endpoints are defined as follows:

$$\underline{Y_4(t)} = \frac{2}{\pi}\int_0^\infty \frac{\underline{R}(\omega,\underline{p})}{\omega}\sin\omega t\,d\omega \tag{5.54a}$$

$$\overline{Y_4(t)} = \frac{2}{\pi}\int_0^\infty \frac{\overline{R}(\omega,\overline{p})}{\omega}\sin\omega t\,d\omega \tag{5.54b}$$

Comparing (5.47) to (5.54) it is seen that the approximate solution $Y_4(t)$ is not guaranteed to enclose the exact solution $X(t)$. On the other hand, detailed (elementary but tedious) analysis shows that $Y_4(t)$ is closer to $X(t)$ than $Y_2(t)$ is to $X(t)$. Indeed, let

$$\underline{e}_{Y_4} = \int_0^\infty \|\underline{X(t)} - \underline{Y_4(t)}\|dt$$

and

$$\overline{e}_{Y_4} = \int_0^\infty \|\overline{X(t)} - \overline{Y_4(t)}\|dt$$

with $\|z\|$ being either z^2 or $|z|$.

Similarly, let

$$\underline{e}_{Y_2} = \int_0^\infty \|\underline{X(t)} - \underline{Y_2(t)}\|dt$$

$$\overline{e}_{Y_2} = \int_0^\infty \|\overline{X(t)} - \overline{Y_2(t)}\|dt$$

It turns out that $e_{Y_4} < e_{Y_2}$ and $e_{Y_4} < e_{Y_2}$.

R e m a r k 5.1. It should be noted that the vectors $\underline{p}$ and $\overline{p}$ involved in formulae (5.54) are in general dependent on the frequency ω i.e. $\underline{p} = \underline{p}(\omega)$ and $\overline{p} = \overline{p}(\omega)$. If this is the case, comparison of (5.26), (5.27) amd (5.54) shows that $Y_4(t) \neq X(t)$ even if $R(\omega, P)$ is the range $r(\omega,P)$. However, in the special case where the endpoints $\underline{r}(\omega,\underline{p})$ and $\overline{r}(\omega, \overline{p})$ are obtained for some constant vectors $\underline{p}$ and $\overline{p}$, respectively, then $Y_4(t)$ may coincide with the exact solution $X(t)$ for some times t. This conclusion follows directly from (5.26), (5.54a) and (2.27), (5.54b).

R e m a r k 5.2. Depending on the accuracy sought various methods for determining the interval extension $R(\omega, P)$ can be used. The simplest approach would be to use the natural interval extension obtained by replacing the real operations in $r(\omega, p)$ by their interval counterparts. More sophisticated methods determine the interval extension by means of appropriate interval mean-value forms. Efficient algorithms for computing the range $r(\omega, P)$ of $r(\omega, p)$ over P have been presented in section 2.3.

Remark 5.3. Assuming (5.33) to hold the errors $\underline{\varepsilon(t)}$ and $\overline{\varepsilon(t)}$ from (5.38) can be estimated on the basis of (5.39) by the quantities

$$\underline{\varepsilon_1(t)} = \frac{2}{\pi}\int_{\varpi}^{\infty} |\underline{R}(\omega,\underline{p})| \frac{|\sin \omega t|}{\omega} d\omega$$

and

$$\overline{\varepsilon_1(t)} = \frac{2}{\pi}\int_{\varpi}^{\infty} |\overline{R}(\omega,\overline{p}) \frac{|\sin \omega t|}{\omega} d\omega$$

In determining the outer solution $Y_2(t)$ or the approximate solution $Y_4(t)$ we are first led to evaluate the endpoints of the interval $R(\omega, P)$ using one of the available methods for determining interval extensions or ranges (Chapter 2). Secondly, we have to determine the bounds $\underline{Y_2(t)}$ and $\overline{Y^2(t)}$ or $\underline{Y^4(t)}$ and $\overline{Y_4(t)}$ by computing the integrals (5.47) to (5.49) or (5.54).

An approximate but rather simple method for evaluating integrals of the above type will be presented now. Unlike the traditional numerical methods this method can (if need be) retain the enclosing property (5.25) for the approximate outer solution $Y_2(t)$. It is based on an appropriate piecewise-linear approximation of the interval extension $R(\omega)$. More specifically, in the context of determining the outer solution $Y_2(t)$, $\underline{R}(\omega)$ is approximated by a piecewise-linear function (PLF) $\varphi_1(\omega)$ such that

$$\varphi_1(\omega) \le \underline{R}(\omega) \tag{5.55a}$$

Similarly, $\overline{R}(\omega)$ is bounded from above by a PLF $\varphi_2(\omega)$, i.e.

$$\varphi_2(\omega) \ge \overline{R}(\omega) \tag{5.55b}$$

Thus, the approximate outer solution to the tolerance problem considered is given by

$$\underline{Y_2(t)} = \frac{2}{\pi}\int_0^{\infty} \frac{\varphi_1(\omega)}{\omega} \sin \omega t \, d\omega + \tilde{S}(t) \tag{5.56a}$$

$$\overline{Y_2(t)} = \frac{2}{\pi}\int_0^{\infty} \frac{\varphi_2(\omega)}{\omega} \sin \omega t \, d\omega - \tilde{S}(t) \tag{5.56b}$$

where

$$\tilde{S}(t) = \frac{2}{\pi}\sum_{\nu \in N_2} \int_{\omega_\nu}^{\omega_{\nu+1}} [\varphi_2(\omega) - \varphi_1(\omega)] \frac{\sin \omega t}{\omega} d\omega \tag{5.56c}$$

It is seen from (5.55) to (5.56) that the approximate outer solution is guaranteed to enclose the exact solution $X(t)$. Since φ_1 and φ_2 are LPFs the integrals (5.56) can be evaluated in a rather simple manner. To simplify notations let us consider the integral

$$y = \frac{2}{\pi}\int_0^\infty \frac{\varphi(\omega)}{\omega}\sin\omega t\, d\omega \tag{5.57}$$

It is assumed that $\varphi(\omega) = 0$ for $\omega \geq \bar{\omega}$ where $\bar{\omega}$ is some final frequency. Now the frequency interval $[0, \bar{\omega}]$ is divided into N subintervals $[0, \omega_1]$, $[\omega_1, \omega_2]$, . . . $[\omega_{N-1}, \bar{\omega}]$. Within an arbitrary subinterval $[\omega_i, \omega_{i+1}]$ $\varphi(\omega)$ is linear.

$$\varphi(\omega) = \varphi_i + k_i\omega$$

with

$$k_i = \frac{\varphi_{i+1} - \varphi_i}{\omega_{i+1} - \omega_i}\,, \quad \varphi_i = \varphi(\omega_i)\,, \quad \varphi_{i+1} = \varphi(\omega_{i+1})$$

So (5.57) becomes (with $\omega_0 = 0$)

$$\begin{aligned} y(t) = \frac{2}{\pi}\sum_{i=0}^{N-1}[\varphi_i\int_{\omega_i}^{\omega_{i+1}}\frac{\sin\omega t}{\omega}d\omega \\ - k_i(\cos(\omega_{i+1}t) - \cos(\omega_i t))] \end{aligned} \tag{5.58}$$

The integrals in (5.58) can be put in the form

$$\int_{\omega_i}^{\omega_{i+1}}\frac{\sin\omega t}{\omega}d\omega = \mathrm{IS}(\omega_{i+1}t) - \mathrm{IS}(\omega_i t)$$

where

$$IS(\omega t) = \int_0^\omega \frac{\sin\omega t}{\omega}d\omega$$

is the so-called integral sinus; its values are tabulated. Finally

$$\begin{aligned} y(t) = \frac{2}{\pi}\sum_{I=0}^{N-1}[\varphi_i(\mathrm{IS}(\omega_{i+1}t) - \mathrm{IS}(\omega_i t)) \\ - k_i(\cos(\omega_{i+1}t) - \cos(\omega_i t))] \end{aligned} \tag{5.59}$$

Thus, $\underline{Y_2(t)}$ and $\overline{Y_2(t)}$ from (5.56) can be found by means of formula (5.59) in which φ_i and k_i must correspond to φ_1, φ_2 and $\varphi_2 - \varphi_1$, respectively. (Note that in evaluating (5.56c) the intervals $[\omega_i, \omega_{i+1}]$ must cover exactly the intervals $[\omega_\nu, \omega_{\nu+1}]$ with $\nu \in N_2$ and ω_ν defined by (5.42), (5.43).

The approximate solution $Y_4(t)$ can be determined in exactly the same way; the piecewise-linear functions φ_1 and φ_2 need not, however, be outer approximations to $\underline{R}(\omega)$ and $\overline{R}(\omega)$, respectively.

The approach of the present section based on the frequency-domain formulation will be illustrated below by way of several examples.

***Example* 5.2.** Consider the circuit shown in Fig. 5.2. It is desired to find an outer solution $Y(t)$ to the step response tolerance problem associated with the output voltage $v(t)$ if R and C belong to certain intervals R^I and C^I.

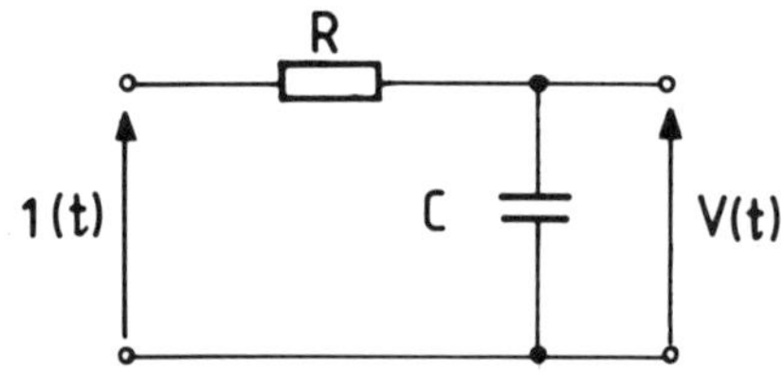

Fig. 5.2. Finding an outer interval solution $Y(t)$ for the output voltage $v(t)$.

For the circuit at hand the frequency response $F(j\omega)$ is

$$F(j\omega) = \frac{b}{b + j\omega}$$

where $b = 1/RC$. Hence, its real part is

$$r(\omega,b) = \frac{b^2}{b^2 + \omega^2} \tag{5.60}$$

Since $R \in R^I$ and $C \in C^I$, $b \in B$ where B is an interval equal to the reciprocal of the product $R^I C^I$. Thus, from (5.60) the natural interval extension of $r(\omega,b)$ over B is

$$R(\omega,B) = \frac{B^2}{B^2 + \omega^2}$$

or equivalently

$$R(\omega,B) = \frac{1}{1 + \left(\dfrac{\omega}{B}\right)^2} \tag{5.61}$$

It is seen from (5.61) that this natural extension yields in fact the range of (5.60) over B with endpoints

$$\underline{r}(\omega, B) = \frac{1}{1 + \left(\dfrac{\omega}{\underline{B}}\right)^2}, \quad \overline{r}(\omega, B) = \frac{1}{1 + \left(\dfrac{\omega}{\overline{B}}\right)^2} \tag{5.62}$$

Furthermore, these endpoints are obtained for constant values of the varying parameter b over the whole frequency interval $[0,\infty)$. Now the outer solution $Y_2(t)$ can be determined by using (5.47) to (5.49) and (5.62). In this example the exact tolerance solution $V(t)$ could be found directly by the formula

$$V(t) = 1 - e^{-Bt} = [1 - \frac{1}{e^{\underline{B}t}}, \; 1 - \frac{1}{e^{\overline{B}t}}] \tag{5.63}$$

Thus, it is easily seen from (5.47) to (5.62), on one hand, and (5.63) on the other, that $V(t) \subset Y_2(t)$ and $V(t) \neq Y_2(t)$.

This example is instructive also in that it shows the possibility that in some cases when the endpoints of the range of $r(\omega,p)$ correspond to two fixed parameter vectors the approximate solution $Y_4(t)$ may provide (within the computation accuracy) the exact solution $X(t)$ to the tolerance problem considered. Indeed, it is easily seen that the approximate solution $Y_4(t)$ obtained by (5.54) and (5.62) yields in fact the exact solution $X(t)$ given by (5.63).

***Example* 5.3.** Consider a two-port network made up of a capacitor C, a resistor R and an inductor L with C being in series with the parallel connection of R and L. It is desired to find an approximate (not necessarily outer) solution $Y(t)$ associated with the output voltage $v(t)$ (across the parallel connection of R and L) when the input voltage is a unit step function and R, L and C belong to some intervals R^I, L^I and C^I, respectively.

The corresponding voltage transfer function is

$$V(s) = \frac{s^2}{s^2 + (1/RC)s + 1/LC}$$

The real part $r(\omega)$ of $V(j\omega)$ is

$$r(\omega) = \frac{-\omega^2(1/LC - \omega^2)}{(1/LC - \omega^2)^2 + \dfrac{\omega^2}{R^2C^2}} \tag{5.64}$$

Using (5.64) an approximate solution $Y_3(t)$ as defined by (5.62) has been determined. The sum $S_n(t, p)$ was obtained applying Simpson's integration rule. The natural interval extension of $S_n(t, p)$ in P was used.

Now let the resistance R, the inductance L and the capacitance C take on values from corresponding intervals given by their nominal (centre) values $C_c = 25$ μF, $R_c = 200$ Ω and $L_c = 2$ H and ± 5% tolerance for C and R and ± 1% tolerance for L. Thus,

$$C^I = [23.75,\ 26.25]\mu\text{F},\quad R^I = [190,\ 210]\Omega,\quad L^I = [1.98,\ 202]\text{H}$$

We shall evaluate the approximation $Y_3(t)$ of the output voltage interval $V(t)$ only for a single time moment $t = t_1 = 0.015$s (approximately at this moment the largest width of the interval $V(t)$ is observed).

From the relation

$$\overline{\omega} = \overline{n}2\pi$$

where $\overline{n}$ is the number of cycles in the interval $[0, \overline{\omega}]$ the final frequency $\overline{\omega}$ for $\overline{n} = 3$ was obtained to be $\overline{\omega} = 1256$ rad/s. The total number n of integration steps was chosen to be $n = 24$ so that $h = \omega/\text{n} = 52.33$ and

$$\omega_0 = 0,\quad \omega_{i+1} = \omega_i + \text{h},\quad 0 \le i \le n-1$$

On the basis of (5.36) and (5.64), after some manipulation

$$a(t,p,\omega_i) = \frac{2}{\pi}\,\frac{-\omega_i \sin\omega_t}{\dfrac{1}{C}\left[\dfrac{1}{L} + \dfrac{\omega_i^2}{R^2\left(\dfrac{1}{L} - \omega_i^2 C\right)}\right] - \omega_i^2} \tag{5.65}$$

Now let $A(t,P,\omega_i)$ be the natural extension of (5.65), i.e.

$$A(t,P,\omega_i) = \frac{2}{\pi}\,\frac{-\omega_i \sin\omega_t}{\dfrac{1}{C^I}\left[\dfrac{1}{L^I} + \dfrac{1}{(R^I)^2\left(\dfrac{1}{L^I\omega_i^2} - C^I\right)}\right] - \omega_i^2} \tag{5.66}$$

On account of (5.50) and (5.52) the approximate solution $Y_3(t)$ is given by the sum

$$Y_3(t_1) = \sum_{i=1}^{n-1} \alpha_i A(t_1,P,\omega_i) \tag{5.67}$$

where $\alpha_i = 4h/3$ for odd i and $2h/3$ for even i. Computation of (5.66), (5.67) yields:

$$Y_3(t_1) = [-0.28679,\ -0.19264]$$

***Example* 5.4.** For two-port network shown in Fig 5.3 find the approximate solution $Y_4(t)$ associated with the worst-case tolerance analysis of the output voltage $v(t)$ when L_1, L_2, R_1 and R_2 belong to some intervals L_1^I, L_2^I, R_1^I and R_2^I.

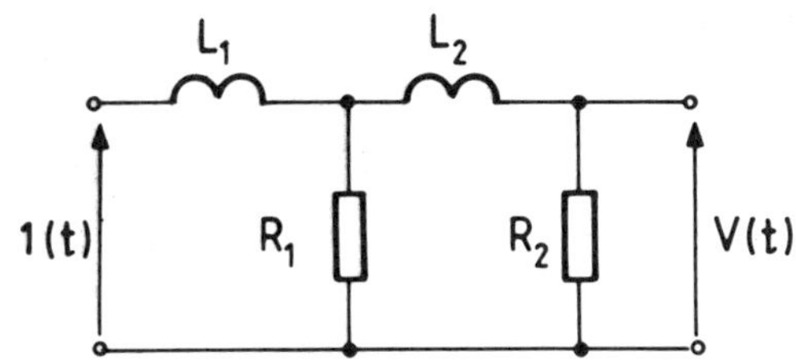

Fig. 5.3. Finding an approximate interval solution $Z(t)$ for the output voltage $v(t)$.

The voltage transfer function is

$$F(s) = \frac{c}{as^2 + bs + c}$$

where

$$a = L_1 L_2 \;, \quad b = (R_1 + R_2) + R_1 L_2 \;, \quad c = R_1 R_2$$

The frequency response is

$$F(j\omega) = \frac{c}{-a\omega^2 + j\omega b + c}$$

so its real part is

$$r(\omega,p) = \frac{c(c - a\omega^2)}{(c - a\omega^2)^2 + \omega^2 b^2} \tag{5.68}$$

where $p = (R_1, R_2, L_1, L_2)$. Let $R(\omega)$ denote some interval extension or the range of (5.68) when R_1, R_2, L_1 and L_2 become intervals (as was shown in the previous example $R(\omega)$ can be evaluated using some appropriate interval analysis method). In practice, $R(\omega)$ will be evaluated for some finite number of discrete frequencies. Suppose (for computational convenience) that for the given intervals R_1^I, R_2^I, L_1^I and L_2^I the upper endpoint $\overline{R}(\omega)$ of $R(\omega)$ is given by the function

$$\overline{R}(\omega) = \frac{10000 - 100\omega^2}{1000 + 700\omega^2 + \omega^4} \tag{5.69}$$

This function was approximated by five linear functions (LF). Each LF was determined by the corresponding points (ω_i, R_i) and (ω_{i+1}, R_{i+1}) with $\omega_0 = 0$, $\omega_1 = 4$, $\omega_2 = 7.5$, $\omega_3 = 13.5$, $\omega_4 = 22.5$ and $\omega = 50$. Using this approximation the upper endpoint $\overline{Y_4(t)}$ was evaluated by means of (5.54b). The graph of $\overline{Y_4(t)}$ is plotted in Fig. 5.4.

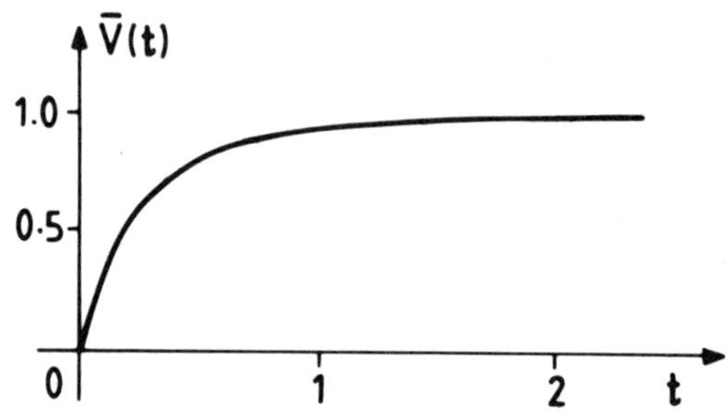

Fig. 5.4. Graph of the upper endpoint of the approximate solution $V(t) = Y_4(t)$ from Example 5.4.

The exact step response $\bar{V}(t)$ corresponding to (5.69) is given (in volts) by

$$\overline{V(t)} = 1 - 1.1725e^{-3.82t} + 0.1725e^{-26.18t} \tag{5.70}$$

It is worth nothing that the graph of the exact solution (5.70) practically coincides with the graph of the approximate solution $\overline{Y_4(t)}$.

5.3. SOLVING TIME-DOMAIN FORMULATION PROBLEMS

In this section we consider the solution of some transient tolerance analysis problems formulated in implicit form in the time-domain. First, two exact methods for solving the auxiliary rather idealized Problems 5.5 and 5.6 (defined in section 5.1.3) are presented in subsection 5.3.1. Using the above solutions, a method for approximate solution of Problems 5.7 and 5.4 is then suggested in section 5.3.2. Finally, outer (enclosing $X(t)$) and inner (enclosed by $X(t)$) solutions for the most general dynamic tolerance Problem 5.3 as well as for the simpler problem 5.4 are presented in section 5.3.3.

5.3.1. Exact method

We shall first solve exactly Problem 5.5. Later on, it will be shown that the solution of Problem 5.6 can be obtained in a similar way.

It is well-known that the solution of the noninterval differential systems (5.11), (5.12)

$$\dot{x} = Ax + b(t), \quad x(0) = c$$

can be determined by formula

$$x(t,c) = W(t)c + W(t)\int_0^t W_1(\tau)b(\tau)d\tau \qquad (5.71a)$$

where $W(t) = e^{At}$ and $W_1(\tau) = e^{-A\tau}$ are $(l \times l)$ matrices (in view of solving Problem 5.5 the notation $x(t,c)$ is used to reflect explicitly the dependence of the solution $x(t)$ on the initial conditions vector c). The above expression can be written in an equivalent form as

$$x(t,c) = W(t)c + f(t) \qquad (5.71b)$$

where the elements of $W(t)$ and $f(t)$ are known functions of time since, according to the formulation of Problem 5.5 the matrix A and the vector $b(t)$ are known exactly.

For fixed t, the relationship (5.71b) defines an affine transformation of the vector c to a corresponding vector $x(t,c)$. Geometrically, the affine transformation involves the following two transformations of c. First, as a result of the multiplication of $W(t)$ and c, the vector c is rotated and lengthened (or shortened). Afterwards, the resultant vector is added to the vector $f(t)$ to obtain the solution vector $x(t, c)$. If now c is allowed to vary within the interval vector C, then (similarly to (5.19) with p being c) the set of vectors $x(t, c)$ defines the reachability set $S(t)$ for Problem 5.5 at time t, i.e.

$$S(t) = \{x(t,c): x(t,c) = W(t)c + f(t),\ c \in C\}$$

It is easily seen from the above definition that for the problem considered $S(t)$ is a hyperparallelopiped: indeed, $S(t)$ is the image of an affine transformation of the hyperbox C. The geometrical transformation of the domain C into the image $S(t)$ is illustrated in Fig. 5.5 for the case of $l = 2$. In the same figure, the corresponding interval vector $X(t)$ defined as the interval hull of $S(t)$ is also shown.

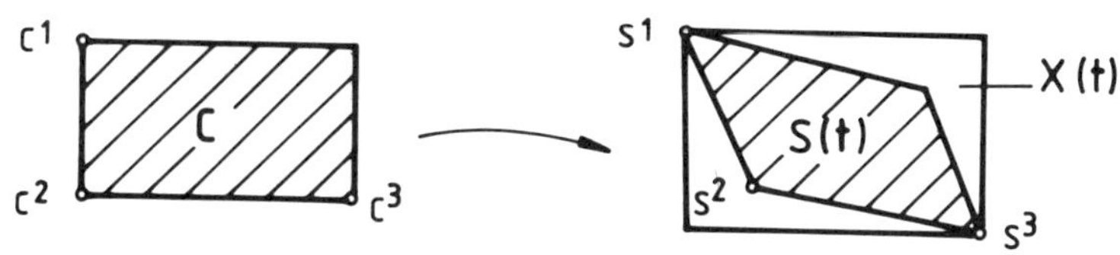

Fig. 5.5. Transformation of the domain C into the image $S(t)$ for the case of $l = 2$.

Since geometrically C is a box in R^l, C can be uniquely determined by fixing (arbitrary) $l+1$ of its vertices c^ν, $\nu = \overline{1, l+1}$. Thus, for example of Fig. 5.5 the shaded region representing C is determined uniquely by (say) its vertices c^1, c^2 and c^3. The remaining $2^l-(l+1)$ vertices can be easily located by means of the basis $l+1$ vertices, the components C_i of the interval vector C and elementary geometrical construction.

In a similar way the hyperparallelopiped $S(t)$ can be uniquely determined if the location of $l+1$ of its vertices $s^\nu(t)$, $\nu = \overline{1, l+1}$ is known; the location of the remaining $2^l - (l+1)$

vertices can be readily found using the basis vertices $s^\nu(t)$. It is natural to assume as basis vertices $s^\nu(t)$ those vertices of $S(t)$ which are the images of the (chosen) basis vertices c^ν of the box C under the affine transformation (5.71b). In this case, the vertices $s^\nu(t)$ are determined by means of the vertices c^ν as follows

$$s^\nu(t) = W(t)c^\nu + f(t), \quad \nu = \overline{1,l+1} \tag{5.72}$$

The above correspondence between c_ν and $s^\nu(t)$ is illustrated by the two-dimensional example of Fig. 5.5

After the basis vertices $s^\nu(t)$ are determined by (5.72) the remaining vertices of $S(t)$ can be found by way of elementary geometrical considerations. Finally, the interval solution $X(t)$ of the dynamic tolerance analysis problem considered can be easily determined as the interval hull of $S(t)$.

Based on the aforegoing the exact solution of Problem 5.5 can be found by the following method. First, l+1 basis vertices c_ν of the box C are chosen. Afterwards l+1 transient analyses

$$\dot{x} = Ax + b(t),\ x(0) = c^\nu, \quad \nu = \overline{1,l+1}$$

(with one and the same circuit and different initial conditions c^ν are carried out whereby the basis vertices $s^\nu(t)$ of $S(t)$ are determined. Using $s^\nu(t)$ the hyperparallelopiped $S(t)$ is constructed. Finally, the interval solution $X(t)$ of Problem 5.5 is determined as the interval hill of $S(t)$. For example, the upper endpoint $\overline{X}_k(t)$ of the kth component $X_k(t)$ of $X(t)$ is found by means of all the vertices of S(t) as the maximum

$$\overline{X_k}(t) = \max_\nu \{s_k^\nu(t), \quad \nu = \overline{1,2^l}\}$$

In a similar way, the lower endpoint $\underline{X}_k(t)$ will be computed as the minimum among all the vertices components $s_k^\nu(t)$.

We now proceed to solving Problem 5.6. First, (for ease of exposition) the input perturbations $b_i(t)$ will be assumed constant, i.e. $b_i(t) = b_i$, $i = \overline{1, l}$. Formula (5.71a) can then be recast in the form

$$x(t,b) = W(t)c + (W(t)\int_0^t W_1(\tau)d\tau)b;$$

hence

$$x(t,b) = W'(t)b + f'(t)$$

where the notation $x(t, b)$ is used to underline the dependence of $x(t)$ on b. Obviously, the latter expression defines an affine transformation of the vector b to the solution vector $x(t, b)$. Therefore similarly to the previous Problem 5.5 if b varies in the box B, the set of all solutions $x(t, b)$ defining the reachability set $S(t)$ is again a hyperparallelopiped.

It is easily verified that $S(t)$ retains this property for input perturbations of the general form $b_i(t) = \varphi_i(t)b_i$, $i = \overline{1, l}$. Therefore, by analogy with Problem 5.5 the exact solution of Problem 5.6 can be obtained by the following method. First, $l + 1$ basis vertices b^ν of the interval box B are chosen Afterwards $l + 1$ transient analysis

$$\dot{x} = Ax + \varphi(t)b^\nu, \quad x(0) = c, \quad \nu = \overline{1, l+1}$$

are carried out (using some of the methods for linear transient analysis) to determine the basis vertices $s^\nu(t)$ of the corresponding hyperparallelopiped $S(t)$. Using $s^\nu(t)$ the remaining vertices of $S(t)$ are then found. Finally, the interval solution $X(t)$ is determined by means of all the vertices of $S(t)$.

5.3.2. Approximate solution

In this section, we shall first solve approximately Problem 5.7. The solution is based on the approach used for approximate solution of a.c. tolerance problems from section 3.3.2, on one hand, and on the exact method for solving Problem 5.6 from the previous section, on the other.

We start by writing the matrix A and the solution of system (5.15) in the form

$$A = A_c + \mathrm{E} \tag{5.73a}$$

$$x = x^c + \eta \tag{5.73b}$$

Substituting (5.7) into (5.15) we have

$$\dot{x}^c + \dot{\eta} = (A_c + \mathrm{E})(x^c + \eta) + b$$

Now taking into account that

$$\dot{x}_c = A_c x^c + b$$

and neglecting the product $\mathrm{E}\eta$ we obtain a linearized system of equations in the increment η

$$\dot{\eta} = A_c\eta + \mathrm{E}x^c(t), \; \eta(0) = 0 \tag{5.74}$$

According to the statement of Problem 5.7 $A \in A^I$ so that the matrix E belongs to the symmetrical matrix $\Delta^I = [-\Delta, \Delta]$ where Δ is the radius of the interval matrix A^I. Thus, the interval equivalent of (5.74) takes on the form

$$\dot{\eta} = A_c\eta + \mathrm{E}x^c(t), \quad \eta(0) = 0, \quad \mathrm{E} \in \Delta^I \tag{5.75}$$

To simplify the exposition it is assumed that we are interested in (approximately) determining only one component $X_k(t)$ of the interval solution $X(t)$ of Problem 5.7. In order to apply the approach from section 3.3.2 we introduce the following notations:

E_i - the ith column of E,
Δ_i - the ith column of Δ^I ,
$D_i(t)$ - diagonal (l x l) matrix, whose nonzero elements are equal to $x_i^c(t)$.

Now, by analogy with section 3.3.2 we solve (5.75) in the following successive manner. First, we set $E_i = 0$ for $i > 1$. Thus (5.75) is reduced to the form

$$\dot{\eta} = A_c\eta + D_1(t)E_1, \quad \eta(0) = 0, \quad E_1 \in \Delta_1^I \tag{5.76}$$

which is obviously a Problem 5.6. Solving (5.76) by the exact method from the previous section with respect to η_k we get two solutions: the lower endpoint $\underline{\eta}_k^{(1)}(t)$ and the upper endpoint $\overline{\eta}_k^{(1)}(t)$ obtained for some for some vertices $\underline{E}_1$ and $\overline{E}_1$ of the box Δ_1^I, respectively. Now, we solve two new Problems 5.6 of the form

$$\dot{\eta} = A_c\eta + D_1(t)\underline{E}_1 + D_2(t)E_2, \quad \eta(0) = 0, \quad E_2 \in \Delta_2^I \tag{5.77a}$$

$$\dot{\eta} = A_c\eta + D_1(t)\overline{E}_1 + D_2(t)E_2, \quad \eta(0) = 0, \quad E_2 \in \Delta_2^I \tag{5.77b}$$

Solving (5.77a) we find an updated estimate $\underline{\eta}_k^{(2)}(t)$ and the corresponding vertex $\underline{E}_2$; similarly, from (5.77b) we get an updated estimate $\overline{\eta}_k^{(2)}(t)$ and the corresponding vertex $\overline{E}_2$. Next, the following two Problems 5.6 are solved

$$\dot{\eta} = A_c\eta + D_1(t)\underline{E}_1 + D_2(t)\underline{E}_2 + D_3(t)E_3, \quad \eta(0) = 0, \quad E_3 \in \Delta_3^I$$

$$\dot{\eta} = A_c\eta + D_1(t)\overline{E}_1 + D_2(t)\overline{E}_2 + D_3(t)E_3, \quad \eta(0) = 0, \quad E_3 \in \Delta_3^I$$

to obtain the corresponding vertices $\underline{E}_3$ and $\overline{E}_3$. We continue the above process until the corresponding vertices $\underline{E}_n$ and $\overline{E}_n$ of the last Δ_n^I are determined. Thus, we have found the lower endpoint $\underline{\eta}_k(t)$ and the upper endpoint $\overline{\eta}_k(t)$ of the interval related to the kth component $\eta_k(t)$ of the increment η. Finally, the approximate solution of Problem 5.7 for $X_k(t)$ is given by the formula

$$\underline{X}_k(t) = x_k^c(t) + \underline{\eta}_k(t)$$
$$\overline{X}_k(t) = x_k^c(t) + \overline{\eta}_k(t)$$

Now we shall show that the method outlined above for approximate solution of Problem 5.7 is also applicable for approximate solution of Problem 5.4. First, we shall consider the following "mixed" problem which is a combination of Problem 5.6 and 5.7.

Problem 5.8 Find an estimate for $X_k(t)$ in the case of the circuit N described by (5.15) with $A \in A^I$, $b \in b^I$ and exactly known vector c.

To solve Problem 5.8 we first observe that if b is written in the form

$$b = b^c + \delta, \quad \delta \in \delta^I$$

then (5.75) takes on the form

$$\dot{\eta} = A_c\eta + \mathrm{E}x^c(t) + \varphi(t)\delta, \quad \eta(0) = 0, \quad \mathrm{E} \in \Delta^I, \quad \delta \in \delta^I \tag{5.78}$$

Obviously, (5.78) can be solved by the above method (indeed, (5.78) differs from (5.75) only in that the system (5.78) contains one more box δ^I along with the boxes Δ_i^I). Let $\underline{\eta}_k(t)$ from (5.78) be obtained for some vertex $(\underline{\mathrm{E}}, \underline{\delta})$ of (Δ^I, δ^I); similarly let $(\overline{\mathrm{E}}, \overline{\delta})$ be the vertex leading to $\overline{\eta}_k(t)$. Now we are in a position to solve approximately Problem 5.4. (when additionally to Problem 5.8 $c \in C$). To do this, we consider the following two Problems 5.5:

$$\dot{x} = A_c x + \underline{\mathrm{E}}x^c(t) + \varphi(t)\underline{\delta}, \quad c \in C \tag{5.79a}$$

$$\dot{x} = A_c x + \overline{\mathrm{E}}x^c(t) + \varphi(t)\overline{\delta}, \quad c \in C \tag{5.79b}$$

Now solving exactly (5.79a) for $\underline{x}_k(t)$ we obtain the lower endpoint of the approximate solution for Problem 5.4 (for its kth component). Similarly, solving exactly (5.79) for $\overline{x}_k(t)$ we find the upper endpoint of the desired approximate solution.

To illustrate the above approach we shall consider the following examples.

***E x a m p l e* 5.5.** We take up the circuit from Example 5.3. The input voltage V = 200V = constant and the inductance L = 2H are given exactly. The interval parameters are R and C and are given by the same nominal values and tolerances as in Example 5.3. It is desired to find the interval of the output voltage $v(t)$ (the voltage across the inductor L) for zero initial conditions $i_L(0) = 0$ and $v_C(0) = 0$.

Since $v = L\, di_L/dt$ and L is noninterval the state vector was chosen to be made up of the variables $x_1 = i_L$ and $x_2 = di_2/t$. The corresponding system for transient analysis of the circuit is then

$$\dot{x}_1 = x_2, \qquad x_1(0) = 0$$

$$\dot{x}_2 = -\frac{1}{LC}x_1 - \frac{1}{RC}x_L, \qquad x_2(0) = \frac{V}{L} = 100$$

When $C \in C^I$ and $R \in R^I$ the dynamic tolerance problem considered is seen to be a Problem 5.3. As mentioned in section 5.1.3 it can however be imbedded (at the cost of obtaining slightly larger interval solution) in the following problem of type 5.7

$$\dot{x}_1 = x_2, \qquad x_1(0) = 0 \tag{5.80a}$$

$$\dot{x}_2 = a_{21}x_1 + a_{22}x_2, \qquad x_2(0) = 100 \tag{5.80b}$$

$$a_{21} \in A_{21}, \qquad a_{22} \in A_{22} \tag{5.80c}$$

with

$$A_{21} = -\frac{1}{LC^I}, \qquad A_{22} = -\frac{1}{R^I C^I}, \tag{5.80d}$$

the intervals A_{21} and A_{22} being treated as independent.

Problem (5.80) was solved on a computer using the approximate method presented above. After determining the interval $X_2(t)$ related to $x_2(t)$ the desired output voltage interval $V(t)$ is found multiplying $X_2(t)$ by L. For $t = t_1 = 0.015$s the approximate solution is given by the interval $[-47.901, \ -35.276]$.

***Example* 5.6.** Again we solve Problem 5.7 for a set of circuits described by the system

$$\dot{x} = Ax, \quad x(0) = x^0, \quad A \in A^I$$

where x^0 is a given initial condition vector while A^I is the interval matrix given in Example 4.5.

It was shown there that A^I is D-stable, so Assumption 5.1 holds.

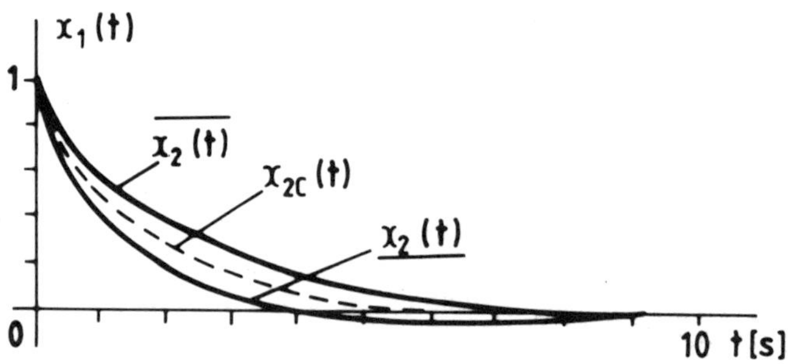

Fig. 5.6. Approximate interval solution for the first component $x_1(t)$ of the dynamic tolerance problem in Example 5.6.

The problem was solved by the approximate method for x^0 = (1, 1, 1, 1, 1) The approximate interval solution related to the first coordinate x_1 is shown in Fig. 5.6.

5.3.3. Outer and inner solution

Two alternate approximate solutions are presented in this section. The first one, $Y(t)$ is an outer solution since it is guaranteed to satisfy condition (5.25):

$$X(t) \subseteq Y(t), \qquad t \in [0,\tau] \tag{5.81}$$

The second solution, $Z(t)$, is called an inner solution since $Z(t)$ is enclosed by the exact interval solution $X(t)$, i.e.

$$Z(t) \subseteq X(t), \qquad t \in [0,\tau] \tag{5.82}$$

An approach for determining an outer solution for the most general Problem 5.3 will be presented now. But first we need to introduce the following auxiliary dynamic tolerance problem.

P r o b l e m 5.9. Find the interval solution of the set of systems

$$\dot{y} = \psi(y,t), \qquad y(0) = c \in C \tag{5.83}$$

where $\psi : R^{m+1} \to R^m$ and C is an m-dimensional vector.

It is seen that in (5.83) only the components of the vector c (the initial conditions $y_i(0)$) are given as intervals, $\psi(y,t)$ being an arbitrary nonlinear function. Problem 5.9 is a standard problem in interval analysis. It has been intensively investigated over the last years; nowadays there exists a general method which yields an outer solution for this problem [61].

Now recall that Problem 5.3 seeks the interval solution of the following set of systems

$$\dot{x} = A(p)x + \varphi(t)b(p) \tag{5.84a}$$

$$x(0) = c(p) \tag{5.84b}$$

$$p \in P \tag{5.84c}$$

The approach adopted here to finding an outer solution to (5.84) is to reduce problem (5.84) to the standard problem (5.83). With this in mind we first replace the set $\{c(p)\colon p \in P\}$ by an interval vector $\tilde{C}$ each component of which may be some interval extension or the range of $c_i(p)$ over P. We next introduce p_i, $i = \overline{1, n}$ as new state variables so that (5.84) takes on the form

$$\dot{x} = A(p)x + \varphi(t)b(p), \qquad x(0) \in \tilde{C} \tag{5.85a}$$

$$\dot{p} = 0, \qquad p(0) \in P \tag{5.85b}$$

Obviously, problem (5.85) is of the form of the standard problem (5.83). Moreover, due to the fact that $\tilde{C}$ contains the set $\{ c(p): p \in P \}$ the original Problem 5.3 given by (5.84) is clearly imbedded in the auxiliary standard problem (5.85). Therefore, if $Y(t)$ is an outer solution to (5.85) then it is an outer solution to Problem 5.3 as well.

It is obvious that the above approach is applicable to treating the simpler Problem 5.4. In this case the equivalent standard problem has the form

$$\begin{aligned} &\dot{x} = Ax + \varphi(t)b, && x(0) \in c^I \\ &\dot{a}_{ij} = 0, && a_{ij}(0) \in a_{ij}^I, \quad i,j = \overline{1,l} \\ &\dot{b}_i = 0, && b_i(0) \in b_i^I, \quad i = \overline{1,l} \end{aligned} \tag{5.86}$$

Now we shall introduce an inner solution, $Z(t)$, to Problem 5.3 (and 5.4). Let each component $X_i(t)$ of the interval solution $X(t)$ be written as $X_i(t) = [\underline{X}_i(t), \overline{X}_i(t)]$. Similarly, the notation $X(t) = [\underline{X}(t), \overline{X}(t)]$ will be used where $\underline{X}(t) = (\underline{X}_1(t), \ldots, \underline{X}_l(t)$ and $\overline{X}(t) = (\overline{X}_1(t), \ldots, \overline{X}_l(t))$. From (5.17), (5.19) and (5.20) it is clear that for a fixed t each component $\underline{X}_k(t)$ can theoretically be determined by finding the global solution to the following problem

$$\underline{X}_k(t) = \min_{p \in P} f(t,p) \tag{5.87a}$$

Similarly

$$\overline{X}_k(t) = \max_{p \in P} f(t,p) \tag{5.87b}$$

Since the global optimization problems (5.87) are in practice intractable it is expedient to seek for some approximate solutions to (5.87). With this in mind, some additional concepts will be introduced.

As has already been mentioned an n-dimensional vector $P_n = (P_1,\ldots,P_n)$ is geometrically represented as some "rectangular box" in R^n defined by the inequalities $\underline{p}_i \leq p_i \leq \overline{p}_i$, $i = \overline{1,n}$. A vertex of P is a real vector $p = (p_1, \ldots, p_n)$ whose components forma a combination of the lower and upper endpoints $\underline{p}_i$ and $\overline{p}_i$. Let the set of all vertices of the n-dimensional interval vector P be denoted by $V(P)$. Clearly, $V(P)$ consists of 2^n vertices (vectors).

Let the interval vector $Z(t) = [\underline{Z}(t, \overline{Z}(t)]$ with $\underline{Z}(t) = (\underline{Z}_1(t), \ldots, \underline{Z}_l(t))$ and $\overline{Z}(t) = (\overline{Z}_1(t), \ldots, \overline{Z}_l(t))$ be an approximation to the interval solution $X(t)$ of the tolerance problem considered. In this section each component $\underline{Z}_k(t)$ of $\underline{Z}(t)$ is defined as the global minimum of the following optimization problem (for fixed t)

$$\underline{Z}_k(t) = \min_{p \in V(P)} f(t,p) \tag{5.88a}$$

Similarly

$$\overline{Z}_k(t) = \max_{p \in V(P)} f(t,p) \tag{5.88b}$$

The interval vector $Z(t)$ whose components are defined by (5.88a) and (5.88b) will be called the inner solution of the worst-case tolerance analysis problem considered. Then the inclusion (5.82) is valid, i.e.

$$\underline{Z}_k(t) \geq \underline{X}_k(t), \quad t \in [0,\tau], \quad k = \overline{1,l} \tag{5.89a}$$

and

$$\overline{Z}_k(t) \leq \overline{X}_k(t), \quad t \in [0,r], \quad k = \overline{1,l} \tag{5.89b}$$

Indeed, clearly $V(P) \subset P$, so comparison of (5.87) and (5.88) leads to (5.89).

Property (5.89) may be useful in some applications. For instance, $X_k(t)$ should not exceed a prescribed threshold value $x_{k\max}$ (typically $x_{k\max}$ is the tolerated overshoot of the dynamic system considered) i.e.

$$\overline{X}_k(t) \leq x_{k\max} \tag{5.90}$$

Now if $\overline{Z}_k(t) > x_{k\,\max}$, it follows from (5.89b) that the system investigated does not meet requirement (5.90).

Now reconsider the solution set $S(0,\tau)$ for the tolerance problem considered. Let us introduce the following subset of $S(0,\tau)$:

$$V(S(0,\tau)) = \{f(t,p): t \in [0,\tau], \;\; p \in V(P)\} \tag{5.91}$$

Obviously, $V(S(0, \tau))$ consists of 2^n real solutions. Each element of $V(S(0,\tau))$ will be called a vertex solution.The set of values

$$V(S(t)) = \{f(t,p): p \in V(P)\} \tag{5.92}$$

of $V(S(0, \tau))$ for fixed time t will also be needed.

Let Ih $V(S(t))$ denote the interval hull of $V(S(t))$. The following theorem states that in fact Ih $V(S(t))$ represents the inner solution $Z(t)$.

T h e o r e m 5.2. The inner solution $Z(t) = (Z_1(t), \ldots, Z_l(t))$ whose interval components $Z_k(t)$ are defined by (5.88) coincides with the interval hull Ih $V(S(t))$, that is,

$$Z(t) = \text{Ih}\, V(S(t)) \tag{5.93}$$

for each time moment $t \in [0, \tau]$.

The proof is obvious and is based on simple geometrical considerations. It is illustrated in Fig. 5.7(a) for the case where $l = 2$. It is assumed for simplicity that only two parameters are allowed to take on values from corresponding intervals giving rise to four vertex solutions $x^1(t)$, $x^2(t)$, $x^3(t)$ and $x^4(t)$.

As is seen in Fig. 5.7(a), generally $Z(t) \subset X(t)$. However, experimental evidence shows that $Z(t)$ is a very close approximation to the exact interval solution $X(t)$. Thus, it is important to know when the inner solution $Z(t)$ represents the interval solution $X(t)$, that is when the inequality in (5.89a) and (5.89b) is replaced by strict equality. The following theorem provides some insight into the issue.

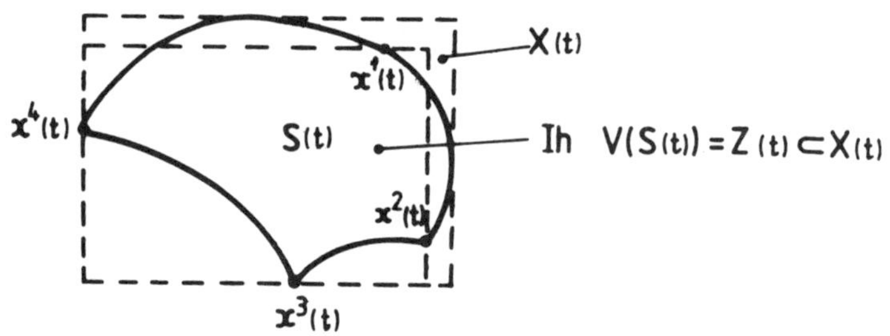

Fig. 5.7(a). Geometrical illustration of Theorem 5.2 for $l = 2$.

Theorem 5.3. The inner solution $Z(t)$ coincides with the interval solution $X(t)$ to the dynamic tolerance problem 5.3 iff

$$S(t) \subseteq \text{Ih}\, V(S(t)) \tag{5.94}$$

The proof is straightforward. It can be visualised using Fig 5.7(b). Indeed, $X(t) = \text{Ih}\, S(t)$. On the other hand if (5.94) is valid then $\text{Ih}\, S(t) = \text{Ih}\, V(S(t))$; hence $X(t) = Z(t)$ which proves the *if* part of the theorem. The *only if* part is obvious since $X(t) = Z(t)$ implies (5.94).

An interesting property of the inner solution is the fact that $Z(t)$ is close to $X(t)$ even when (5.94) is violated. Indeed, for reasons of continuity and smoothness of the tolerance problem considered it is not to be expected that $\text{Ih}\, S(t)$ containing $\text{Ih}\, V(S(t))$ should be much larger than $Z(t)$. This is a heuristic explanation of the experimentally confirmed fact that the inner solution $Z(t)$ is a very good approximation to $X(t)$.

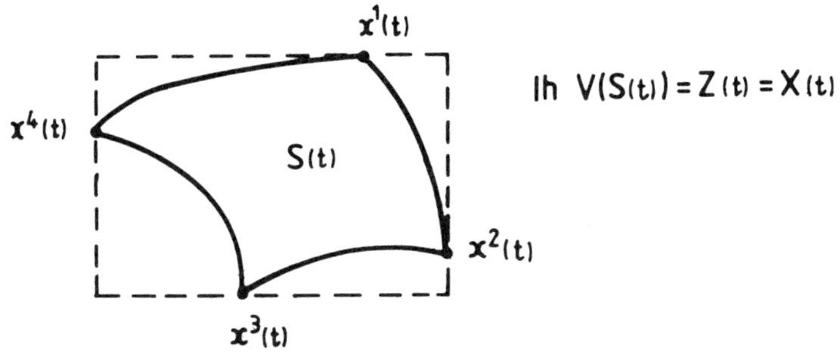

Fig. 5.7(b). Geometrical illustration of Theorem 5.3 for $l = 2$.

It is clear that definition (5.88) of the inner solution remains also valid in the case of Problem 5.4 if the parameter vector p includes the elements of A, b and c.

Based on Theorem 5.2 the inner solution $Z(t)$ can be determined by the following straightforward method.

Method 5.1

The transient analysis (5.84a), (5.84b) is carried out 2^n times for all vertices $p \in V(P)$. Let $x^q(t) = (x_1^q(t), \ldots, x_l^q(t))$ denote a vertex solution related to a given qth vertex from $V(P)$. All solutions $x^q(t)$ are stored. Then obviously $\underline{Z}_k(t)$ is obtained by the formula

$$\underline{Z}_k(t) = \min_q \{x_k^q(t), \quad q = \overline{1, 2^n}\} \tag{5.95a}$$

Similarly

$$\overline{Z}_k(t) = \max_q \{x_k^q(t), \quad q = \overline{1, 2^n}\} \tag{5.95b}$$

When implementing the method on a computer it should be borne in mind that each vertex solution $x^q(t)$ is computed with some error. Therefore, if a traditional numerical integration method is used, the validity of the inclusion (5.82) may be violated. In order to ensure (5.82) notwithstanding all possible computation errors it is recommended to use the interval integration method from [61]. Using this method each component $x_k^q(t)$ of a vertex solution $x^q(t)$ is obtained as an interval $x_k^q(t) = [\underline{x}_k^q(t), \overline{x}_k^q(t)]$. Thus, in practice, $\underline{Z}_k(t)$ is finally found by the formula

$$\underline{Z}_k(t) = \min_q \{ \overline{x}_k^q(t), \quad q = \overline{1, 2^n} \}$$

similarly

$$\overline{Z}_k(t) = \max_q \{ \underline{x}_k^q(t), \quad q = \overline{1, 2^n} \}$$

Although very simple as an algorithm the present method has a computational complexity of exponential type. Thus, its efficient application to higher dimension problems is questionable.

Now we shall focus on Problem 5.8. For ease of exposition we shall assume additionally that

$$b_i(t) = b_i, \quad i = \overline{1, l} \tag{5.96}$$

The inner solution $Z(t)$ for this problem could be found by the combination method (5.95). We shall however show that for this particular problem an alternate approximate solution $W(t)$ can be determined in a much easier manner. The new solution $W(t)$ being very close to $Z(t)$ is also a fairly good approximation to the exact interval solution $X(t)$ of Problem 5.8.

The approximation $W(t)$ is based on Method 5.1 and the implicit Euler method for integrating ordinary differential equations and can be determined by the following method.

Method 5.2

Let b^I and A^I from Problem 5.8 be written in the form

$$b^I = [b^c - \delta, \ \ b^c + \delta] \tag{5.97a}$$

$$A^I = [A^c - \Delta, \ \ A^c + \Delta] \tag{5.97b}$$

where b^c and A^c are the centres and δ and Δ are the radii of b^I and A^I, respectively. Using (5.97a) each real vector $b \in b^I$ can be written as

$$b = b^c + u \tag{5.98a}$$

where

$$u \in u^I = [-\delta, \delta]$$

thus

$$b^I = b^c + u^I$$

In a similar way the real matrix A can be written as

$$A = A^c + U \tag{5.98b}$$

where

$$U \in U^I = [-\Delta, \Delta]$$

Hence

$$A^I = A^c + U^I$$

Consider (5.15a) written in the form (with $\varphi(t) = E$, E being the identity matrix)

$$\dot{x} = (A^c + U)x + b^c + u \tag{5.99}$$

where $U \in U^I$ and $u \in u^I$. Using the implicit Euler integration method the discrete approximation of (5.99) is

$$x^{\nu+1} - x^{\nu} = h(A^c + U)x^{\nu+1} + h(b^c + u), \quad \nu \geq 0 \tag{5.100}$$

where h is the integration step size and ν is the current step number. From (5.100)

$$(E - hA^c - hU)x^{\nu+1} = x^{\nu} + hb^c + hu$$

or equivalently

$$(B^0 + B)x^{\nu+1} = c^{\nu} + v, \quad \nu \geq 0 \tag{5.101a}$$

with

$$B^0 = E - hA^c, \quad c^{\nu} = x^{\nu} + hb^c \tag{5.101b}$$

$$B = -hU, \quad v = hu \tag{5.101c}$$

Bearing in mind that $U \in U^I$ and $u \in u^I$ it is seen from (5.101c) that $B \in B^I$ and $v \in v^I$ with

$$B^I = -hU^I, \quad v^I = hu^I \tag{5.102}$$

Consider again (5.101) and (5.102). If B and v are some vertices of B^I and v^I, then the sequence x^{ν}, $\nu \geq 0$ resulting from the recursive solution of (5.101a), is a discrete approximation of the corresponding vertex solution $x(t)$. Neglecting the inaccuracy of Euler integration method we see that $x^\nu = x(t_\nu)$

Suppose that we are interested in determining the upper endpoint $\overline{W}_k(t)$ of an approximate solution $W(t)$ having the property that it remains close to the inner solution $Z(t)$. It will be shown below that $W_k(t)$ can be found using the following procedure.

Procedure 5.1. Solve iteratively the interval systems

$$(B^0 + B^I)w^{\nu+1} = c^\nu + v^I \tag{5.103}$$

where B^I and v^I are given by (5.102). Initially, for $\nu = 0$,

$$c^0 = x^0 + hb^c$$

where x^0 is the initial condition $x(0)$. Afterwards ($\nu > 0$) c^ν is obtained as follows. Solve the interval system

$$(B^0 + B^I)W^1 = c^0 + v^I$$

for the interval solution W^1. We get $2l$ l-dimensional vectors whose ith components determine the lower endpoint $\underline{W}_i^1$ and the upper endpoint $\overline{W}_i^1$ of the ith component W_i^1 of the interval solution W.

Let the vector which represents the vertex corresponding to $\overline{W}_k^1$ be denoted as $\overline{w}^1$. Put

$$c^1 = \overline{w}^1 + hb^c$$

and proceed to solve the interval system

$$(B^0 + B^I)w^2 = c^1 + v^I \tag{5.104}$$

Now (5.104) is solved for W^2 and the new vector $\overline{w}_2$ (corresponding to $\overline{W}_k^2$) is found. We put

$$c^2 = \overline{w}^2 + hb^c$$

and the iterative process continues according to (5.103).

The lower endpoint $\underline{W}_k(t)$ of the component $W_k(t)$ for the discrete time t_ν can be determined by a similar way.

Procedure 5.2. Solve iteratively the linear interval system

$$(B^0 + B^I)w^{\nu+1} = c^\nu + v^I, \quad \nu \geq 0 \tag{5.105}$$

with

$$c^0 = x^0 + hb^c, \quad c^v = \underline{w}^v + hb^c$$

where $\underline{w}^v$ represents the vertex corresponding to the lower endpoint $\underline{W}_k^v$ of the kth component of the current interval solution W^v of (5.105).

Now we shall show that for the simplified Problem 5.8 $W(t)$ is close to $Z(t)$ and, hence, to $X(t)$. First, the following result will be proven.

Theorem 5.4. The inner solution $Z(t)$ and the exact solution $X(t)$ to the dynamic tolerance problem 5.8 (with the specific condition (5.96) coincide over some time interval $[t_1, \infty)$, $t_1 \geq 0$.

Proof. Owing to Assumptions 5.1 and (5.96) $\dot{x}(\infty) = 0$ and (5.15a) (with $A \in A^I$ and $b \in b^I$) reduces at infinity to the following linear interval system

$$A^I x_\infty = -b^I \tag{5.106}$$

where x_∞ stands for $x(\infty)$. Thus, the reachability set at infinity $S_\infty = S(\infty)$ of the problem considered is in fact the solution set S of (5.106). From the properties of S (cf. section 3.2.1) it follows that S_∞ is always contained in the interval hull of its vertices, i.e.

$$S_\infty \subseteq \mathrm{Ih}\, V(S_\infty) \tag{5.107}$$

Thus, on account of Theorem 5.3

$$Z_\infty = X_\infty$$

Due to the continuity and smoothness of the tolerance problem considered it is clear that its reachability set $S(t)$ is varying continuously and smoothly as t changes. Therefore, it follows from (5.107) that

$$S(t) \subseteq \mathrm{Ih}\, V(S(t)) = Z(t)$$

for some time interval $[t_1, \infty)$ which completes the proof of the theorem.

Now suppose that

$$W(\infty) = X(\infty) \tag{5.108}$$

On account of Theorem 5.4 and (5.108) the approximation $W(t)$ can be different from the exact solution $X(t)$ only over a bounded time interval. Therefore, using the same argument of continuity and smoothness as in the case of the inner solution $Z(t)$ it could be expected that $W(t)$ should be a good approximation to $X(t)$ provided (5.108) is fulfilled. Experimental evidence show that this is the case for all the numerical examples considered so far.

The methods for solving the linear interval system (5.103), (5.105) were given in section 3.2.

In order to compare the computational efficiency of the above two methods assume the first method is implemented by the implicit Euler integration method using the same integration step h as in Method 5.2. The computation efforts needed in each method can be estimated by the number N of multiplications per step. Let N_1 denote this quantity for Method 5.2. It is easily seen that if all the components of A^I and b^I are intervals then

$$N_1 = l^2 2^{l^2+l} \tag{5.109}$$

Let N_2 denote the amount of multiplications per step for the second transient tolerance analysis method when it uses the general Rohn's method from section 3.2. Then

$$N_2 = 2l^4 2^l \tag{5.110}$$

It is seen from (5.109) and (5.110) that Method 5.2 is vastly superior over Method 5.1 for large l.

The computational superiority of the second method is further enhanced whenever $C^I = B^0 + B^I$ from (5.103) is an inverse-stable matrix (section 3.2). In this case

$$N_2 = 4l^5$$

and Method 5.2 is preferable computationwise to Method 5.1 even for moderate l.

We shall now consider several numerical examples. They were solved on an IBM personal computer by both methods. In implementing Method 5.1 each vertex solution was obtained as an interval by means of the general interval integration method from [61].

***Example* 5.7.** In this example the dynamic tolerance problem is that of Example 5.5. The corresponding system of differential equations is

$$\begin{aligned} \dot{x}_1 &= x_2 , & x_1(0) &= 0 \\ \dot{x}_2 &= -\frac{1}{LC}x_1 - \frac{1}{RC}x_2 , & x_2(0) &= \frac{V}{L} = 100 \end{aligned} \tag{5.111}$$

with $C \in C^I$ and $R \in R^I$. As underlined in Example 5.5 this is a problem of type 5.3. It was first solved approximately determining the corresponding inner solution $Z(t)$ by means of Method 5.1. The interval for the output variable $v(t)$ at $t = t_1 = 0.015$ obtained by this method is

$$V(t) = [-47.33, \ -35.754] \tag{5.112}$$

To apply Method 5.2, Problem (5.111) was embedded in the problem (5.80) of type 5.4. Application of Procedures 5.1 and 5.2 yielded the following interval for the same output variable $v(t)$ at the same time $t = t_1$

$$V(t) = [-47.5820, \ -35.1332] \tag{5.113}$$

As expected interval (5.113) is wider than (5.112) because the solution set of problem (5.111) is contained in the solution set of problem (5.80) (in the latter problem the interval A_{21} and A_{22} from (5.80d) are assumed to be independent which is not the case).

***Example* 5.8.** Consider the worst-case tolerance problem associated with system

$$\dot{x} = A^I x + b \ , \quad x(0) = x^0 \tag{5.114}$$

where

$$b = (2, 2)^T, \quad x^0 = (0, 0)^T, \quad A^I = [A^c - \Delta, \ A^c + \Delta]$$

(the symbol T denotes transpose) with

$$A^c = \begin{bmatrix} -2 & 0 \\ 1 & -3 \end{bmatrix}, \quad \Delta = \begin{bmatrix} 0.1 & 0 \\ 0.1 & 0 \end{bmatrix}$$

We seek the approximate tolerance solutions within the interval $[0, t_\infty]$; here t_∞ is some large enough time at which the steady-state solution of (5.114) has already been attained. It was found that for the problem considered t_∞ can be estimated to be $t_\infty = 6$s. An integration step $h = 0.1$s was assumed when using Method 5.2.

Since each component of the approximate solutions $Z(t)$ and $W(t)$ is monotonously increasing with t we only give data for the solutions at the time $t = t_\infty$ where they have its largest width. The following results have been obtained by both methods used:

$$Z_1(t_\infty) = Z_2(t_\infty) = W_1(t_\infty) = W_2(t_\infty) = [0.952381, 1.052632]$$

To show the improved accuracy of the present methods as compared with that of the method from section 5.3.2 we also give the values of $Y_1(t_\infty)$ and $Y_2(t_\infty)$ obtained through the latter method:

$$Y_1(t_\infty) = Y_2(t_\infty) = [0.950000, 1.050000]$$

It is seen from (5.115) that both methods lead to the same intervals at t_∞, that is, $Z(t_\infty)$ obtained by Method 5.1 is equal to $W(t_\infty)$ obtained by Method 5.2. Furthermore, since $W(t_\infty) = Z(t_\infty)$ the condition (5.108), $W(t_\infty) = X(t_\infty)$, is also fulfilled on account of Theorem 5.4. Thus, $W(t)$ can be expected to be reasonably close to $X(t)$. In fact, it was found that both methods yield the exact interval solution $X(t)$.

Example **5.9.** Again we consider a system of the form (5.114) with

$$b = (2,2,2,2,2)^T, \quad x^0 = (1,1,1,1,1)^T$$

while the interval matrix A^I is the matrix from Example 4.5 and is stable, so the interval solution to the tolerance problem considered is guaranteed to be bounded for the whole time interval $[0, \infty)$.

We provide data about the 5th component $W_5(t)$ of $W(t)$ for two time moments: $t = t_1 = 4.256$s ($\underline{W}_5(t)$ has minimum value at t_1) and $t = t_\infty$ (when the steady-state solution of the tolerance problem is reached). The following results have been obtained:

$$Z_5(t_1) = [-0.57063, \ 1.50094]$$

by Method 5.1,

$$W_5(t_1) = [-0.60077, \ 1.49906]$$

by Method 5.2 (with $h = 0.04256$s),

$$W_5(t_\infty) = [-0.557924, \ 1.765511]$$

by Method 5.2 (with the above h).

It is interesting to note that the same result (5.116) was obtained (as well as for all the remaining components of $W(t_\infty)$) by solving the corresponding static problem $A^I x = -b$. Thus, by Theorem 5.4 the approximations $Z(t)$ and $W(t)$ are expected to be reasonably close to the exact solution $X(t)$. Moreover, the interval matrix $B^0 + B^I$ associated with this example turned out to be inverse-stable, thus enabling the more efficient version of Method 5.2 to be used.

Comments

Section 5.1 Tolerance analysis is an important stage in the design of linear electrical circuits and systems. However, the overwhelming majority of known results is confined to the case of steady-state tolerance analysis.

On the other hand, there are numerous situations where the prime concern is whether the circuit (system) studied will meet some specified dynamic performance requirements for all possible variations in a set of parameters. This dynamic extension of tolerance analysis is usually referred to (especially in control engineering literature) as the performance robustness problem (e.g. [54]). A well-known example of performance robustness is the problem where a state variable $x_k(t)$ should not exceed some prescribed threshold value $x_{k\max}$ under all admissible parameter variations (typically, $x_{k\max}$ is the tolerated overshoot of the dynamic system considered). A similar problem arises in the setting of relay protections: the relay should not react to all responses of the circuit protected due to normal parameter variations but should operate under abnormal conditions. Once again determination of the maximum value of the corresponding circuit response under all possible parameter changes is of paramount importance.

Transient tolerance analysis of linear electric circuits gives rise to a greater variety of problems as compared with the steady-state tolerance analysis studied in Chapters 2 and 3. In section 5.1 of this chapter an attempt is made to introduce a classification scheme for the dynamic tolerance analysis problems and the methods for their solution. Three basic approaches to formulating transient tolerance problems are presented: explicit form formulation, frequency-domain formulation and time-domain formulation.

The explicit formulation (subsection 5.1.1) is a rather general form of stating dynamic tolerance problems since it allows for arbitrary input parameters and nonzero initial conditions depending on the input parameters. Using this approach the dynamic tolerance problem is, in fact, reduced to an explicit static tolerance problem and can be solved approximately or exactly by means of the methods developed in Chapter 2. However, since the relationship between the output variable and the input parameters must be known explicitly, the explicit form formulation is possible only for circuits of low order of complexity.

The frequency-domain formulation (subsection 5.1.2) is an alternative explicitly form formulation. This formulation applicable only for tolerance analysis of step responses of linear circuits with zero initial conditions is based on the relationship between the time step response and the real part $r(\omega, p)$ of the frequency response of the circuit investigated. Unlike the explicit formulation it can be used for tolerance analysis of circuits of higher complexity order.

The time-domain formulation (subsection 5.1.3) is an implicit form of stating dynamic tolerance problems. To prepare the ground for introduction of tolerance analysis methods based on this approach five problems of different complexity are therein stated.

It should be noted that only a small part of the problems formulated in section 5.1 has been treated in tolerance analysis literature. Thus, for a special case where the varying parameters p_i are elements of the matrix A a method for enclosing e^{At}, $A \in A^I$, is suggested in [63]. This method could be used for tolerance analysis of the natural responses of linear circuits (when there is no excitation). For the same case again, under the additional rather restrictive assumption that the elements of A depend on a single (determining) parameter a method for enclosing e^{At}, $A \in A^I$ arbitrarily sharply is proposed in [64]. The special problem 5.5 is considered in [65].

Section 5.2. In this section approximate solutions for the dynamic tolerance Problem 5.2 (i.e. the worst-case analysis of the step response of circuits with zero initial conditions in its frequency-domain formulation) are suggested in [60].

Two outer solutions $Y_1(t)$ and $Y_2(t)$ (having the property to enclose the exact interval step response $X(t)$) are presented in subsection 5.2.1. They are based on two different approaches from interval analysis to evaluating integrals. For fixed t, the former solution $Y_1(t)$ is obtained as the solution of an associated static tolerance problem which can be solved by some of the methods from Chapter 2. It should be stressed that this outer solution can be made rather close to the exact solution $X(t)$ at the cost of greater computational effort by increasing the limit frequency $\overline{\omega}$ and the number of integration steps n. The latter solutions $Y_2(t)$ is easier to compute (especially if the range of $r(\omega, p)$ over P is already known) but leads to relatively larger intervals than $X(t)$.

Two alternate approximate solutions $Y_3(t)$ and $Y_4(t)$ are introduced in subsection 5.2.2. Although they do not guarantee the enclosing property of the outer solutions $Y_1(t)$ and $Y_2(t)$ they prove to be rather good approximation to the exact solution $X(t)$. At the same time they require less computational effort than their counterparts $Y_1(t)$ and $Y_2(t)$. Therefore, they should be preferred to the outer solutions if the enclosing property is not essential for the dynamic tolerance problem at hand (for example, at an early stage of the system (circuit) design).

Section 5.3. The last section of Chapter 5 deals with approximate solutions of tolerance problems formulated in implicit time-domain form.

In subsection 5.3.1 the two auxiliary Problems 5.5 and 5.6 (formulated in subsection 5.1.3) are solved exactly (of course, the term "exactly" does not account for computational errors). The method for solving Problem 5.5 was first suggested in [66].

In subsection 5.3.2 the exact solutions of Problems 5.5 and 5.6 are used to solve approximately Problems 5.7 and 5.4. Based on an appropriate linearization of the original nonlinear problem, first an approximate solution for Problem 5.7 is obtained. This approach is then extended to the "mixed" Problem 5.8 (a combination of Problems 5.6 and 5.7). Finally, an approximate solution for the general Problem 5.4 (in the class of linear differential equations with independent interval coefficients and interval initial conditions) is suggested. It should be noted that similarly to the approximate solutions $Y_3(t)$ and $Y_4(t)$ from section 5.2.2 the approximate solutions of this section do not possess the enclosing property of the outer solutions $Y_1(t)$ and $Y_2(t)$ from subsection 5.2.1. Furthermore, (being only a linearization of the respective original problem) they provide good approximation to the exact interval solution $X(t)$ only for relatively narrow input parameter intervals.

The outer solution $Y(t)$ suggested in subsection 5.3.3 is obtained by first transforming the original tolerance Problems 5.3 or 5.4 into an equivalent nonlinear initial value problem for which only the initial conditions are intervals (Problem 5.9). The solution $Y(t)$ is then found by solving Problem 5.9 using general interval integration methods (e.g. the method developed in [61]). However, apart from the computational difficulties arising from the increased dimension of the equivalent Problem 5.9 this approach suffers from the drawback that the enclosing bounds on $X(t)$ may be rather conservative for larger t (especially in the case of the general Problem 5.3) if the input parameter intervals are relatively wider. This is due to the fact that the equivalent differential systems (5.85) and (5.86) are nonlinear (bilinear in the case of (5.86)) and that their integration by interval methods is associated with the so-called "wrapping effect" [2], [66]. Nevertheless, the outer solution $Y(t)$ seems to be the only available enclosing solution for the general transient tolerance Problem 5.3.

In order to assess the closeness of the outer solution $Y(t)$ to the exact solution $X(t)$ an inner solution $Z(t)$ for Problems 5.3 and 5.4 is also suggested in the same section. Indeed, if $Y(t)$ and $Z(t)$ do not differ very much they provide a good approximation to $X(t)$ since

$$Z(t) \subseteq X(t) \subseteq Y(t)$$

The inner solution turns out to be a very good approximation to the exact solution and in some cases $Z(t)$ may coincide with $X(t)$ (Theorems 5.3 and 5.4). The solution $Z(t)$ may be computed exactly (not accounting for the computational errors) by a combinational method (Method 5.1) or approximated (by $W(t)$) in a rather efficient manner (Method 5.2) even in the case of circuits of increased size. Experimental evidence shows that the approximate solutions $Z(t)$ and $W(t)$ are a very good approximation to $X(t)$. At the same time they require (especially $W(t)$) considerably less computation time (by an order) than the statistical methods (if the accuracy of the latter methods is to be comparable with the above interval methods).

CHAPTER 6

ANALYSIS OF NONLINEAR ELECTRIC CIRCUITS

6.1. DETERMINATION OF ALL OPERATING POINTS OF RESISTIVE CIRCUITS

The problem of finding the set of all d.c. operating points of nonlinear electric circuits has received a great deal of attention among circuit theoreticians in view of its numerous applications. This problem has two versions depending on whether the nonlinear resistors are modelled by piecewise-linear functions (PLF problem) or by continuously differentiable functions (CDF problem). In the class of traditional (noninterval) methods only the former problem has been solved in the case where the resistive circuit equations are written in the known hybrid-representation form [71]. The traditional methods are not capable of solving the CDF problem since they do not guarantee that all the operating points will be located.

In contrast, the existing interval methods [67] to [70] for solving the *CDF* problem are capable of finding infallibly all operating points within prescribed accuracy in a finite number of steps. Such methods suggested for the general case where the nonlinear resistive circuit is described by a system of nonlinear equations of general form are presented in section 6.1.1. The case where the circuit equations are in the hybrid-form representation is exposed in section 6.1.2.

6.1.1. General form description

In this case the nonlinear resistive circuit is described by the vector equation

$$f(x) = 0 \tag{6.1}$$

where f: $R^n \rightarrow R^n$ is a C^1 (continuously differentiable) function. The components x_i of x (branch currents, branch or nodal voltages) are in practice bounded within some admissible intervals, i.e.

$$x_i \in X_i^0,\ i = \overline{1,n} \tag{6.2}$$

or in vector notation

$$x \in X^0 \tag{6.3}$$

where X^0 is a vector with components X_i^0 .

The d.c. analysis problem herein considered may be formulated as follows: given the C^1 vector function f find all the real solutions of (6.1) within the prescribed box X^0.

This problem can be efficiently solved by some of the interval Newton method exposed in section 1.4. (The reader is strongly advised to go over subsections 1.4.1 and 1.4.2). Recall that the Newton method is associated with repeatedly solving the linear interval system (1.79)

$$J(X)(y - x) = -f(x) \tag{6.4}$$

with respect to y for different subboxes $X \subseteq X^0$ ($J(X)$ is the interval extension of the Jacobian matrix $J(x)$ of $f(x)$ with element $J_{ij} = \partial f_i / \partial x_j$ and x is the centre of X). The existing interval methods for solving problem (6.1), (6.3) differ essentially in the way the interval linear system (6.4) is solved.

In this section we shall present three versions of the Interval Newton method suitable for nonlinear d.c. circuit analysis. The first version appeals to Hansen's method and the second one implements Krawczyk's method (cf. section 1.4.2). The last version is based on a method suggested by Allefeld and Herzberger in [10].

First method

Recall that according to Hansen's method system (6.4) is first premultiplied by the matrix $B = [J(x)]^{-1}$ to give the equivalent linear interval system (1.81)

$$A(X)(y - x) = b(x) \tag{6.5a}$$

where

$$A(X) = BJ(X),\ b(x) = -Bf(x) \tag{6.5b}$$

As the computation process proceeds the initial box X^0 is dynamically subdivided into subboxes X. For each X, system (6.5) is solved in a "succesive iteration" mode by formula (1.90). The interval element $J_{ij}(X)$ is defined as follows [4]

$$J_{ij}(X) = I_{ij}(X_1,\dots,X_j,\ x_{j+1},\dots,x_n) \tag{6.6a}$$

where X_k ($k = \overline{1, j}$) is the kth component of X, and x_m ($m = \overline{j + 1, n}$) is the corresponding element of x. If the circuit considered contains only uncoupled two-terminal resistors, then (6.6a) is simplified to

$$J_{ij}(X) = J_{ij}(X_j) \tag{6.6b}$$

The real vector b and the interval matrix A are now computed by (6.5b). According to (1.90) a new set X' with components X'_i is obtained as follows:

$$Y_i = x_i - [b_i + \sum_{j=1}^{i-1} A_{ij}(X_j' - x_j) + \sum_{j=i+1}^{n} A_{ij}(X_j - x_j)]/A_{ii} = x_i - C_i/A_{ii} \tag{6.7a}$$

$$X' = Y_i \cap X_i \tag{6.7b}$$

As each new component Y_i, $i = \overline{1, n}$, is computed, it is immediately intersected with X_i so that the newest result X_i' is used in finding $Y_{i+1}, \ldots, Y_n$.

Since the interval A_{ii} may contain zero for one or more values of i, extended interval arithmetic is used to compute Y_i from (6.7a). If $0 \in A_{ii}$, first the quotient C_i/A_{ii} from (6.7a) is obtained by formula (1.25); then its negative is formed and finally Y_i is computed by adding x_i. As is seen from (1.25) Y_i may consist of one or two infinite segments of R. However, since Y_i is intersected with the finite interval X_i the resultant set X_i' is always finite and may consist of:

(i) one single interval

(ii) two dispoint intervals X_i^L and X_i^R, the subscripts L and R meaning left and right,

(iii) empty set

In the first case the new interval X_i'(renamed X'_j) is directly used in (6.7a).

In case (ii) there are two disjoint interval X_i^L and X_i^R and a gap G_i between them. One possible approach to tackle this case is to use X_i^L and X_i^R (renamed X_j^L and $\underline{X_j^R}$) separately in (6.7a). However, since case (ii) may occur for all components X_i', $i = \overline{1, n}$, such an approach would lead to a rather complicated algorithm. For simplicity (following the recommendation from [8]) we have adopted the following approach. As can be easily seen

$$X_i^L \cup G_i \cup X_i^R = X_i$$

Thus, whenever case (ii) occurs the whole interval X_i (renamed as X_j') is used in (6.7a) rather than X_i^L and X_i^R. However, we store all the gaps G_i, $i \in \overline{1, n}$, and the corresponding intervals X_i^L and X_i^R. Finally, we only retain the index i_0 and the corresponding intervals $X_{i_0}{}^L$ and $X_{i_0}{}^R$ with the largest gap between them. Thus, the new set X' will be composed of only two boxes: one whose i_0th component is $X_{i_0}{}^L$ and one whose i_0th component is $X_{i_0}{}^R$. The other components of the two boxes are the same as those of X for all $i \neq i_0$. When each X_i' is a single interval ($i = \overline{1, n}$) then the new set X' represents one single box. If X' is smaller than X the above procedure is applied to X'(renamed X). If $X \subseteq X'$ then no reduction of the size of X takes place and the current box X is divided in half (along its largest component) and each subbox is processed separately.

If the intersection X_i' is empty (case (iii)) for at least one i, the current box X cannot contain a solution, so X is deleted.

Whenever a current box is divided into two boxes we put one of these new boxes in a list L to be processed later on and work on the other. Subsequent boxes may also have to be subdivided, thus adding to the list L of boxes yet to be processed. So the number of boxes in L tends to grow initially. Eventually, the number of boxes in the list L finally decreases to the number of all real solutions of (6.1) contained in the initial box X^0; if (6.1) has no solution in X^0 the list L becomes finally empty.

According to [4], [8] it is guaranteed that if the width w of a box from L at the final stage of the algorithm is small enough, it contains a solution of (6.1). The width w of a box with components $X_i = [\underline{X}_i, \overline{X}_i]$ is defined to be (1.27):

$$w = \max_i ((\overline{X}_i - \underline{X}_i), \quad i = \overline{1,n}) \tag{6.8}$$

The computation process is terminated when the width w of each box in L becomes less than a prescribed accuracy ε. Then each solution is assumed to be midpoint of the corresponding box.

The method outlined above has the following algorithm [67].

Algorithm 6.1.

Initially the list L of boxes to be processed consists of a single box X^0. The subsequent steps are to be done in the following order except as indicated by branching.

Step 1. Let $X = X^0$, $l = 1$ and $\nu = 0$ (l is the current length of the list L, and ν denotes the number of solutions found so far).

Step 2. Compute the centre x, $J(x)$, $J(X)$, B, b and A corresponding to X. If $J(x)$ happens to be singular and hence B does not exist, go to Step 8.

Step 3. Put $i = 1$.

Step 4. Calculate the interval X_i' using (6.7). If X_i' is empty, eliminate X from the list L and reset $l = l - 1$. If the new value $l = 0$ go to Step 12; otherwise choose the most recent box from L, rename it X and go to Step 11.

Step 5. If X_i' consists of a single interval put $i = i + 1$. If $i \le n$ go to Step 4; otherwise go to Step 7.

Step 6. If X_i' consists of two subintervals X_i^L and X_i^R, find the width of the gap G_i:

$$\text{gap}_i = \underline{X}_i^R - \overline{X}_i^L$$

where $\underline{X}_i^R$ is the lower end of the right-hand subinterval $\overline{X}_i^R$ and X_i^L is the upper end of the left-hand subinterval X_i^L. Store the index i as i_0 and the corresponding intervals $X_{i_0}^L$ and $X_{i_0}^R$ if $\text{gap}_{i_0} > \text{gap}_{i_0-1}$; put $X' = X$. Set $i = i + 1$ and go to step 4 if $i \le n$; otherwise skip to Step 9.

Step 7. If $X \subseteq X'$ (no reduction of the size of X occurs), go to Step 8. Otherwise put $X = X'$ and skip to Step 11.

Step 8. Find the index j of the largest component X_i of X. Divide X_j into two subintervals $X_j^L = [\underline{X}_j, (\underline{X}_j + \overline{X}_j)/2]$ and $X_j^R = [(\underline{X}_j + \overline{X}_j)/2, \overline{X}_j]$. Form the boxes $X^L = (X_1, \ldots, X_j^L, \ldots, X_n)$ and $X^R = (X_1, \ldots, X_j^R, \ldots, X_n)$. Go to Step 10.

S t e p 9. Form the boxes $X^L = (X_1, \ldots, X_{i_0}^L, \ldots X_n)$ and $X^R = (X_1, \ldots, X_{i_0}^R, \ldots X_n)$.

S t e p 10. Add X^R to the list L. Put $l = l + 1$ and $X = X^L$.

S t e p 11. Compute w for the current X using (6.8). If $w > \varepsilon$ go to Step 2. If $w \leq \varepsilon$, set $\nu = \nu + 1$. Print ν, X and its midpoint. Delete X from L and put $l = l - 1$.

If $l = 0$ the algorithm is terminated and ν is the number of solutions contained in X^0, if $l > 0$ choose the most recent box from L, rename it X and go to Step 2.

S t e p 12. In this case $\nu = 0$ and hence there is no d.c. solution of the nonlinear circuit considered in the initial region X^0.

R e m a r k 6.1. According to Step 7 of the algorithm we go to Step 11 whenever X' is smaller than X. However, if the reduction of the size of X' with regard to X is negligible this would result in slow convergence so that it is better, in this case, to go to Step 8.

R e m a r k 6.2. In order to ensure that the initial box X^0 contains all the d.c. solutions of the circuit considered one can choose X^0 as large as possible. However, this approach will result in greater computer time. Therefore it is expedient to find limits on each variable with the help of the no-gain property [72] or any other techniques.

Second method

This method for d.c. nonlinear analysis is based on Krawctyk's version of the Newton method (cf. section 1.4.2)). According to (1.87) and (1.89) the iterative procedure of the method is

$$K(x^{(k)}, X^{(k)}) = b(x^{(k)}) + x^{(k)} + [I - A(X^{(k)})]\,(X^{(k)} - x^{(k)}) \tag{6.9a}$$

$$X^{(k+1)} = X^{(k)} \cap K(X^{(k)}, X^{(k)}), \quad k \geq 0 \tag{6.9b}$$

(k is the iteration number).

This procedure is based on vector operations. However, better results can be obtained if componentwise operations are used. Similarly to (6.7) whenever a reduction of a component $X_i^{(k+1)}$ of the current box $X^{(k)}$ occurs, the reduced component is immediately used in trying to reduce the remaining components $X_j^{(k+1)}$, $j = \overline{i+1, n}$. Thus, the componentwise procedure is based on the following formulae

$$\begin{aligned} Y_i^{(k)} = b_i(x^{(k)}) + x_i^{(k)} &+ \sum_{j=1}^{i-1} C_{ij}(X^{(k)})(X_j^{(k+1)} - x_j^{(k+1)}) \\ &+ \sum_{j=i}^{n} C_{ij}(X^{(k)})(X_j^{(k)} - x_j^{(k)}) \end{aligned} \tag{6.10a}$$

$$X_i^{(k+1)} = X_i^{(k)} \cap Y_i^{(k)}, \quad k \geq 0 \tag{6.10b}$$

where $C_{ij}(X^{(k)})$ is the corresponding element of the interval matrix $I - A(X^{(k)})$. Based on the componentwise procedure (6.10) an algorithm for the second method has been developed.

Algorithm 6.2.

It has essentially the same structure as Algorithm 6.1. Some obvious simplifications occur due to the fact that now there is no division by an interval containing zero. Hence, the new set X' is always a single box and Steps 6 to 9 of Algorithm 6.1 must be skipped. Of course, the elements $X_i' = X_i^{(k+1)}$ of X' are now calculated by (6.10).

Third method

Now we shall sketch an alternate version of the interval Newton method suggested in [10] and applied for d.c. analysis in [68]. It is based on the following approach. The interval Jacobian matrix $J(X)$ from (6.4) is represented as the sum of two matrices as follows

$$J(X) = D(X) - B(X) \tag{6.11}$$

where $D(X)$ is formed by the diagonal elements of J and $B(X)$ includes the remaining elements of J (with changed sign). Then (6.4) can be written as

$$[D(X) - B(X)](x - y) = f(x) \tag{6.12}$$

or equivalently as

$$y = x - D^{-1}(X)[B(X)(x - y) + f(x)] \tag{6.13}$$

Based on (6.13) and the consideration that the zero(s) y of (6.1) in X must be in X the following iterative procedure has been proposed in [10]

$$Y^{(k)} = x^{(k)} - D^{-1}(X^{(k)})[B(X^{(k)}(x^k - X^{(k)}) + f(x^k)] \tag{6.14a}$$

$$X^{(k+1)} = X^{(k)} \cap Y^{(k)}, \quad k \geq 0 \tag{6.14b}$$

The above procedure is based on vector operations. Similarly to the second method better results are obtained if componentwise operations are used. Thus, by analogy with (6.10) the componentwise procedure is now based on the following formulae

$$Y_i^{(k)} = x_i^{(k)} - D_{ii}^{-1}(X^{(k)}\left[\sum_{j=1}^{i-1} B_{ij}(X^{(k)})(x_j^{(k+1)} - X_j^{(k+1)})\right. \tag{6.15a}$$
$$\left. + \sum_{j=i+1}^{n} B_{ij}(X^{(k)})(x_j^{(k)} - X_j^{(k)}) + f_i(x^{(k)}\right]$$

$$X_i^{(k+1)} = X_i^{(k)} \cap Y_i^{(k)}, \quad k \geq 0 \tag{6.15b}$$

Based on the componentwise procedure (6.15) an algorithm has been developed for the third version of the Newton method.

Algoritm 6.3.

It has the same structure as Algorithm 6.1. Again, extended interval arithmetic is used because $D_{ii}(X^{(k)})$ may contain zero. Thus, the only difference between the two algorithms refer to the following steps:

Step 2 is changed to read: compute x, $f(x)$ and $J(X)$ corresponding to X;

In Step 4 the interval $X_i' = X_i^{(k+1)}$ is computed using (6.15).

Unlike other known (traditional) methods for solving the *CDF* problem of nonlinear resistive circuit analysis the three versions presented of the interval Newton method guarantee that all solutions contained in an initial bounded hyper-rectangular region will be found within a prescribed accuracy. At the same time this method requires comparatively lesser computational efforts for low- and medium-size problems. For high-dimensional problem, however, the memory requirements of the methods in its present form may be excessive.

Two more efficient interval methods for nonlinear resistive circuit analysis will be presented in the next section. The improved efficiency of the methods is, however, achieved at the cost of narrowing the class of circuits to that described by the known hybrid form represenation [71].

6.1.2. Hybrid form representation

In this section we are concerned with the same d.c. nonlinear circuit analysis problem as that studied in section 6.1.1. However, it is now assumed that the circuit investigated allows the so-called hybrid representation [71]:

$$\begin{bmatrix} \varphi_a(v_a) \\ \varphi_b(i_b) \end{bmatrix} = \begin{bmatrix} H_{aa} & H_{ab} \\ H_{ba} & H_{bb} \end{bmatrix} \begin{bmatrix} v_a \\ i_b \end{bmatrix} + \begin{bmatrix} s_a \\ s_b \end{bmatrix} \tag{6.16}$$

where $v_a = (v_1, \ldots, v_l)$, $i_b = (i_{l+1}, \ldots, i_n)^T$ are the hybrid variables. If we introduce the column-vector $x = (v_a|i_b)^T$ and the vector function $\varphi(x) = (\varphi_1(x_1), \ldots, \varphi_n(x_n))^T$ Eq.(6.16) can be recast in the form

$$\varphi(x) = Hx + s \tag{6.17a}$$

where

$$\varphi_i(x) = \varphi_i(x_i), \quad i = \overline{1,n} \tag{6.17b}$$

The problem is to find all the operating points of the circuit studied, that is, to find all the real solutions of (6.17) contained in some (large enough) interval box X^0.

The problem formulated could be solved by the general methods from the previous section. Its computational efficiency, however, seems to be limited to circuits of low dimensionality since it involves recursive splitting of the initial region X^0 into subregions X^v, and the number of X^v and hence the computational effort needed to locate all the real solutions of (6.1) in X^0 tend to grow exponentially with the dimensionality n of the problem. On the other hand, the specific form of the hybrid representation (6.17) permits the elaboration of two specialized, more efficient interval methods for analysis of nonlinear circuits of larger size [69], [70]. They will be referred to as Method 4 and Method 5, respectively.

Fourth method

In order to expose the new method some additional notions from interval analysis are needed.

So far, we have considered the notion of interval extension of a scalar function in several variables, i.e. $f : R^n \to R$. If $f(x)$ is a map $f : R^n \to R^n$, then the interval extension $F(X)$ of $f(x)$ in X is an interval vector whose components are the interval extensions $F_i(X)$ of the corresponding components $f_i(x)$ of $f(x)$.

Let $l : R^n \to R^n$ be a linear (affine) map, i.e.

$$l(x) = Ax + b \tag{6.18}$$

where A is a constant (noninterval) matrix. Let $x \in X$ where X is an interval vector. The image of X under l will be denoted by Z. It can be easily seen that the interval extension $L(X)$ of $l(x)$ is the interval hull of Z (i.e. the smallest interval vector which contains the set Z). The geometrical interpretation of X, Z and $L(X)$ for $n = 2$ is given in Fig. 6.1

Let $f : R^n \to R^n$ be a continuous map with an interval extension $F(X)$. Then $F(X)$ is called inclusion monotonic in X^0 if $Y \subset X$ entails $F(Y) \subset F(X)$ for any $X, Y \subseteq X^0$.

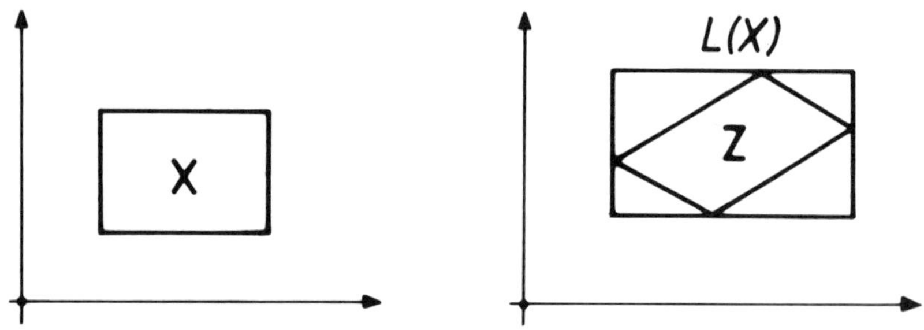

Fig. 6.1. A geometrical interpretation of the domain X, the image Z and the interval extension $L(X)$.

Finally, consider the equation

$$x = f(x) \tag{6.19}$$

If there exists an x^* such that (6.19) holds, then x^* is called a fixed point of the map f. Now consider the equation

$$X = F(X) \tag{6.20}$$

where X is an interval vector and $F(X)$ is the interval extension of $f(x)$. By analogy to the fixed point of $f(x)$, X^* is called a fixed interval of the map f if X^* satisfies (6.20).

To make the new method easier to understand it will be initially assumed that the functions $\varphi_i(x_i)$, $i = \overline{1, n}$, describing the nonlinear resistors are strictly monotone. Later on, this restriction will be removed.

A. Monotonic caracteristics

*A*1. *Basic results*

Assume that Eq.(6.17) has N solutions x^s, $s = \overline{1, N}$, in a box $X \subseteq X^0$. Rewrite (6.10) in the form

$$\varphi(x) = l(x) \tag{6.21}$$

First we shall prove the following lemma.

Lemma 6.1. The images of all N solutions to (6.21) are contained in the intersection of $\Phi(X)$ and $L(X)$, i.e.

$$\varphi(x^s) = l(x^s) \in \Phi(X) \cap L(X) \tag{6.22}$$

where $\Phi(X)$ is the interval extension of $\varphi(x)$ and $L(X)$ is the interval extension of $l(x)$.

P r o o f. Since $x^s \in X$, $s = \overline{1, N}$ then $\varphi(x^s) \in \Phi(X)$ and $l(x^s) \in L(X)$. But x^s is a solution, so $\varphi(x^s) = l(x^s)$. Thus $\varphi(x^s) \in \Phi(X)$ and at the same time $\varphi(x^s) \in L(X)$; hence (6.22) holds.

C o r o l l a r y 6.1. Let $A = \Phi(X)$, $B = L(X)$ and $C = (A \cup B) \setminus (A \cap B)$. Then

$$\varphi(x^s) = l(x^s) \notin C, \quad s = \overline{1,N} \tag{6.23}$$

or, in other words, the solutions x^s cannot be outside of $\Phi(X) \cap L(X)$.

P r o o f. Assume that $\varphi(x^s) \in C$. Then $\varphi(x^s) \notin \Phi(X) \cap L(X)$. But according to Lemma 6.2 this is a contradiction since x^s is a solution.

C o r o l l a r y 6.2. If

$$\phi(X) \cap L(X) = \varnothing \tag{6.24}$$

then equation (6.21) has no solutions in X.

P r o o f. The proof of this corollary follows immedietly from Lemma 6.1 and Corollary 6.2.

Let for brevity $A = \Phi(X)$ and $B = L(X)$. We shall assume that

$$A \neq B$$

Let X^s denote the smallest box containing all solutions x^s to (6.10) in X. It will be shown that under appropriate conditions X can be reduced to a smaller box X^1 containing X^s.

First we shall consider the case where

$$(A \cap B) \subset A \tag{6.25}$$

We shall show that whenever (6.25) holds we are able to reduce X. Indeed, according to Corollary 6.1, $\varphi(x^s) \notin C = A \setminus (A \cap B)$ for each solution $x^s \in X$ and we can remove C. Thus, we retain only the intersection $A \cap B$ for further inspection. Since each component $\varphi_i(x_i)$ of $\varphi(x)$ is strictly monotonic, the inverse map φ^{-1} exists. Therefore we are now able to find a new box X^1 by the formula

$$X^1 = \varphi^{-1}(A \cap B) \tag{6.26}$$

The box X^1 is reduced in size as compared to X. Indeed, we first note that $\varphi^{-1}(Y)$ is inclusion monotonic in A. Besides,

$$X = \varphi^{-1}(A) \tag{6.27}$$

Now it follows from (6.25) to (6.27) and the inclusion monotonicity property of $\varphi^{-1}(Y)$ that $X^1 \subset X$. Furthermore, all the solutions x^s to (6.10) are still in X^1 since X^1 is the inverse image of $A \cap B$ under φ^{-1}, while on the other hand, according to Lemma 6.1 $A \cap B$ contains all the images $\varphi(x^s)$. Thus we have proved the following theorem.

Theorem 6.1. If the condition (6.25) is fulfilled (with $A = \Phi(X)$ and $B = L(X)$) and the box $X^s \subset X$, then the box X^1 obtained by (6.26) contains X^s and $X^1 \subset X$.

Consider the interval equation

$$Y = \varphi^{-1}(\varphi(Y) \cap L(Y)) \tag{6.28}$$

and let the fixed interval of (6.28) be denoted by X^*. Consider the following iterative procedure:

$$X^{(k+1)} = \varphi^{-1}(\varphi(X^{(k)}) \cap L(X^{(k)}) , \quad k \geq 0 \tag{6.29}$$

On the basis of Theorem 6.1 it is easily seen that the iterative procedure (6.29) converges to the fixed interval X^* of (6.28) whenever $X^0 \supset X^*$ with $X^0 = X$. It should be noted that generally $X^* \neq X^s$ and $X^s \subset X^*$.

Now we shall consider the case where

$$(A \cap B) \subset B \tag{6.30}$$

Here again it is possible to reduce the size of X. With this in mind we first rewrite (6.17a) in the equivalent form

$$Hx = \varphi(x) - s \tag{6.31}$$

Let $H^{-1}(X)$ be the interval extension of $H^{-1}(x)$ when $x \in X$ (H is assumed invertible). Using the same arguments as in case (6.25), the following results are readily obtained [70].

Lemma 6.2. All N solutions of (6.17) are also in the intersection

$$X \cap H^{-1}((\Phi(X) \cap L(X)) - s)$$

Corollary 6.3. No solution of (6.17) is outside of

$$X \cap H^{-1}((\Phi(X) \cap L(X)) - s)$$

Corollary 6.4. If

$$X \cap H^{-1}((\Phi(X) \cap L(X)) - s) = \varnothing \tag{6.32}$$

Eq.(6.17) has no solution in X.

On the basis of Lemma 6.2 and Corollary 6.3 and using the same arguments as in case (6.25) the following theorem is easily seen to be valid.

Theorem 6.2. If the condition (6.30) is fulfilled and $X^s \subset X$, then the box X^1 obtained by the formula

$$X^1 = X \cap (H^{-1}(A \cap B) - s) \tag{6.33}$$

contains X^s and $X^1 \subseteq X$.

It should be stressed that unlike Theorem 6.1, Theorem 6.2 does not guarantee that X^1 is always smaller in size than X. However, if $X \supset X^*$ where X^* is the fixed interval of the interval equation

$$X = X \cap (H^{-1}(\varphi(X) \cap L(X)) - s) \tag{6.34}$$

then $X^1 \subset X$.

Similarly to (6.29), consider the iterative procedure

$$X^{(k+1)} = X^{(k)} \cap (H^{-1}(\varphi(X^{(k)}) \cap L(X^{(k)}) - s), \; k \geq 0 \tag{6.35}$$

On the basis of Theorem 6.2 it is readily seen that the procedure (6.35) converges to the fixed interval X^* of (6.34) whenever $X^0 \supset X^*$ with $X^0 = X$. Here again $X^* \neq X^s$ in general and $X^s \subset X^*$.

Finally, we shall consider a third case where both (6.25) and (6.30) hold, i.e.

$$(A \cap B) \subset A \quad \text{and} \quad (A \cap B) \subset B \tag{6.36}$$

In this general case it is reasonable to exploit the capability of both (6.29) and (6.35) in trying to reduce X. This gives rise to the following iterative procedure:

$$\begin{aligned} Y^{(k)} &= \varphi^{-1}(\varphi(X^{(k)}) \cap L(X^{(k)}) \\ X^{(k+1)} &= Y^{(k)} \cap (H^{-1}(\varphi(Y^{(k)}) \cap L(Y^{(k)}) - s) \,, \quad k \geq 0 \end{aligned} \tag{6.37}$$

Now we are in a position to expose the basic ideas behind the current method.

A2. Outline of the method

In order to find all the solutions to (6.17) we shall start with a large enough initial region (box) $X^{(0)}$. Then the iterative procedure (6.37) is applied and a fixed interval $X^* \subset X^{(0)}$ is obtained. (Actually, the iterative process is stopped whenever the reduction of the size of the current box $X^{(k+1)}$ as compared to that of the peceding box $X^{(k)}$ is smaller than a constant ε_1.)

At this stage the resulting box X^* is split along its widest side into two boxes X^L and X^R (left and right). Let the widest side of X^* have the coordinate number i_0. Clearly

$$X^L = (X_1^*,\ldots,[\underline{X}_{i_o}^*, m(X_{i_o}^*)], X_{i_o+1}^*,\ldots X_n^*)$$
$$X^R = (X_1^*,\ldots,[m(X_{i_o}^*), \overline{X}_{i_o}^*], X_{i_o+1},\ldots X_n^*) \tag{6.38}$$

The right box X^R is put into a list L for further processing.

The left box X^L is now renamed $X^{(0)}$ and the iterative procedure (6.37) is again applied. If X^L contains more than one solution, a new fixed interval X^* with nonzero width will be found. The new box X^* will be again split into two subboxes and the resulting right box X^R stored in the list L.

If the left box X^L generated at some stage contains only one solution, the fixed interval X^* obtained by (6.37) will eventually reduce to a point. (Actually, the iterative process is stopped whenever the width of $X^{(k+1)}$ becomes smaller than a constant ε_2.) Owing to Theorems 6.1 and 6.2, this point is a solution x^s to (6.17).

Whenever a solution x^s is found, the last box stored in the list L is retrieved from L and renamed $X^{(0)}$. The iterative procedure (6.37) is again applied to $X^{(0)}$.

The process of splitting the fixed interval vectors may result in a generation of boxes which do not contain a solution. Each of these boxes will be deleted in a finite number of iterations on account of rule (6.24) or (6.32). Whenever this occurs, the last box stored in L is retrieved and processed by the procedure (6.37).

The described process of generating, storing and retrieving boxes will terminate when list L becomes empty. Owing to the theoretical results obtained and the fact that $\varepsilon_1 > 0$ and $\varepsilon_2 > 0$, the termination of the iterative process will occur in a finite number of steps.

A3. A componentwise procedure

The iterative procedure suggested by (6.37) for reducing the size of the current box is based on vector operations. Its convergence rate can be improved if componentwise operations are introduced. Thus whenever a reduction of a component $X_i^{(k+1)}$ of the current box $X^{(k)}$ occurs, this will be used immediately for reducing the remaining components $X_j^{(k+1)}$, $j = \overline{i+1, n}$. The new componentwise procedure has the following structure.

Procedure 6.1.

Step 1. Let $k = 0$, $X_j = X_j^{(k)}$ and compute $A_j = \varphi_j(X_j)$, $j = \overline{1, n}$.
Step 2. Let $i = 1$.
Step 3. Compute

$$B_i = L_i(X) = \sum_{j=1}^{n} h_{ij} X_j + s_i \tag{6.39}$$

and $C_i = A_i \cap B_i$. If $C_i = A_i$ go to Step 5; othewise ($C_i \subset A_i$) proceed to next step.
Step 4. Evaluate $X_i = \varphi_i^{-1}(C_i)$ and recompute $A_i = \varphi_i(X_i)$.
Step 5. For $\nu = 1$ to n calculate

$$B_\nu = \sum_{j=1}^{n} h_{\nu j} X_j + s_\nu \quad \text{and} \quad C_\nu = A_\nu \cap B_\nu$$

S t e p 6. Let $R = H^{-1}$ with $R = (r_{ij})$. Compute

$$Y_i = X_i \cap \left[\sum_{j=1}^{n} r_{ij}(C_j - s_j) \right]$$

If $Y_i = X_i$ go to Step 7; otherwise ($Y_i \subset X_i$) put $X_i = Y_i$ and recompute $A_i = \varphi_i(X_i)$.

S t e p 7. Put $i = i + 1$ and go back to Step 3 until $i \le n$.

S t e p 8. Put $k = k + 1$. Let $X_i^{(k)} = X_i$, $i = \overline{1, n}$ and go back to Step 2 untill the reduction of the current box size become smaller than a preset constant ε_1; that is, until the condition

$$w(X_i^{(k)})/w(X_i^{(k-1)}) < 1 - \varepsilon_1 \tag{6.40}$$

is fulfilled for each i.

A4. Algorithm for monotonic characteristics

On the basis of the aforegoing, the following algorithm of the present method is suggested for the case when the nonlinear elements are described by monotonic characteristics.

A l g o r i t h m 6.4a.

S t a g e 1. Choose ε_1, ε_2 and $X^{(0)}$. Let $X = X^{(0)}$. Put $k = 0$ (k is the iteration number from Procedure 6.1), $m = 0$ (m is the current length of the list L) and $s = 0$ (s is the number of the solutions so far found).

S t a g e 2. Call Procedure 6.1. If at some step an intersection $A_\nu \cap B_\nu = \varnothing$ ($\nu \in [1, n]$) go to Stage 5; otherwise proceed to the next stage.

S t a g e 3. If the width of the box X obtained on exit from Procedure 6.1 is smaller than ε_2 go to Stage 6; othewise proceed to the next stage.

S t a g e 4. Split X along its widest side into X^L and X^R according to (6.38). Put X^R into the list L and let $m = m + 1$. Rename X^L as X and go to Stage 2.

S t a g e 5. Remove X and put $m = m - 1$. If $m = 0$ go to Stage 8; othewise go to Stage 7.

S t a g e 6. Put $s = s + 1$ and print the current solution found in interval form $X = (X_1, \ldots, X_n)$. As an approximation to the exact solution the midpoint of X is taken. Put $m = m - 1$. If $m = 0$ go to Stage 9; othewise proceed to the next stage.

S t a g e 7. Retrieve the last box X^R from the list L, rename it X and go back to Stage 2.

S t a g e 8. Termination 1: in this case Eq.(6.17) has no solution in $X^{(0)}$.

S t a g e 9. Termination 2: all the solutions to (6.17) in $X^{(0)}$ have already been found.

B. Nonmonotonic characteristics

Here the method suggested above will be generalized to encompass the case where the functions $\varphi_i(x_i)$ are not monotonic. Only the distinctions from the previous section will be stressed. To highlight the underlying ideas we shall restrict ourselves to a simple example.

Consider the function $\varphi_i(x_i)$ whose plot is given in Fig. 6.2. (Similarly to the monotonic characteristics case, here the subscript i denotes the coordinate number of the widest side of the current box X). Let $A = \varphi_i(X_i)$ and $B_i = L_i(X)$ (computed by (6.39)). Assume that $A_i \supset B_i$.

From the preceding theoretical results and the obvious geometrical considerations (see Fig. 6.2) it is clear that the subintervals X_i' and X_i'' do not contain a solution to (6.17). Thus we can reduce the initial interval X_i by deleting X_i''. Then we can remove the subinterval X_i'(referred to as a gap) to obtain two new intervals X_i^L and X_i^R. Using X_i^L and X_i^R while keeping the remaining intervals X_j ($j \neq i$) unchanged, two new boxes X^L and X^R, respectively, are now produced. Similarly to the monotonic characteristics case, the box X^R is stored in the list L for subsequent treatment. We rename the box X^L as X and try to reduce or split it in just the same way as above by the use of $\varphi_i(X_i)$ corresponding to the widest side of X.

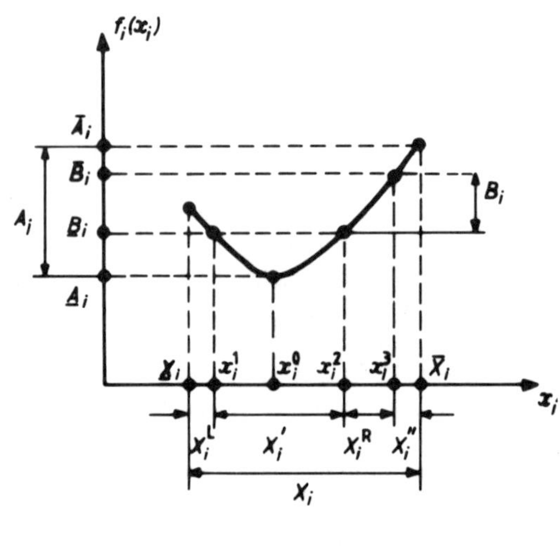

Fig. 6.2. Nonmonotonic characteristic of the ith nonlinear element.

Consider again Fig. 6.2. In order to find X_i' and X_i'' we have to solve the following two equations:

$$\varphi_i(x_i) = \overline{B}_i \tag{6.41a}$$

$$\varphi_i(x_i) = \underline{B}_i \tag{6.41b}$$

These are nonlinear equations of a single variable and can easily be solved by various iterative techniques. Care should be taken, however, to ensure that the corresponding approximate solutions for x_i^1 and x_i^3 on one hand and for x_i^2 on the other are obtained from the right and from the left, respectively. This will guarantee that the remaining intervals X_i^L and X_i^R are computed with some excess, so that there is no risk of omitting a solution to (6.17) when removing X_i' and X_i''.

In order to find A_i we have to compute the range $\varphi_i(X_i)$ of $\varphi_i(x_i)$ over X_i. Since this is a one-dimensional problem it can be solved by various noninterval or interval methods [2]. Similarly to X_i^L and X_i^R, the approximation to A_i should be outward.

If the functions $\varphi_i(x_i)$ are complex enough, several gaps may occur simultaneously along each X_i. If we remove all the gaps this will result in an increased number of newly generated boxes (especially for higher n). In an attempt to keep the number of stored boxes as low as possible, we have preferred to adopt the following simpler scheme: only the widest gap is removed from the current interval X_i (just as we did in implementing the general method from section 6.1.1). This approach has enabled us to adopt readily Algorithm 6.4a to the general case of nonmonotonic characteristics. This modified algorithm will be referred to as Algorithm 6.4b. The technical detailes involved in the modification are straightforward and are therefore omitted.

It is seen that in the general case the present method reduces essentially to finding repeatedly the range and solutions of nonlinear functions in a single variable over an interval. This fact accounts for the relative computational simplicity of the method.

Fifth method

This method is, in fact, an improvement of the third method from the previous section related to the case where the circuit studied is described by system (6.17).

The nonlinear system (6.1) to be solved is now of the form

$$f(x) = \varphi(x) - Hx - s = 0 \tag{6.42}$$

The corresponding Jacobian is

$$J(X) = \Phi'(X) - H \tag{6.43}$$

Hence $D(X)$ from (6.11) is the diagonal matrix

$$D(X) = \text{diag}\left\{\phi_i'(X_i) - h_{ii},\ i=\overline{1,n}\right\} \tag{6.44}$$

while the remaining part $B(X)$ is now a constant matrix

$$B = \left\{h_{ij},\ i \neq j,\ \ i,j = \overline{1,n}\right\} \tag{6.45}$$

where h_{ij} are the elements of H.

On account of (6.42) to (6.45) the iterative procedure (6.15) takes on the form

$$\begin{aligned} Y_i^{(k)} = x_i^{(k)} &- D_{ii}^{-1}(X_i^{(k)})[\varphi_i(x_i) - h_{ii}x_i - s_i \\ &- \sum_{j=1}^{i-1} h_{ij}X_j^{(k+1)} - \sum_{j=i+1}^{n} h_{ij}X_j^{(k)}] \end{aligned} \tag{6.46a}$$

$$X_i^{(k+1)} = X_i^{(k)} \cap Y_i^{(k)},\ \ k \geq 0 \tag{6.46b}$$

In the previous methods x is the centre m of the current box X. The present method (similarly to the modified mean-value forms from section 2.2.1) appeals to two new vectors x' and x'' distinct from m when computing the next box X' in an attempt to reduce the size of X'. It should be underlined straight away that this modification is applied only at those iterations and for those components $Y_i^{(k)}$ for which the following condition holds

$$0 \notin D_{ii}(X_i^{(k)}) \tag{6.47}$$

(if (6.47) is not fulfilled the present method makes use of the extended division as in methods M1 and M3 from section 6.1.1).

To explain the basic idea behind this section's method, we need formula (6.46a) rewritten for convenience as

$$Y_i = x_i - A_i[t_i(x_i) - C_i] \tag{6.48}$$

where x_i is now treated as an unknown from the interval X_i,

$$t_i(x_i) = \varphi_i(x_i) - h_{ii}x_i \tag{6.49}$$

while A_i and C_i are independent intervals of obvious expressions.

From (6.48) the lower endpoint $\underline{Y}_i$ and the upper endpoint $\overline{Y}_i$ of Y_i are clearly functions of x_i. Now we shall introduce two points x_i^L and x_i^U such that the former moves $\underline{Y}_i$ as high as possible while the latter shifts $\overline{Y}_i$ as low as possible. These poles will be referred to as the lower pole and upper pole of Y_i.

First, we shall give a rigorous definition of the poles. With this in mind, note that (6.48) is in fact short notation of the following equality

$$y_i = x_i - a_i[t_i(x_i) - c_i] \tag{6.50a}$$

with

$$a_i \in A_i, \quad c_i \in C_i \tag{6.50b}$$

Indeed, a_i and c_i appear only once to the first power in (6.50a) and no division by an interval containing zero occurs in (6.48) because of

$$A_i = 1/D_{ii}(X_i^{(k)}) \tag{6.51}$$

and condition (6.47). Therefore, by Theorem 1.4 $\underline{Y}_i$ for fixed x_i is in fact the lower endpoint of the range $y_i(x_i, A_i, C_i)$. Similarly, $\overline{Y}_i$ for fixed x_i is the upper endpoint of the range. Now, if we free x_i within X_i, y_i becomes a function of all three variables x_i, a_i and c_i belonging to their respective intervals. Thus, x_i^L can be defined as the solution of the following minmax problem:

$$x_i^L = \min_{\substack{a_i \in A_i \\ c_i \in C_i}} \max_{x_i \in X_i} \{x_i - a_i[t_i(x_i) - c_i]\} \tag{6.52}$$

In a similar way, x_i^U is the solution of the following maxmin problem

$$x_i^U = \max_{\substack{a_i \in A_i \\ c_i \in C_i}} \min_{x_i \in X_i} \{x_i - a_i[t_i(x_i) - c_i]\} \tag{6.53}$$

Based on Theorem 1.4, it is easily seen that an alternative way to define the poles of Y_i is to solve the problems

$$x_i^L = \min_{\nu} \max_{x_i \in X_i} \{y_i^{(\nu)}(x_i), \quad \nu = \overline{1,4}\} \tag{6.54}$$

$$x_i^U = \max_{\nu} \min_{x_i \in X_i} \{y_i^{(\nu)}(x_i), \quad \nu = \overline{1,4}\} \tag{6.55}$$

where

$$y_i^{(1)}(x_i) = x_i - \underline{a}_i[t_i(x_i) - \underline{c}_i] \tag{6.56a}$$

$$y_i^{(2)}(x_i) = x_i - \underline{a}_i[t_i(x_i) - \overline{c}_i] \tag{6.56b}$$

$$y_i^{(3)}(x_i) = x_i - \overline{a}_i[t_i(x_i) - \underline{c}_i] \tag{6.56c}$$

$$y_i^{(4)}(x_i) = x_i - \overline{a}_i[t_i(x_i) - \overline{c}_i] \tag{6.56d}$$

Remark 6.1. It should be stressed that the operations min and max in (6.54) and (6.55) are to be carried out in the order (from left to right) given in the respective problem since (as is easily verified) these operations are not commutative.

The exact determination of the poles x_i^L and x_i^U from (6.52), (6.53) or (6.54) to (6.56) in, in general, a very difficult, time-consuming task. That is why it is reasonable to try and find some approximations x_i' and x_i'' to x_i^L and x_i^U, respectively, that are simple to compute and are at the same time accurate enough. One way to do this is to calculate $Y_i(x_i)$ from (6.48) for three points, namely $\underline{x}_i$, m_i and $\overline{x}_i$ (m_i being the centre of X_i). Then the lower endpoint of the (hopefully) reduced interval Y_i corresponding to x_i' is determined as follows

$$\underline{Y}_i = \max\{\underline{Y}_i(\underline{x}_i), \underline{Y}_i(m_i), \underline{Y}_i(\overline{x}_i)\} \tag{6.57}$$

In a similar way, the upper endpoint of the narrowed interval Y_i corresponding to x_i'' is chosen as follows

$$\overline{Y}_i = \min\{\overline{Y}_i(\underline{x}_i), \overline{Y}_i(m_i), \overline{Y}_i(\overline{x}_i)\} \tag{6.58}$$

This simple approach turns out to be rather efficient in the case of diode-transistor circuits all transistors of which have been represented by the Ebers–Moll model. Indeed, now

$$\varphi_i(x_i) = \alpha_i(e^{\beta x_i} - 1) \tag{6.59}$$

$$\varphi_i'(x_i) = \alpha_i \beta_i e^{\beta x_i} \tag{6.60}$$

We need to calculate the derivative

$$t_i'(x_i) = \varphi_i'(x_i) - h_{ii} \tag{6.61}$$

of $t_i(x)$ for $x_i = \underline{x}_i$ and $x_i = \overline{x}_i$ in order to calculate A_i. Thus, as is readily seen from (6.59) to (6.60) it is sufficient to calculate only $e^{\beta \underline{x}_i}$ and $e^{\beta \overline{x}_i}$ to be able to compute $Y_i(\underline{x}_i)$ and $Y_i(\overline{x}_i)$ in a most economical way. Thus, the new approach (6.57), (6.58) requires roughly speaking, the same amount of computation as the traditional approach when only $Y_i(m_i)$ is calculated, providing however most often a narrower interval Y_i.

6.1.3. Numerical examples

To illustrate the applicability of the above methods we shall consider several examples.

***Example* 6.1.** The circuit to be analysed contains two tunnel diodes, a linear resistor and a voltage source connected in series. The voltage-current characteristics of the tunnel diodes are

$$i_1 = (2.5v_1^3 - 10.5v_1^2 + 11.8v_1)\times 10^{-3}$$
$$i_2 = (0.43v_2^3 - 2.69v_2^2 + 4.56v_2)\times 10^{-3}$$

while the source voltage e is constant with $e = 30$V and $r = 13.3$ kΩ. If the voltage across the first and second diode are denoted as x_1 and x_2, the circuit equations are

$$f_1(x_1,x_2) = 30 - 13.3(2.5x_1^3 - 10.5x_1^2 + 11.8x_1 - x_1 - x_2 = 0$$
$$f_2(x_1,x_2) = 2.5x_1^3 - 10.5x_1^2 + 11.8x_1 - 0.43x_2^3 + 2.69x_2^2 - 4.5x_2 = 0$$

We have to find all the operating points of the circuit studied. This problem was solved by methods M1 and M2.

To apply the above methods we need the interval Jacobian matrix $J(X)$ where X is a two-dimensional interval vector. The real Jacobian matrix $J(X)$ has the following elements J_{ij}:

$$J_{11}(x_1,x_2) = -13.3(7.5x_1^2 - 21x_1 + 11.8) - 1$$
$$J_{12}(x_1,x_2) = -1$$
$$J_{21}(x_1,x_2) = 7.5x_1^2 - 21x_1 + 11.8$$
$$J_{22}(x_1,x_2) = -1.29x_2^2 + 5.38x_2 - 4.56$$

It is seen that J_{ij} are functions of the variable x_j only (if at all), i.e.

$$J_{ij}(x_i,x_j) = J_{ij}(x_j)$$

since the circuit studied contains only two-terminal uncoupled resistors. Now, if x_j is replaced by X_j we shall have the corresponding natural interval extension $J_{ij}(X_j)$. Thus, we have shown the validity of formula (6.6b). For example, the natural interval extension for J_{22} is

$$J_{22}'(X_2) = -1.29X_2^2 + 5.38X_2 - 4.56$$

It is expedient to use the nested form of the above polynomial (section 1.2.2)

$$J_{22}(X_2) = X_2(-1.29X_2 + 5.38) - 4.56$$

since according to the subdistributivity property $J_{22}(X_2)$ is, in general, a narrower interval than $J_{22}'(X_2)$. For this reason the remaining elements of $J(X)$ are determined as follows:

$$J_{11}(X) = -13.3[X_1(7.5X_1 - 21) + 11.8] - 1, \quad J_{12}(X) = -1$$
$$J_{21}(X) = X_1(7.5X_1 - 21) + 11.8$$

We chose $X_1^0 = X_2^0 = [0, 4]$ for the components of X^0 and $\varepsilon = 10^{-3}$ for the accuracy.

Computer programs implementing algorithms A1 and A2 were developed ([67], [78]). Using these programs all nine operating points of the circuit were found infallibly within the prescribed accuracy. The following table shows the results obtained by A1 (the components x_1 and x_2 of the corresponding operating point OP_i are given as the centres of the components X_1 and X_2 of the respective interval solution).

Table 6.1

	OP_1	OP_2	OP_3	OP_4	OP_5	OP_6	OP_7	OP_8	OP_9
x_1	0.221	0.215	0.198	1.661	1.701	1.819	2.303	2.289	2.189
x_2	0.817	1.715	3.746	0.723	1.842	3.667	0.678	1.856	3.666

To assess the computational efficiency of the present algorithms, the switching-parameter algorithm [73] was also programmed. It was applied to the example considered using the same four starting points as in [73]. It was observed that the present algorithms requires far less computer time as compared to the method from [73].

As regards the computer memory requirements of the present algorithms the bulk of the needed memory volume is determined by a two-dimensional array $V = l_m \times 2n$ where l_m is the maximum length of the list L and n is the number of nonlinear equations. It should be borne in mind that l_m may be a very large number for high-dimensional problems.

***Example* 6.2.** The circuit investigated is shown in Fig.6.3.

Using the Ebers–Moll model for the transistor the following description of the circuit in the form (6.17) was obtained (with x_i being the corresponding voltage v_i)

$$\varphi_1(x_1) = 10^{-9}(e^{38x_1} - 1)$$
$$\varphi_2(x_2) = 1.98 \times 10^{-9}(e^{38x_2} - 1)$$
$$\varphi_3(x_3) = 10^{-9}(e^{38x_3} - 1)$$

$$H = \begin{bmatrix} -0.6689 & 1.6722 & -0.6689 \\ -0.6622 & -1.3445 & -0.6622 \\ -1 & 1 & -4 \end{bmatrix}$$

$$s = (8.0267, \ -4.0535, \ 6)^T$$

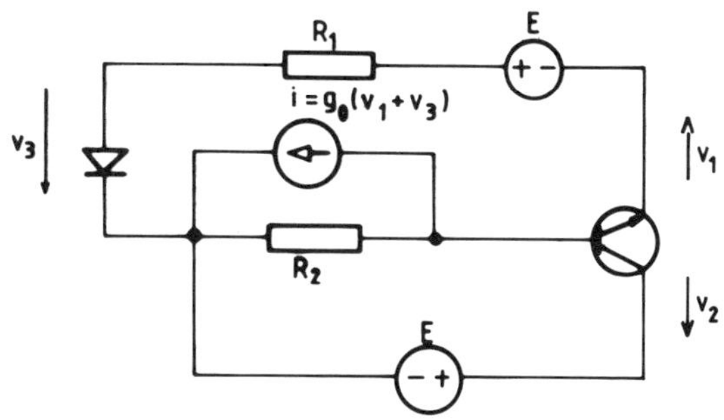

Fig. 6.3. Nonlinear circuit investigated in Example 6.2.

The problem is to find all the solutions of the circuit in the initial region

$$X^{(0)} = ([0, 1], [-5, 0], [0, 1])$$

with $\varepsilon = 0.01$. It has been solved by methods M2, M3, M4 and M5 [78]. There is only one solution in $X^{(0)}$ located with the desired accuracy within the interval X^s

$$X^s = ([0.555, 0.556], [-3.519, -3.517], [0.468, 0.469])$$

The number of iterations N_i needed to obtain X^s by the different methods is given in Table 6.2.

Table 6.2

Method	M2	M3	M4	M5
N_i	92	18	43	18

Example **6.3.** The circuit to be analyzed is described by system (6.17) with $n = 4$. The nonlinear elements are zener diodes while the active multiport corresponding to the RHS of (6.17a) may have arbitrary structure. It is assumed that the circuit equations are:

$$
\begin{aligned}
2(e^{x_1}-1) &= 3.86548x_1-0.38126x_2-0.14836x_3-0.17986x_4\\
3(e^{x_2}-1) &= 14.9484x_1+0.00764x_2-0.97901x_3+11.568x_4-9\\
2(e^{x_3}-1) &= 13.3092x_1-4.99094x_2-4.25872x_3+9.76315x_4-8\\
5(e^{x_4}-1) &= -8.91431x_1+3.34286x_2+4.17818x_3+2.61661x_4+5
\end{aligned}
\tag{6.62a}
$$

where x_i denotes the diode voltage v_i, $i = \overline{1, 4}$.

We want to find the set of all real solutions to (6.62a) in the "rectangular" region defined by the inequalities

$$-1 \le x_i \le 2, \quad i = \overline{1,4} \tag{6.62b}$$

The problem (6.62), was solved using the following algorithms:

A1: Algorithm 6.1 from section 6.1.1.

A3: Algorithm 6.3 from section 6.1.1.

A4a: Algorithm 6.4a from section 6.1.2 (for monotonic characteristics).

A4b: The algorithm for nonmonotonic characteristics based on A4a.

In order to apply algorithm A4b the original system (6.62a) was transformed into the equivalent form

$$
\begin{aligned}
&2(e^{x_1}-1)-3.86548x_1=-0.3812x_2-0.14836x_3-0.17986x_4\\
&3(e^{x_2}-1)-0.00764x_2+9=149484x_1-0.97901x_3+11.568x_4\\
&2(e^{x_3}-1)-4.25872x_3+8=13.3092x_1-4.99094x_2+9.76315x_4\\
&5(e^{x_4}-1)-2.61661x_4-5=-8.91431x_1+3.34286x_2+4.17818x_3
\end{aligned}
\tag{6.63}
$$

The following values were chosen for the parameters involved in the algorithms: $\varepsilon_1 = 0.01$ (equation (6.40)), $\varepsilon_2 = 0.001$ (width of the interval solution).

The nonlinear equations of the type (6.41) involved in algorithm A4b were solved as follows. First, the point x_i^o corresponding to $\underline{A}_i$: is found (see Fig.6.2). Thus the interval X_i is divided into two subintervals: $Y_1 = [\underline{X}_i, x_i^0]$ and $Y_2 = [x_i^0, \overline{X}_i]$. Then the appropriate approximation (from the right or left side) to each of the solutions x_i^2, x_i^2 and x_i^3 is evaluated using the bisection method for the corresponding subinterval Y_1 or Y_2 until it is reduced 100 times. This accuracy will be denoted as ε_3. It should be mentioned that in the present version of algorithm A4b the accuracy ε_3 remains constant at each step. Obviously a lot of iterations could be saved if ε_3 is coarsened each time a solution to (6.55) is approached (since Y_1 and Y_2 are then on the order of ε_2).

The problem (6.62) has six solutions:

$$x^1 = (\ 0.2 \quad 0.2 \quad 0.5 \quad 1.0\ )$$
$$x^2 = (\ 1.0 \quad 0.5 \quad 1.0 \quad 0.5\ )$$
$$x^3 = (\ 0.5 \quad 1.0 \quad 0.5 \quad 1.0\ )$$
$$x^4 = (\ 1.0 \quad 1.0 \quad 0.2 \quad 0.1\ )$$
$$x^5 = (-0.1966 \quad -0.82741 \quad -0.82176 \quad 0.19116)$$
$$x^6 = (\ 0.8457 \quad 0.8596 \quad 0.9193 \quad 0.8184\ )$$

All these solutions were found approximately (as the midpoint of the corresponding interval solutions) by each of the interval algorithms A1, A3 to A4b implemented on a personal computer. However, the algorithms A4a and A4b have better convergence rates as compared to A1 and A3. Indeed, the computer time needed by A4a and A4b to find the set of all solutions to the example considered was 10% and 50% respectively less than that of A1 while A3 was superior to A1 by only 8%.

***Example* 6.4.** We take up the circuit investigated in [73]. It contains four transistors and is described by the following system of equations

$$T\varphi(x) + Gx + b = 0$$

where

$$x = (v_1,\ v_2,\ v_3,\ v_4)^T$$

is the vector of the unknown voltages across the diodes in the Ebers–Moll model of the transistors. The components $\varphi_k(x_k)$ of $\varphi(x)$ are

$$\varphi_k(x_k) = 10^{-9}(e^{40x_k} - 1), \quad k = \overline{1,4}$$

while

$$T = \begin{bmatrix} 6103.168 & 2863.168 & 0 & 0 \\ 3580 & 6620 & 700 & 500 \\ 0 & 0 & 6103.168 & 2863.168 \\ 700 & 500 & 3580 & 6620 \end{bmatrix}$$

$$G = \begin{bmatrix} 0 & 4.36634 & 0 & 0 \\ 5.4 & 0 & 1 & 0 \\ 0 & 0 & 0 & 4.36634 \\ 1 & 0 & 5.4 & 0 \end{bmatrix}, \quad b = \begin{bmatrix} -12 \\ -22 \\ -12 \\ -20 \end{bmatrix}$$

The description (6.17) is now easily obtained with

$$H = -T^{-1}G, \quad s = -T^{-1}b$$

We seek all solutions in the region X^0 defined by

$$X_1^0 = [-1.1,\ 0.4],\quad X_2^0 = [-5,\ 0.4]$$

$$X_3^0 = [-1.6,\ 0.4],\quad X_4^0 = [-4,\ 0.4]$$

with $\varepsilon = 0.01$. The solutions have been found by the methods M2, M4 and M5 [78]. The corresponding numbers of iterations N_i are given in Table 6.3.

Table 6.3

Method	M2	M4	M5
N_i	207	143	79

R e m a r k 6.3. The data for N_i referring to column M5 from Table 6.2 and 6.3 are obtained using a somewhat different implementation of the fifth method [78].

E x a m p l e **6.5.** In this example it will be shown that the general interval method from section 6.1.1 can be successfully used for solving problems related to electric circuit synthesis. The problem to be solved herein is to realize the following voltage transfer function

$$V(s) = \frac{0.186s^2 + 2.474}{0.327s^3 + 2.640s^2 + 4.945s + 4.949}$$

by means of the circuit shown in Fig. 6.4. It is desired to determine the component values G_1, C_1, C_2, C_3, G_L and $\Gamma_1 = 1/L_1$.

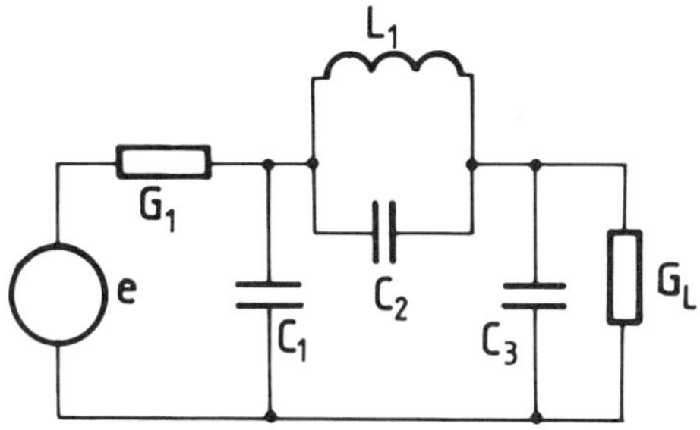

Fig. 6.4. Electric circuit considered in Example 6.5.

The parameters C_3 and G_L must satisfy the inequalities

$$C_3 \geq 1 \ , \quad G_L \geq 1 \tag{6.64}$$

It is shown in [74] that the following equations (the so-called component equations) are valid:

$$\begin{aligned}
&G-1/L_1 = 2.474\\
&G_1 C_2 = 0.186\\
&(G_1 + G_L)/L_1 = 4.949\\
&(C_1 + C_3)L_1 + G_1 G_L = 4.945\\
&G_1(C_1 + C_3) + G_L(C_1 + C_2) = 2.640\\
&C_1(C_2 + C_3) + C_2 C_3 = 0.327
\end{aligned}$$

After manipulation this system of six equations is transformed equivalently into a system of only 3 nonlinear equations

$$\begin{aligned}
&a_1 x_1 + a_2(X_2 + x_3) = 0\\
&x_1[a_3 x_2 + (a_4 + x_1 x_3)x] + a_5 = 0\\
&a_6(x_2 + x_3) + x_1(x_2 x_3 - a_7) = 0
\end{aligned} \tag{6.65a}$$

where

$$x_1 = G_1, \quad x_2 = C_1 \quad \text{and} \quad x_3 = C_3$$

Taking into account (6.64) the following restrictions on x_i were chosen:

$$0.1 \le x_1 \le 2, \quad 0.01 \le x_2 \le 1, \quad 0.1 \le x_3 \le 5 \tag{6.65b}$$

The system (6.65) was solved by Hansen's method. Two solutions for x_1, x_2 and x_3 and hence two solutions for the component values Γ_1, G_1, C_1, C_2, C_3 and G_L were obtained. They are given (as centres of corresponding intervals) in the columns S_1 and S_2 of Table 6.4.

Table 6.4

	S_1	S_2	
G_1	1.7829	1.3824	x_1
C_1	0.1775	0.0532	x_2
C_3	1.0945	1.6541	x_3
C_2	0.1043	0.1400	–
G_L	1.7836	1.8624	–
Γ_1	1.3876	1.8624	–

It is worthwhile nothing that using a traditional (noninterval) method only S_1 was obtained in [74].

***Example* 6.6.** This example serves to show that the general methods for circuit analysis can be used for solving nonlinear problems arising in areas other than circuit theory.

The problem herein considered stems from field theory [78]. It is desired to identify (locate) an impurity in a conductive medium. The problem is two-dimensional and the impurity is approximated by a rectangular region. Given the conductivity and the size of the medium, the unknown parameters are the centre and the widths of the rectangular impurity. Based on experimental design the problem is equated to that of solving a system of four equations in four variables (two for the coordinates of the centre and two for the sides of the impurity region). Each equation is of the form

$$b_0 + \sum_{i=1}^{n} b_i x_i + \sum_{i=1}^{n} b_{ii} x_i^2 + b_{12}x_1x_2 + b_{13}x_1x_3$$

$$+ b_{14}x_1x_4 + b_{23}x_2x_3 + b_{24}x_2x_4 + b_{34}x_1x_4 = 0$$

The system has been solved by Krawctyk's method. All the solutions contained in the interval box

$$X^0 = ([-2, 2], \ [-2, 2], \ [-2, 2], [-2, 2])$$

have been infallibly located with an accuracy $\varepsilon = 10^{-3}$.

6.1.4. Underdetermined systems

In this section, we shall touch upon the problem of solving underdetermined systems, that is, systems that have more unknowns that equations.

In the context of nonlinear d.c. circuit analysis, systems of n equations in $(n + 1)$ variables [73] or $(n - 1)$ equations in n variables [79] arise when homotopy methods are used for circuit analysis, n being the number of unknowns from the general form description (6.1) or the hybrid form representation (6.16).

Here we shall consider systems of two nonlinear algebraic equations in n variables with $n > 2$. As was shown in sections 2.5.1 such systems arise in connection with tolerance analysis in probabilistic formulation.

Let the system under consideration be

$$f_1(x) = 0 \qquad (6.66a)$$

$$f_2(x) = 0 \qquad (6.66b)$$

where $x = (x_1, \ldots, x_n)$ and f_1 and f_2 are assumed of class C^1. The problem is to verify whether the above system has at least one solution or no solution in a given box X^0, i.e. when

$$x \in X^0 \qquad (6.66c)$$

An interval method for solving this problem will be now presented. With this in mind we first introduce extensions of $f_1(x)$ and $f_2(x)$ in X

$$F_1(X) = f_1(x) + \sum_{j=1}^{n} G_{1j}(X)\,[X_j - x_j] \qquad (6.67a)$$

$$F_2(X) = f_2(x) + \sum_{j=1}^{n} G_{2j}(X)\,[X_j - x_j] \qquad (6.67b)$$

where X is any subbox of X^0 (i.e. $X \subseteq X^0$), x is a fixed point in X (usually x is the centre of X) while $G_{ij}(X)$ are the interval extensions of the derivatives $g_{ij} = \partial f_i/\partial x_j$ $(i = 1, 2; j = 1, n)$ in X. Next we form the two-dimensional interval vector

$$F(X) = (F_1(X), F_2(X))^T = ([\underline{F}_1, \overline{F}_1], [\underline{F}_2, \overline{F}_2])^T.$$

A necessary condition for (6.66) to have a solution is the inclusion

$$0 \in F(X) \qquad (6.68)$$

A sufficient condition for (6.66) not to have a solution is the exclusion

$$0 \notin F_1(X) \tag{6.69a}$$

or

$$0 \notin F_2(X) \tag{6.69b}$$

The present method for checking the compatibility of (6.66) is based on a procedure for verifying (6.69) or (6.68). It has, essentially the same algorithm as the interval methods from section 6.1.1.

Initially we set $X = X^0$ and evaluate (6.67) (using natural interval extension for $G_{ij}(X)$). If (6.69) is fulfilled Problem (6.66) has no solution and the computation process is terminated.

If (6.69) is not fulfilled (i.e. (6.68) holds) then we try to reduce the size of the current box X by deleting parts of X that are guaranteed not to contain a solution of (6.66a), (6.66b). If we fail to reduce X we then split it into two halves X^1 and X^2 along the largest side of X. One of the halves (say X^2) is stored in a list L of subboxes to be processed later. The other half X^1 is renamed as X and is processed as before: $F(X)$ is evaluated again and the exclusion (6.69) is checked for the new box X. This process of reducing the size of the current box or generating (by splitting) new subboxes and deleting part or all of them may end in only two possible outcomes:

a) the list L becomes empty after a finite number of steps;

b) the list L contains at least one box whose size tends to decrease until some accuracy condition $\varepsilon_4 > 0$ is met.

In the former case we have established that (6.66) has no solution. In the latter case we have enclosed at least one solution of (6.66).

Now we shall describe a procedure for reducing (if possible) the size of the current box X.

Procedure 6.2.

We first try to delete points from the intervals X_1 and X_2 that are not solutions of (6.66a), (6.66b). With this in mind (6.67) is written as a system of equations in the following form

$$G_{11}(X)(y_1 - x_1) + G_{12}(X)(y_2 - x_2) = B_1 \tag{6.70a}$$

$$G_{21}(X)(y_1 - x_1) + G_{22}(X)(y_2 - x_2) = B_2 \tag{6.70b}$$

where

$$B_i = f_i(x) + \sum_{j=3}^{n+1} G_{ij}(X)(X_j - x_j),\ i = 1{,}2 \tag{6.71}$$

are known intervals. Since $G_{ij}(X)$, $i = 1, 2$ are also known intervals, (6.70) is in fact a system of linear interval equations (cf. section 3.2.1). Based on their theory it will be shown that the sets of points along X_1 and X_2 which should be retained (because some of them might be solutions of (6.66a), (6.66b) in X) can be found in the following way.

Case 1.

In this case

$$0 \notin G_{11}G_{22} - G_{12}G_{21} = D \tag{6.72}$$

where, for simplicity, the argument X from G_{ij} is omitted. In this case the interval solution $W = (W_1, W_2)$ of the interval system

$$\begin{aligned} G_{11}w_1 + G_{12}w_2 &= B_1 \\ G_{21}w_1 + G_{22}w_2 &= B_2 \end{aligned} \tag{6.73}$$

with

$$w_1 = y_1 - x_1^0,\quad w_2 = y_2 - x_2^0$$

can be found by Rohn's method (section 3.2.2). First the following four real (noninterval) linear systems are solved:

$$\begin{aligned} \underline{G}_{11}w_1 + \underline{G}_{12}w_2 &= \underline{B}_1 \\ \underline{G}_{21}w_1 + \underline{G}_{22}w_2 &= \underline{B}_2 \end{aligned}$$

$$\begin{aligned} \overline{G}_{11}w_1 + \overline{G}_{12}w_1 &= \overline{B}_1 \\ \underline{G}_{21}w_1 + \underline{G}_{22}w_2 &= \underline{B}_2 \end{aligned}$$

$$\begin{aligned} \underline{G}_{11}w_1 + \underline{G}_{12}w_2 &= \underline{B}_1 \\ \overline{G}_{12}w_1 + \overline{G}_{22}w_2 &= \overline{B}_2 \end{aligned}$$

$$\begin{aligned} \overline{G}_{11}w_1 + \overline{G}_{12}w_2 &= \overline{B}_1 \\ \overline{G}_{12}w_1 + \overline{G}_{22}w_2 &= \overline{B}_2 \end{aligned}$$

Let the components w_1 and w_2 of the corresponding solutions be denoted as

$$w_1^{(k)},\ w_2^{(k)},\quad k = \overline{1,4}$$

Then the endpoints $\underline{W}_1, \overline{W}_1$ of W_1 and $\underline{W}_2, \overline{W}_2$ of W_2 are determined as follows

$$\underline{W}_i = \min(w_i^{(k)},\ k = \overline{1,4}),\ i=1,2$$
$$\overline{W}_i = \max(w_i^{(k)},\ k = \overline{1,4}),\ i=1,2$$

Now, from (6.74) the interval vector $Y = (Y_1, Y_2)$ is formed

$$Y_1 = x_1 + W_1,\ \ Y_2 = x_2 + W_2$$

Finally, the new intervals X_1' and X_2' are determined:

$$X_1' = X_1 \cap Y_1,\ \ X_2' = X_2 \cap Y_2 \tag{6.75}$$

If $X_1' \subset X_1$ and/or $X_2' \subset X_2$ then some parts of X_1 and/or X_2 not containing solutions of (6.66a), (6.66b) in X have actually been deleted. If

$$X_i \cap Y_i = \emptyset,\ \ i = 1 \text{ or } i = 2$$

then the current box X cannot have solutions of (6.66a), (6.66b) (this assertion is based on Theorem 1.18).

Case 2.

Now (6.72) is not fulfilled. Nevertheless, the components X_1' and X_2' can be determined in the following way. First, from (6.73) the components W_1 and W_2 can be expressed as

$$W_1 = (B_1G_{22}-B_2G_{12})/D = A_1/D$$

$$W_2 = (B_2G_{11}-B_1G_{21})/D = A_2/D$$

Since in this case $0 \in D$ one is led to use extended interval arithmetic (formula (1.25)) to find W_1 and W_2. It follows from (1.25) that the components W_1 and W_2 and hence Y_1 and Y_2 are now infinite. However, after intersecting with the finite intervals X_1 and X_2 in (6.75) the new components X_1' and X_2' will always be finite. As has been explained in section 6.1.1 the sets X_1' and/or X_2' may be:

i) empty set (no solution in X)
ii) one single interval
iii) two disjoint subintervals

If the last case occurs we proceed in exactly the same manner as in algorithm A1 from section 6.1.1.

Next, we try to reduce the size of X_3 and X_4 using the same approach as for X_1 and X_2. (An alternative possibility is to try to reduce once again X_2 and X_3. This is to be done at least once if n is odd.) Now, system (6.70) will have the form

$$G_{13}(y_3 - x_3^0) + G_{14}(y_4 - x_4^0) = B_1'$$
$$G_{23}(y_3 - x_3^0) + G_{24}(y_4 - x_4^0) = B_2'$$

However, here

$$B_i' = f_i(x^0) + G_{i1}(X_1' - x_1^0) + G_{i2}(X_2' - x_2^0) + \sum_{j=5}^{n} G_{ij}(X_j - x_j^0), \quad i = 1,2$$

The above procedure continues until the last pair X_{n-1}, X_n has been tried for possible reduction.

If no (or negligible) reduction has been achieved the current box X is split into two subboxes, one of them is stored in the list L while the other one (renamed as X) is processed anew.

Remark 6.4. The above method for checking whether (6.66) has a solution or not differs from the general methods of section 6.1.1 since the latter methods have been developed to solve n systems of nonlinear equations in n unknown while (6.66a), (6.66b) is a system of 2 equations in n unknowns ($n > 2$). However, most of the steps involved in the present method: generating and entering boxes in the list L, retrieving and deleting boxes from L, etc. are the same (or slightly altered) as in the previous methods from section 6.1.1.

Experimental results about the applications of the present method for tolerance analysis have been given in section 2.5.3, Example 2.14a. Some technical details concerning the two algorithms A11a and A11b used there will be briefly considered here.

In checking condition (6.69a) the mean-value form (1.44) was used for the interval extension of (2.148a) while the natural extension was used for (2.148b). Furthemore, only Case 1 of Procedure 6.2 has been incorporated in the present version (whenever Case 2 of the procedure occurs the current box is split into two subboxes). Lastly, Procedure 6.2 for reducing the current box X was implemented in two versions:

a) after the attempt to reduce the size of the pair X_1 and X_2 we try to reduce the pair X_3 and X_4;

b) after the attempts to reduce the size of X_1 and X_2 we try to reduce consecutively the pair X_2, X_3 and X_3, X_4.

Thus, algorithms A11a and A11b are essentially the same except that A11a uses Procedure 6.2a while A11b implements Procedure 6.2b.

Finally, the following remark is pertinent. When system (2.148) from Example 2.14a has a solution it is to be located with a given accuracy. Since the order of the nominal parameters values x_1 to x_4 differ enormously ($x_1 = x_2 = 10^5$, $x_3 = 10^{-9}$, while $x_4 = 10^{-7}$) the usual bisection criterion to split the current box X along its widest side and the stopping criterion

$$\max(w(X)) \geq \varepsilon$$

are not adequate. Indeed, even for $\varepsilon = 10^{-6}$ the above criteria will leave x_3 and x_4 totally unchanged. That is why, the following approach was introduced. The intervals X_1^0 to X_4^0 were scaled to one and the same (unit) width only for the purposes of bisecting and stopping. Then the former bisection and stopping rules were applied with $\varepsilon = 10^{-2}$. This approach ensures that the solution of system (2.148) will be located in a small volume ∇V^s which, for the example considered and the accuracy chosen, is 10^{-8} times smaller than the volume of the initial box X^0, each side X_i^0 of X^0 having been reduced at least hundredfold.

6.2. ANALYSIS OF DYNAMIC CIRCUITS

6.2.1. Finding all the periodic steady-states

In this section we consider the problem of finding all the periodic steady-states of a given period that are established in a nonlinear electric circuit exited by periodic sources of the same period. More exactly, given the system of nonlinear ordinary differential equations (ODE's):

$$\dot{x} = \psi(x, t) \tag{6.76}$$

where ψ: R^n x $R \rightarrow R^n$ is asumed to be a T-periodic function in t ensuring the existence of $x(t)$ for $t \in [0,\infty)$ and the continuous dependance of $x(t)$ on the initial condition point $x(0) = x_0$, we seek all the T-periodic solutions of (6.46) contained in a bounded region $D \subset R$. The problem considered here numerous important applications in various fields of science and engineering.

An interval method [75] for solving the problem formulated will be now presented.

Let an arbitrary solution of (6.76) starting from the initial point x_0 at the time t_0 be denoted as:

$$x(t) = f(x_0, t_0)$$

Assuming $t_0 = 0$ we have equivalently

$$x(t) = f(x_0) \tag{6.77}$$

Hence for $t = T$

$$x(T) = f(x_0) \tag{6.78}$$

When $x(t)$ is a T-periodic solution to (6.76), then

$$x(T) = x(0) = x_0 \tag{6.79}$$

In this case, from (6.78) and (6.79) we get

$$x_0 = f(x_0) \tag{6.80}$$

Thus, it is seen that the original problem of determining the T-periodic solutions of (6.76) is equated to that of finding the fixed points of the map $f: R^n \to R$ from (6.80). It is worth noting that f is not known explicitly. However, every image of x_0 under f can be found as $x(T)$ by solving (6.76) with $x(0) = x_0$ for t varying from $t = 0$ to $t = T$.

Let X^0 be a given interval vector (an n-dimensional box). What we seek is the determination of all the fixed points of the map f and, hence, the determination of all the T-periodic solutions of (6.76) when $x_0 \in X^0$. Indeed, if a fixed point x_0 to (6.80) is known, then the corresponding periodic solution can be found by integrating (6.76) for the period $[0,T]$ using the fixed point as a starting point for integration.

In this section, an interval method is suggested for solving the problem of finding all T-periodic solutions of (6.76) contained in a bounded region of R^n. The method suggested is based on the equivalent transformation of the original problem into the fixed point problem (6.80), from one hand, and on a well-known interval scheme [10] for finding all the solutions of (6.80), on the other. It involves dynamically dividing the initial box X^0 into subboxes X^v some of which are entered into a list L of subboxes to be processed subsequently. For each of the arising X_v the following steps are carried out.

S t e p 1. Find an outer solution $Y(t)$ of the interval transient analysis problem

$$\dot{x} = \psi(x,t), \quad x_0 \in X^v \tag{6.81}$$

for $t \in [0,T]$ using some interval methods from Chapter 5. (As is seen from (6.81) the arising transient analysis problem is in fact the standard problem 5.9 since only the components of the initial conditions vector x_0 are given as intervals.) Let Y^v denote the outer interval solution of (6.81) for $t = T$ (i.e. Y^v is an interval enclosure of the set of all "point" solutions $x(t)$ of (6.76) at $t = T$ when $x_0 \in X^v$).

S t e p 2. If

$$Y^v \cap X^v = 0 \tag{6.82}$$

the subbox X^v does not contain a solution of (6.80) and is, therefore, deleted (not entered into the list L).

S t e p 3. If

$$(Y^v \cap X^v) \subset X^v \tag{6.83}$$

(the inclusion is strict), then put

$$X^{v+1} = Y^v \cap X^v \tag{6.84}$$

and replace X^v by X^{v+1} in L.

Step 4. If

$$Y^{v} \supseteq X^{v} \tag{6.85}$$

(no reduction of the size of X^v takes place), then X^v is split along its widest size into two boxes that are entered into the list L for further processing.

Remark 6.3. The relations of intersection and inclusion in (6.82) to (6.85) are meant componentwise.

Remark 6.4. It is seen from Steps 1 to 4 that the fixed points problem (6.80) is infallably solved by means of a zero-order interval method (not using derivatives of the function $f(x_0)$ with respect to the components x_{0i} of x_0).

The above process continues until all the fixed points of f are located within a prescribed accuracy ε, that is, until

$$Y^{v} \subseteq X^{v} \tag{6.86}$$

and

$$w(X^{v}) \leq \varepsilon \tag{6.87}$$

where $w(X^v)$ is the width of the box X^v .

As regards the computational efficiency of the method suggested the most crucial is Step 1. In our implementation we have used the interval method due to R. Lonher [61] to solve repeatedly the nonlinear interval transient problem (6.81).

A computer program implementing the method suggested has been developed. To test its applicability (at least for low-dimensional problems) the following example was solved.

***Example* 6.7.** The system of differential equations is

$$\begin{aligned} \dot{x}_1 &= x_2 \\ \dot{x}_2 &= -0.04448x_2 - x_1^3 + 2\cos(2.345t) \end{aligned} \tag{6.88}$$

For $T = 2.6793967$s it has three T-periodic solutions. The corresponding fixed points to the associated equation (6.80) are:

$$\begin{aligned} x_{10} &= 2.904; \quad x_{10} = -0.371; \quad x_{10} = -2.519 \\ x_{20} &= 1.492; \quad x_{20} = 0.017; \quad x_{20} = 1.023 \end{aligned} \tag{6.89}$$

The initial box X^0 was first chosen to be:

$$X_1^0 = [-2.6,\ 3.0];\quad X_2^0 = [0.0,\ 1.5] \tag{6.90a}$$

It contains all the fixed points (6.89). Having chosen $\varepsilon = 0.05$ all three T-periodic solutions were found in 72 iterations (integrations of system (6.81)). The maximum length l_m of the list L needed to solve the problem considered was $l_m = 9$.

Next, a smaller initial box X^0

$$X_1^0 = [-2.6,\ 0.0];\quad X_2^0 = [0.0,\ 1.5] \tag{6.90b}$$

containing the second and the third fixed point from (6.89) was chosen. Using the same accuracy ε the two T-periodic solutions starting from the corresponding fixed points were found in 39 iterations requiring a maximum list length $l_m = 8$.

***Example* 6.8.** [78]. A parametric current source is considered in this example. Its behaviour is modelled by the following system of three nonlinear differential equations:

$$\begin{aligned}
\frac{d\varphi_1}{d\tau} &= \sin(\tau + \psi) - A_1[(\alpha_1 + \alpha_2)\varphi_1 - \alpha_3\varphi_2] \\
\frac{d\varphi_2}{d\tau} &= A_2 v_c + A_3[\alpha_3\varphi_1^* - (\alpha_2 + \alpha_3)\varphi_2^*] \\
\frac{dv_c}{d\tau} &= A_4[\alpha_3\varphi_1 - (\alpha_2 + \alpha_3)\varphi_2]
\end{aligned} \tag{6.91a}$$

where φ_1, φ_2 (flux-linkages) and v_c (capacitor voltage) are normalized values of the state variables and $\tau = 314$ t is the normalized time. Furthermore,

$$\begin{aligned}
A_1 &= 0.835\times10^{-3},\ A_2 = 1.11 \\
A_2 &= 0.97\times10^{-2},\ \ A_4 = 0.128
\end{aligned}$$

while

$$\begin{aligned}
\alpha_1 &= \alpha_1(\varphi_1) = 7.4\times10^{-4} + 2.9\times10^{-5}\varphi_1^{10} \\
\alpha_2 &= \alpha_2(\varphi_2) = 7.4\times10^{-4} + 2.9\times10^{-5}\varphi_1^{10} \\
\alpha_3 &= \alpha_3(\varphi_1,\varphi_2) = 1 + 1.4\times10^{-5} + 0.78\times10^{-5}(\varphi_1 - \varphi_2)
\end{aligned} \tag{6.91b}$$

and the nonlinear functions (6.91b) are obtained if the curve $H = f(B)$ of the material used is approximated by a polynomial

$$H = 3.76B + 0.244B^{11}$$

of eleventh degree. The system (6.91) has been solved for various values of the initial phase ψ. For $\psi = 1.5$ a unique periodic solution with initial conditions (computed as the centres of respective interval solutions)

$$\varphi_1^0 = -0.0613, \quad \varphi_2^0 = 0.0113, \quad v_c = -0.1503$$

in the box

$$X^{(0)} = ([-0.1,\ 0.1],\ [-0.1,\ 0.1],\ [-0.2,\ 0.2])$$

has been found.

6.2.2. Uniqueness of the periodic steady-state

In this subsection we shall touch upon the challenging problem of establishing the uniqueness of periodic steady-states in nonlinear electric circuits. One possible approach to handling this problem may be to appeal to the method presented in the previous subsection; if the fixed point solution of (6.80) is unique in a very large box X^0 (theoretically for an infinitely large X^0) then obviously so is the corresponding periodic steady-state. This approach is, however, limited (at least in its present numerical implementation) to problem of low size (with n not exceeding 2 or 3).

Here an alternate approach will be presented which is based on a sufficient condition for uniqueness of the periodic solution of a class of nonlinear differential equation system [76], namely the so-called separable systems. A system of nonlinear differential equations is called separable if it is of the form:

$$\dot{x} = f(x) + b(t) \tag{6.92}$$

Unlike the general form equation (6.76)

$$\dot{x} = \psi(x,t)$$

now the function $\psi(x,t)$ (the right-hand side) of (6.92) is a sum of two functions each one depending only on x or t, respectively (the argument x and t are separated, hence the term "separable system").

A large class of nonlinear electric circuits can be described by the separable form system (6.92). For simplicity we shall only consider circuits whose nonlinear (two-terminal) elements are the inductors and capacitors, the resistors being linear. Introduce the column-vectors

$$v_L = (v_{L_1},\dots,v_{L_l})^T, \quad i_C = (i_{C_1},\dots,i_{C_{n-l}})^T \tag{6.93}$$

Assuming that the resultant active linear multiport obtained after extracting the nonlinear elements has a hybrid form representation we can write

$$\begin{bmatrix} v_L \\ i_c \end{bmatrix} = H \begin{bmatrix} i_L \\ v_c \end{bmatrix} + \begin{bmatrix} b_1(t) \\ b_2(t) \end{bmatrix} \tag{6.94}$$

where H is constant (n x n) matrix. Now suppose the nonlinear characteristics of the inductors and capacitors are given as the continuously differentiable functions

$$i_{L_k} = i_k(\varphi_k), \quad k = \overline{1,l} \tag{6.95a}$$

$$v_{C_p} = v_p(q_p), \quad p = \overline{1,n-l} \tag{6.95b}$$

where φ_k is the flux linkage of the kth inductor and q_p is the charge on the pth capacitor. On introducing the vector φ with components φ_k, the vector q with components q_p and the vector functions $i(\varphi)$ and $v(q)$ with components (6.95) we have the following relations in vector form

$$\begin{gathered} i_L = i(\varphi) \quad , \quad v_C = v(q) \\ v_L = \frac{d\varphi}{dt} = \dot{\varphi} \ , \quad i_C = \frac{dq}{dt} = \dot{q} \end{gathered} \tag{6.95c}$$

Hence, (6.94) can be written as

$$\begin{bmatrix} \dot{\varphi} \\ \dot{q} \end{bmatrix} = H \begin{bmatrix} i(\varphi) \\ v(q) \end{bmatrix} + \begin{bmatrix} b_1(t) \\ b_2(t) \end{bmatrix} \tag{6.96}$$

Finally, if the vectors

$$x = \begin{bmatrix} \varphi \\ q \end{bmatrix}, \quad \dot{x} = \begin{bmatrix} \dot{\varphi} \\ \dot{q} \end{bmatrix}$$

and the vector nonlinear function

$$\psi(x) = \begin{bmatrix} i(\varphi) \\ v(q) \end{bmatrix} \tag{6.97}$$

are introduced, system (6.96) takes on the separable form

$$\dot{x} = H\psi(x) + b(t) \tag{6.98a}$$

It should be noted that the function $\psi(x)$ from (6.98a) has the special form

$$\psi_k(x) = \psi_k(x_k) \tag{6.98b}$$

which follows from (6.95) and (6.97). It is assumed that all the sources of the circuit described by (6.98) are periodic with one and the same period T. Hence the vector $b(t)$ is also T-periodic.

Now we shall present the following sufficient condition for the periodic solution of (6.92) to be unique (in the large, i.e. for any initial conditions vector $x_0 \in R^n$) [76].

Theorem 6.3. Let $f(x)$ be a C^1 function and $b(t)$ a T-periodic function. Furthermore, let $J(x)$ denote the Jacobian matrix of $f(x)$ with components $J_{ij}(x) = \partial f_i / \partial x_j$. If all the eigenvalues of the matrix

$$C(x) = \frac{1}{2}[J(x) + J^T(x)] \tag{6.99}$$

(T denoting transpose) have negative real parts for any $x \in R^n$ then the system of nonlinear differential equation (6.92) has a unique T-periodic solution.

Since $\psi(x)$ and $b(t)$ from (6.98) are a C^1 function and a T-periodic function, respectively, Theorem 6.3 can be obviously applied to system (6.98). Let $d(x)$ be a diagonal matrix whose nonzero element $d_k = d_{kk}(x_k) = \partial \varphi_k / \partial x_k$. So the matrix $C(x)$ defined by (6.99) will, in the present case, be

$$C(x) = \frac{1}{2}[Hd(x) + d(x)H^T] \tag{6.100}$$

Recall that a matrix is stable iff all its eigenvalues have negative real parts (section 4.2.2). Thus, we have the following corollary.

Corollary 6.5. If the matrix $C(x)$ defined by (6.100) is stable for any $x \in R^n$, the nonlinear circuit described by (6.98) has a unique T-periodic steady-state.

Finding all the eigenvalues of $C(x)$ even in the case (6.100) for all $x \in R^n$ is computationnaly an intractable problem.

To be able to use Corollary 6.5 we shall assume (which is most often the case in practice) that each characteristic of the nonlinear elements given by (6.95) has a finite slope. Then each diagonal element d_k of the matrix $d(x)$ will belong to some interval D_k, that is,

$$d_k = d_{kk}(x) \in D_k, \quad k = \overline{1,n} \tag{6.101}$$

for any $x_k \in R$. Now we shall introduce the notations:

d – a real diagonal matrix whose nonzero elements are defined by (6.101)

D – an interval diagonal matrix whose nonzero elements are the intervals D_k from (6.101).

It follows from (6.101) and (6.100) that $C(x)$ will be stable for all $x \in R^n$ iff the following set of matrices

$$C = \frac{1}{2}[Hd + dH^T] \tag{6.102a}$$

$$d \in D \tag{6.102b}$$

is stable. Thus we have the following theorem.

Theorem 6.4. The nonlinear circuit described by (6.98) has a unique T-periodic steady-state if the set of matrices C defined by (6.102) is stable.

The advantage of Theorem 6.4 over (the equivalent) Corollary 6.5 lies in the fact that the assertion of Theorem 6.4 can be verified by some of the methods from section 4.2 for checking the stability of matrices with interval data. It should however be stressed that the matrix C defined by (6.102a) is a matrix whose elements C_{kj} are not independent since C is a symmetric matrix which follows from

$$C^T = \frac{1}{2}[Hd + dH^T]^T = \frac{1}{2}[d^T H^T + (H^T)^T d^T] =$$
$$= \frac{1}{2}[dH^T + Hd] = C$$

Moreover, the elements C_{kj} are all functions of the elements d_k and d_j of the diagonal matrix d. Indeed, it is seen from (6.102a) that

$$c_{kj} = \frac{1}{2}[c_{kj}' + c_{kj}''] = \frac{1}{2}[h_{kj}d_j + d_k h_{jk}] = c_{jk}$$

Since Theorem 6.4 provides only a sufficient condition for uniqueness of the periodic steady-state of the nonlinear circuit studied it is preferable to use a necessary and sufficient condition test for checking the stability of the set of matrices (6.102). Therefore, it is recommended to first transform the matrix stability problem associated with (6.102) into an equivalent polynomical formulation stability problem. The resultant (characteristic) polynomial is a polynomial with dependent coefficients. Its stability or instability can be established by the criteria from sections 4.3.1.

The above approach to proving the uniqueness of T-periodic steady-states in nonlinear circuits of the separable class considered will be illustrated by the following examples.

***Example* 6.9.** The circuit studied is shown in Fig. 6.5. The only nonlinear element is the inductor with a given nonlinear characteristic

$$i = i(\varphi) \tag{6.103}$$

To obtain the circuit equations in the separable form (6.92) we first write down a system of equations in the "usual" form

$$\begin{aligned} R_1 i_1 + v_C + v_L &= v(t) \\ R_2 i_2 = v_L &= \dot{\varphi} \\ i_1 &= i_2 + i \end{aligned} \tag{6.104}$$

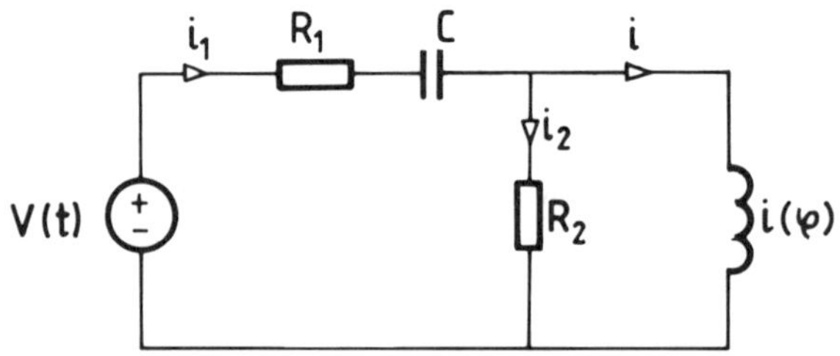

Fig. 6.5. Circuit studied in Example 6.9.

Using (6.103) and the relations

$$v_C = \frac{q}{c}, \quad i_1 = \dot{q} \tag{6.105}$$

system (6.104) is then transformed equivalently into the system

$$\begin{aligned} \dot{\varphi} &= -\frac{R_1}{\alpha} i(\varphi) - \frac{1}{\alpha C} q + \frac{1}{\alpha} v(t) \\ \dot{q} &= \frac{1}{\alpha} i(\varphi) - \frac{1}{\alpha C R^2} q + \frac{1}{\alpha R_2} v(t) \end{aligned} \tag{6.106a}$$

where

$$\alpha = (R_1 + R_2)/R_2 \tag{6.106b}$$

The system (6.106) is of the separable form (6.98):

$$\begin{aligned} \dot{\varphi} &= h_{11} i(\varphi) + h_{12} q + b_1(t) \\ \dot{q} &= h_{21} i(\varphi) + h_{22} q + b_2(t) \end{aligned} \tag{6.107}$$

where

$$h_{11} = -\frac{R_1R_2}{R_1+R_2}, \quad h_{12} = -\frac{R_2}{(R_1+R_2)C}$$
$$h_{21} = \frac{R_2}{R_1+R_2}, \quad h_{22} = -\frac{1}{(R_1+R_2)C} \tag{6.108}$$

Let $d = d(\varphi)$ denote the derivative of i in φ. Then

$$J = \begin{bmatrix} h_{11}d & h_{12} \\ h_{21}d & h_{22} \end{bmatrix}$$

and hence

$$C = \begin{bmatrix} h_{11}d & 0.5h_{12}+0.5h_{21}d \\ 0.5h_{12}+0.5h_{21}d & h_{22} \end{bmatrix} = \begin{bmatrix} c_{11}(d) & c_{12}(d) \\ c_{21}(d) & c_{22}(d) \end{bmatrix} \tag{6.109}$$

where the notation $c_{kj}(d)$ is used to underline the dependence of the corresponding elements on d. Now suppose that $d = d(\varphi)$ is bounded for all $\varphi \in (-\infty, \infty)$. Then d belongs to some interval D, i.e.

$$d \in D \tag{6.110}$$

Therefore, according to Theorem 6.4 the circuit considered has a unique periodic steady-state (when the supply voltage $v(t)$ is periodic) if the set of matrices (6.109), (6.110) is stable.

To verify the stability of (6.109), (6.110), first the associated characteristic polynomial is formed

$$det \begin{bmatrix} c_{11}(d)-\lambda & c_{12}(d) \\ c_{12}(d) & c_{22}-\lambda \end{bmatrix} =$$
$$= \lambda^2 - [c_{11}(d) + c_{22}]\lambda + c_{11}(d)c_{22} - c_{12}^2(d) = 0 \tag{6.111}$$

By Theorem 4.13 the set of polynomials (6.111), (6.110) is stable iff

$$-c_{11}(d) - c_{22} > 0 \tag{6.112a}$$

and

$$c_{11}(d)c_{22} - c_{12}^2(d) > 0 \tag{6.112b}$$

with $d \in D$. To get simpler conditions (6.112) system (6.106a) will be modified as follow. First the variable q is replaced by a new variable u:

$$q = ku \tag{6.113}$$

where k is an unknown constant. Using (6.113) the system (6.107) takes on the form

$$\begin{aligned} \dot{\varphi} &= h_{11}i(\varphi) + h_{12}ku + b_1(t) \\ \dot{u} &= \frac{h_{21}}{k}i(\varphi) + h_{12}u + b_2(t)\frac{1}{k} \end{aligned} \tag{6.114}$$

The matrix C related to (6.114) is then

$$C = \begin{bmatrix} h_{11}d & \frac{1}{2}(h_{12}k + \frac{h_{21}}{k}d) \\ \frac{1}{2}(h_{12}k + \frac{h_{21}}{k}d) & h_{12} \end{bmatrix} \tag{6.115}$$

so

$$c_{12} = c_{21} = \frac{1}{2}(h_{12}k + \frac{h_{21}}{k}d) \tag{6.116}$$

Now let interval D be put in the centered form

$$D = d_0 + [-\Delta, \Delta] \tag{6.117}$$

Then d can be written as

$$d = d_0 + \delta \tag{6.118a}$$

where

$$\delta \in [-\Delta, \Delta] = \Delta^I \tag{6.118b}$$

Using (6.118a) the expression (6.116) for c_{12} can be rewritten as

$$c_{12} = \frac{1}{2}[h_{12}k + \frac{h_{21}}{k}d_0) + \frac{1}{2}\frac{h_{21}}{k}\delta] \tag{6.119}$$

In order to obtain simpler (and less conservative) conditions (6.112) the constant k is chosen such that the first term in (6.119) be zero, i.e.

$$h_{12}k + \frac{h_{21}}{k}d_0 = 0 \tag{6.120}$$

It is seen from (6.108) and (6.120) that

$$k = \sqrt{Cd_0} \tag{6.121}$$

Now on the basis of (6.119) and (6.120) the conditions (6.112) corresponding to the modified matrix (6.115) are

$$-h_{11}(d_0 + \delta) - h_{22} > 0, \quad \delta \in \Delta^I \tag{6.122a}$$

$$h_{11}(d_0 + \delta)h_{22} - \frac{h_{21}\delta^2}{4k^2} > 0, \quad \delta \in \Delta^I \tag{6.122b}$$

It is easily seen that on account of (6.108) and (6.121) conditions (6.122) reduce to

$$R_1R_2C(d_0 + \delta) + 1 > 0, \quad \delta \in \Delta^I \tag{6.123a}$$

$$R_1d_0 + R_1\delta - \frac{R_2}{4d_0}\delta^2 > 0, \quad \delta \in \Delta^I \tag{6.123b}$$

We shall check conditions (6.123) for the following data:

$$R_2 = 1.6\ \text{k}\Omega, \quad C = 1.69\,\mu\text{F}, \quad d_0 = 0.40563\ \text{H}^{-1} \tag{6.124a}$$

$$\Delta^I = [-0.37562, \quad 0.37562]\ \text{H}^{-1} \tag{6.124b}$$

The problem is to find (approximately) the smallest value of R_1, for which the circuit investigated is still guaranteed to have a unique periodic steady-state.

Since for the above data

$$d = d_0 + \delta > 0, \quad \delta \in \Delta^I$$

(the characteristic (6.103) of the nonlinear inductor is strictly increasing) and R_1, R_2 and C are positive numbers, the first condition (6.123a) is always satisfied (for any R_1). Thus, we need only to check the second condition (6.123b). To do so we have to find the global minimum $\underline{y}$ of the function

$$y = y(\delta) = R_1d_0 + R_1\delta - \frac{R_2}{4d_0}\delta^2 \tag{6.125a}$$

when

$$\delta \in \Delta^I \tag{6.125b}$$

Using a very simple technique based on testing the sign of the derivative

$$R_1 - \frac{R_2}{2d_0}\delta > 0, \quad \delta \in \Delta^I \tag{6.126}$$

of (6.125a) it has been found that $\underline{y} > 0$ for

$$R_1 \geq 4.64\,\mathrm{k}\,\Omega \tag{6.127}$$

Thus, the circuit investigated is guaranteed to have a unique periodic steady-state if (6.127) is fulfilled since for such values of R_1 the conditions (6.123) are satisfied.

***Example* 6.10.** This example serves to show that the present sufficient condition for uniqueness can be applied (unlike other known criteria) even in the case of nonlinear elements with nonmonotonic characteristics. We shall consider the circuit shown in Fig. 6.6. As compared with the previous example (Fig. 6.5.) now the circuit contains additionally a nonlinear current-controlled resistor R_3 with volt-ampere characteristic

$$v_3 = v_3(i) \tag{6.128}$$

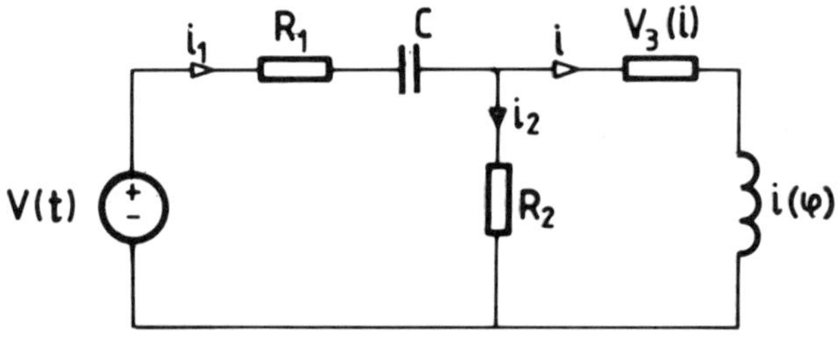

Fig. 6.6. Circuit considered in Example 6.10.

It can be easily verified that choosing the magnetic flux and the capacitor charge q as state variables the circuit equations are

$$\frac{d\varphi}{dt} = -\frac{R_1}{\alpha}i(\varphi) - v_3(i) - \frac{q}{\alpha C} + \frac{1}{\alpha}v(t) \tag{6.129a}$$

$$\frac{dq}{dt} = \frac{1}{\alpha}i(\varphi) - \frac{v_3(i)}{R_2} - \frac{q}{\alpha C R_2} + \frac{1}{\alpha R_2}v(t) \tag{6.129b}$$

with α given by (6.106b). Substituting (6.103) into (6.128) and (6.129) we finally get

$$\frac{d\varphi}{dt} = a_{11}(\varphi) + a_{12}q + b_1(t) \tag{6.130a}$$

$$\frac{dq}{dt} = a_{21}(\varphi) + a_{22}q + b_2(t) \tag{6.130b}$$

where

$$a_{11}(\varphi) = -\frac{R_1}{\alpha}i(\varphi) - v_3(i(\varphi)) \tag{6.131a}$$

$$a_{21}(\varphi) = \frac{1}{\alpha}i(\varphi) - \frac{1}{R_2}v_3(i(\varphi)) \tag{6.131b}$$

Now system (6.130) is of separable form and the uniqueness of its T-periodic solution may be checked by Theorem 6.3. It will be shown that this is possible even if the function (6.128) is nonmonotonic. Indeed, let $d_{11}(\varphi)$ and $d_{21}(\varphi)$ denote the derivative of $a_{11}(\varphi)$ and $a_{21}(\varphi)$, respectively, in φ. Combining (6.130) and (6.131) the Jacobian matrix related to (6.130) is seen to be

$$J = \begin{bmatrix} d_{11}(\varphi) & a_{12} \\ d_{21}(\varphi) & a_{22} \end{bmatrix}$$

so that the matrix C is

$$C = \begin{bmatrix} d_{11}(\varphi) & 0.5[d_{21}(\varphi) + a_{12}] \\ 0.5[d_{21}(\varphi) + a_{12}] & a_{22} \end{bmatrix} = \begin{bmatrix} c_{11}(\varphi) & c_{12}(\varphi) \\ c_{12}(\varphi) & c_{22} \end{bmatrix} \tag{6.132}$$

The stability of matrix C from (6.132) when $\varphi \in (-\infty, \infty)$ can be checked in a similar way as this was done in the previous example. The sufficient conditions (6.112) are now replaced by the inequalities

$$y_1(\varphi) = -c_{11}(\varphi) - a_{22} > 0 \tag{6.133a}$$

$$y_2(\varphi) = c_{11}(\varphi)a_{22} - c_{12}(\varphi) > 0 \tag{6.133b}$$

with

$$\varphi \in (-\infty, \infty) \tag{6.133c}$$

or equivalently

$$\underline{y_1(\varphi)} > 0 \tag{6.134a}$$

$$\underline{y_2(\varphi)} > 0 \tag{6.134b}$$

where the symbol $\underline{y(\varphi)}$ denotes the lower endpoint of $y(\varphi)$ when φ belongs to the corresponding domain. Thus, the problem of assessing the uniqueness of the periodic steady-state of the circuit considered has been reduced to that of finding the global minimum of two functions in one variable. The latter problem can be solved infallibly by some of the available interval methods for global optimization (e.g. [80]). Obviously, conditions (6.134) may be satisfied even in the case where the characteristic (6.128) of the nonlinear resistor R_3 is nonmonotonic.

Comments

Section 1. In this section, the problem of determining all operating points of resistive circuits whose nonlinear two-terminal resistors are modelled by continuously differentiable functions (CDF problem) has been considered in the framework of the interval analysis approach. Such an approach was, seemingly for the first time, suggested in [67]. It differs favorably from the traditional noninterval approach in that all the operating points are infallibly located within a prescribed accuracy in a finite number of iterations.

In the case (subsection 6.1.1) where the nonlinear equations describing the circuit are of the general form (6.1) three interval methods for solving the d.c. nonlinear analysis problem have been presented: Hansen's method, Krawctyk's method and Alefeld–Herzberger's method. Detailed componentwise algorithms implementing these methods have been developed.

It should be borne in mind that nowadays there exist two improved versions of Hansen's method. The first one is referred to as the Hansen–Greenberg realization. It has been applied in [81] to find the solution to a model of a bipolar transistor having nine equations in nine variables. The second one (due to Kearfott [82]) is based on a special (optimal) preconditioning of (6.4) (premultiplying it by a matrix different from the matrix B used in subsection 6.1.1) and has been reported to yield (in some cases) substantially better results than the original Hansen's method.

Subsection 6.1.2 covers the case where the nonlinear circuit permits the hybrid form representation (6.16). Two methods for solving the associated d.c. analysis problem have been presented. The former method (M4) is, in fact, a modification of the known interval zero-order fixed-point method for solving a system of nonlinear algebraic equations which takes into account the specific diagonal form (6.17b) of the nonlinearities involved. The latter one (M5) is an improvement of Alefeld–Herzberger method and is based on the introduction of two (suboptimal) points x_i' and x_i'' in an attempt to reduce the interval Y from (6.48). It should be noted that the idea of using two vectors x' and x'' different from the centre m has been suggested in [83] in the context of Krawctyk's method for the general case of system (6.1) and in [69] for the special case of the hybrid representation (6.17). In the latter case it is possible to find the optimal points x_i^L and x_i^U at a reasonable computational cost provided the nonlinear functions (6.49) are monotonic in the current intervals X_i.

Numerical examples illustrating the applicability of the above methods are given in section 6.1.3. It is very difficult (at this stage even impossible) to draw any conclusion regarding the relative numerical efficiency of a particular method – so much depends on technical details in the algorithmic implementation of the method considered. For instance, the nonlinear equations (6.41) from method M4 can be solved in a far more efficient way than it was done in the present algorithm A4b. The experimental evidence available so far seems to show that for the hybrid form representation case, Algorithms A4b and A5 (at least in their present implementation) are more efficient than the remaining interval methods considered in sections 6.1.1, 6.1.2. It will be interesting to apply Kearfott's method for solving the hybrid system (6.17) and to compare its efficiency with the above two algorithms.

Examples 6.5 and 6.6 show that the general methods from section 6.1.1 can be applied successfully for solving problems other than d.c. circuit analysis.

In the last section 6.1.4 a method for solving underdetermined systems of two-equations of n variables ($n > 2$) has been presented. Such systems occur in method M2 for tolerance analysis in probabilistic formulation from section 2.5.1. Its present algorithmic implementation allows for certain improvements (incorporation of Case 2 from Procedure 6.2 and more efficient transition from a pair X_i, X_{i+1} to the next pair X_{i+1}, X_{i+2} in solving the linear interval systems of type (6.70)). It should be noted that the tolerance method based on the above method for solving underdetermined systems (even in the present imperfect form) shows a rather high convergence rate.

Section 6.2. In the first subsection of this section, the problem of finding all the periodic steady-states of a given period arising in a nonlinear electric circuit has been considered. An interval method for solving this problem has been suggested. It reduces the original problem of determining the T-periodic solutions of system (6.76) to that of finding the fixed points of Eq. (6.80). The latter problem is then solved using a zero-order interval method.

It should be stressed that the method in its present implementation is very time-consuming because of the need to find (in Step 1 of the method) an outer solution $Y(t)$ of the interval transient analysis problem (6.81) at each iteration. The latter problem has been solved using Lohner's integration method. Depending on the integration step size, the number of Taylor's terms used in the expansion of the right-hand side of (6.81) the size of the initial region X^0 and the accuracy ε chosen the solution of Example 6.7 took from 1 to 10 minutes on a VAX 9000 computer. Therefore, the present method can be applied to circuits of low dimension ($n \leq 3$). Improving the numerical efficiency of the interval methods for integrating nonlinear ODE's, and more specifically overcoming the wrapping effect might lead in the near future to better methods for solving the global T-periodic solution considered in section 6.2.1.

In the last subsection of the chapter the challenging problem of establishing the uniqueness of a T-periodic steady-state in nonlinear electric circuits has been touched upon. A new result has been obtained for the special case of circuits for which the system of ODE's is of the separable form (6.92). Theorem 6.4 provides a sufficient condition for uniqueness in this class of circuits. It reduces the original uniqueness problem to that of checking whether an associated set of matrices is stable. The latter problem can be handled by some of the methods from section 4.2.

Two numerical examples illustrate the application of Theorem 6.4.

The problem of determining the uniqueness of a T-periodic solution for a system of nonlinear ODE's in the general case (6.76) is extremely difficult. Sufficient conditions (in noninterval form) for some special cases can be found in, among others, [84] and [85]. It should be noted that the approach from [84] can be implemented using interval global optimization techniques from sections 2.3 to 2.4, thus reducing the conservativeness of the known results.

CONCLUSIONS

In conclusion, a few brief remarks of general character will be made here.

The present book is the first monograph to deal with interval analysis applications in circuit theory. It covers a limited number of topics that have been mainly in the author's areas of research. Many other applications are conceivable in the domain of circuit analysis, both in the case of linear and nonlinear circuits. Indeed, so far only real interval arithmetic has been used for the purposes of circuit analysis. It is expected that the use of complex interval arithmetic [10] will lead to new interesting interval methods for analyzing a.c. circuits. Further application of the interval analysis approach seems especially promising in the domain of nonlinear circuits. It suffices to note here that already interval methods have been designed to rigorously verify the existance of chaos in dynamic systems [86]. Finally, it should be noted that the interval methods available to date treat circuit analysis problems only: the area of developing interval methods for solving electric circuit (or control system) synthesis problems has not been investigated as yet.

The interval methods for circuit analysis presented in the book cover essentially two major topics: robust analysis of linear circuits (static and dynamic tolerance analysis, robust stability) and some aspects of the global analysis of nonlinear circuits (finding all d.c. or periodic steady-states, uniqueness of the periodic steady-state). They are essentially based on the present state of the art of those interval analysis techniques that are related to the topics considered. It should, however, be borne in mind that interval analysis is presently undergoing a period of rapid development and it is to be expected that new improved mathematical tools (methods for obtaining narrower interval extensions for solving more efficiently linear and nonlinear equations, global optimization problems as well as more efficient hardware and software realizations) will be soon available. Therefore, it is belived that the numerical efficiency of the interval methods for circuit analysis may be substantially improved. However, even in their present (far from being perfect) implementation these methods are superior to the existing traditional (point) methods in many respects: guaranteed global convergence and required accuracy, reliable stopping criteria (unlike the noninterval methods based on global optimization or global nonlinear analysis for which one always faces the risk of terminating the computation process prematurely before the global solution(s) is (are) reached or continuing it uselessly in the hope to find new solutions), lesser computation times in most of the cases studied so far.

The interval methods suggested in the book have been designed to solve primarily electric circuit analysis problems. Indeed, some of them exploit advantageously the specific structure of the electric circuit class considered in an effort to devise algorithms of improved numerical efficiency. On the other hand many of the electric circuit analysis problems herein considered – tolerance analysis via global optimization, robust stability and performance analysis, global nonlinear analysis in the case of equations of arbitrary form – are of more general nature. The methods proposed for solving these latter problems can, therefore, be applied (directly or after minor modifications) to tackling similar problems arising in systems of arbitrary physical constituency. For this reason, it is hoped that these more general interval methods will be useful not only to electrical or electronics engineers but also to systems analysts, control engineers and other specialists striving to use modern computer methods in their respective research or application areas. Finally, some of the specific electric circuit analysis problems formulated in the book constitute challenging mathematical problems and might draw the attention of applied mathematicians and, more specifically, of interval analysis specialists.

If the present monograph arouses interest among all those who develop or apply modern computer methods in their special fields and encourage them to consider the possibility of including interval analysis methods in their work, the author's main objective would be largely fulfilled.

REFERENCES

[1] R.E. Moore, *Interval Analysis* (Englewood Cliffs, NJ: Prentice-Hall, 1966).

[2] R.E. Moore, *Methods and Applications of Interval Analysis* (Philadelphia: SIAM, 1979).

[3] S.M. Markov, *C.R. Acad. Bulgare des Sciences* **30** (1977) 1239.

[4] E.R. Hansen and S. Sengupta, *BIT* **21** (1981) 203.

[5] S. Scelboe, *BIT* **14** (1974) 87.

[6] S.M. Markov, in *Fundamentals of Numerical Computation* (*Computer-Oriented Numerical Analysis*), ed. G. Alefeld and R.D. Grigorieff (Wien-New York: Springer-Verlag, 1980)

[7] E.R. Hansen, in *Topics in Interval Analysis*, ed. E. Hansen (Clarendon Press, Oxford, 1966).

[8] E.R. Hansen, *Num. Math.* **34** (1980) 247.

[9] E.R. Hansen, in *Topics in Interval Analysis,* ed. E. Hansen (Clarendon Press, Oxford, 1966).

[10] G. Alefeld and J. Herzberger, *Introduction to Interval Computation* (Academic Press, New York, 1983) p. 333.

[11] K. Reichmann, *Computing* **22** (1979) 355.

[12] A. Neumaier, in *Interval Mathematics*, ed. K. Nickel (Academic Press, New York, 1985).

[13] J. Rohn, *Freiburger Intervall-Berichte* **7** (1984) 1.

[14] J. Rohn, *Freiburger Intervall-Berichte* **4** (1985) 15.

[15] J. Rohn, *Freiburger Intervaal-Berichte* **4** (1985) 93.

[16] R.E. Moore, *SIAM J. Numer. Anal.* **14** (1977) 611.

[17] R. Krawczyk, *Computing* **4** (1969) 187.

[18] N. Asaithambi and R. Moore, *Computing* **28** (1982) 225.

[19] E.R. Hansen and S. Seugupta, in *Interval Mathematics*, ed. K. Nickel (Academic Press, New York, 1980).

[20] Y. Fuji and K. Tochida, in *Interval Mathematics*, ed. K. Nickel (Academic Press, New York, 1985).

[21] S. Scelboe, *IEEE Trans. Circuits Systems* **26** (1979) 874.

[22] E.R. Hansen, in *Interval Mathematics*, ed. K. Nickel (Springer-Verlag, Berlin, Heidelberg, 1975).

[23] R. Krawczyk and A. Neumaier, *SIAM J. Numer. Anal.* **22** (1985) 604.

[24] A. Neumaier, *Interval Methods for Solving Equations* (Cambridge University Press, London, 1990).

[25] K. Reinschke, *Numerische Verfahren zur Analyse passiver Linear Netwerke unter Berucksichtingung des Einflußes es der Toleranzen der Schaltelemente* (Dissertation, Fakultat für Elektrotechnik, Tech. Univ., Dresden, 1966).

[26] K. Reinschke, *Zuverläßichkeit von Systemen* (Verlag Technik, Bd. 2, Berlin, 1974).

[27] K. Geher, *Theory of Network Tolerances* (Akademial Kiodo, Budapest, 1971).

[28] J.K. Fidler and C. Nightingale, *Computer-aided Circuit Design* (Nelson, 1978).

[29] P.R. Adby, *Applied Circuit Theory* (Horwood, Chichester, 1980).
[30] E.R. Hansen, *Math. Com.* **22** (1978) 153.
[31] L.V. Kolev, V. Mladenov and S. Vladov, *IEEE Trans. Circuits Syst.* **CAS-35** (1988) 967.
[32] S. Skelboe, *True Worse-case Analysis of Linear Electrical Circuits by Interval Arithmetic* (Rep 1T, II, Inst. of Circuit Theory and Telecommunication, Tech. Univ. of Denmark, Lyngby, 1977).
[33] K. Madsen and H. Jacobsen, *IEEE Trans. Circuits Syst.* **CAS-26** (1979) 775.
[34] E. Baumann, *Freiburger Intervall-Berichte* **6** (1986) 1.
[35] J. Rohn, in *Int. Conf. on Interval Methods for Numerical Computation* (Oberwolfach, West Germany, March 1990).
[36] L.V. Kolev and S. Vladov, in *Proceedings on the Sixth Int. Symp. on Networks, Systems and Signal Processing, ISYNT 89* (Zagreb, Yugoslavia, June 1989).
[37] J. Rohn, *Linear Algebra and its Applications* **126** (1989) 39.
[38] C. Jansson, *Computing* **46** (1991) 265.
[39] J. Pinel and K. Roberts, *IEEE Trans. Circuit Theory* **19** (1972) 612.
[40] M. Glesner and A. Blum, in *European Conf. on Circuit Theory and Design* (London, 1974).
[41] J.W. Bandler, P.C. Lin and H. Tromp, *IEEE Trans. Circuits Syst.* **CAS-23** (1976) 155.
[42] V.L. Kharitonov, *On a Generalization of a Stability Criterion* (Izv. Akad. Nauk. Ser.: Phys.-Mat., **1**, Kazakh. SSR, 1978).
[43] S. Bialas and J. Garloff, *IEEE Trans. Auto. Control* **AC-30** (1985) 310.
[44] N. Balabanian and T.A. Bickart, *Electrical Network Theory* (Robert E. Kreiger Publishing Company, Florida, 1985).
[45] J.A. Heinen, *Int. J. Control* **39** (1984) 1323.
[46] D. Xu, *Int. J. Control* **41** (1985) 289.
[47] R.K. Yedavalli, *Int. J. Control* **43** (1986) 767.
[48] A. Rachid, *Int. J. Control* **50** (1989) 1563.
[49] Y.–T. Juang and C.–S. Shao, *Int. J. Control* **49** (1989) 1401.
[50] G. Dudic *et al.*, *Analysis and Optimal Synthesis of Control Systems Using Computers* (Science, Moscow, 1984).
[51] A. Neumeier, Private communication (1990).
[52] C. Soh and S. Berger, *IEEE Trans. Auto. Control* **AC-33** (1988) 351.
[53] K. Wei and R. Yedavalli, *IEEE Trans. Auto. Control* **AC-32** (1987) 907.
[54] R.E. De Gaston and M.E. Safronov, *IEEE Trans. Auto. Control* **AC-33** (1988) 156.
[55] A. Sideris and R. Sauchez Pena, *IEEE TRans. Auto. Control* **AC-34** (1989) 1272.
[56] F. Kraus and W. Truol, *IEEE Trans. Auto. Control* **AC-53** (1991) 967.
[57] R. Frazer and W. Duncan, in *Proc. Royal Society A*, **124** (1929) 642.
[58] A. Vicino, A. Tesi and M. Milanese, *IEEE Trans. Auto. Control* **35** (1990) 835.
[59] J.C. Doyle, J. Wall and G. Stein, *Proc. IEEE Conf. Decision Control* (Orlando, FL, Dec. 1982).

[60] L.V. Kolev, *Int. J. Circuit Theory Appl.* **20** (1992) 649.

[61] R. Lohner, Einschliessung der Lösungen gewöhnlicher Anfangs-und Randwertaufgaben und Anwendungen (Dissertation, Karlsruhe (TH), Deutschland, 1988)

[62] S.–S. Wang and W.–G. Lin, Control Theory Advanced Tech. 7 (1991) 271.

[63] P. Botchev, *Int. Conf. on Interval Analysis* (Basel, 1989).

[64] E.P. Oppenheimer and A.N. Michel, *IEEE Trans. Circuits Syst.* **35** (1988) 1230.

[65] L.V. Kolev and S. Vladov, *Scientific Session of the Higher Institute for Mech. and Elect. Eng.* (Sofia, Bulgaria, Oct. 1989).

[66] K. Nickel, in *Interval Mathematics*, ed. K. Nickel (Academic Press, New York, 1980).

[67] L.V. Kolev, *Int. J. Circuit Theory Appl.* **12** (1984) 175.

[68] L.V. Kolev and V. Mladenov, *Scientific Session of the Higher Institute for Mech. and Elect. Eng.* (Sofia, Bulgaria, Oct. 1989).

[69] L.V. Kolev, *Int. Conf. on Interval Methods for Numerical Computations* (Oberwolfach, West Germany, March 1990).

[70] L.V. Kolev and V. Mladenov, *Int. J. Circuit Theory Appl.* **18** (1990) 257.

[71] L.O. Chua and P.M. Lin, *Computer Aided Analysis of Electronic Circuits: Algorithms and Computational Techniques* (Prentice–Hall, Englewood Cliffs, NJ, 1975).

[72] L.O. Chua, Y.F. Lam and K.S. Stromsoe, *Proc.* 1976 *IEEE Int. Symp. on Circuits and Systems* (TUM, Munich, West Germany, April, 1976).

[73] L.O. Chua and A. Ushida, *Int. J. Circuit Theory Appl.* **4** (1976) 215.

[74] A.A. Lanne, *Optimal Synthesis of Linear Electric Circuits* (Communication, Moscow, 1969).

[75] L.V. Kolev and V. Mladenov, *Int. Conf. on Scientific Computation* (Albena, Bulgaria, Oct. 1990).

[76] B.A. Pliss, *Nonlocal Problems of Oscillation Theory* (Science, Moscow, 1964).

[77] S. Malan, M. Milanese, M. Taragna and J. Garloff, *Int. Conf. on Intelligent Control and Instrumentation* (Singapore, Feb. 1992).

[78] V. Mladenov, *On Some Problems of the Global Analysis of Nonlinear Circuits* (PhD dissertation, Sofia, 1993)

[79] L.V. Kolev, *Int. J. Circuit Theory Appl.* **15** (1987) 181.

[80] H. Ratschek and J. Rokne, *New computer Methods for Global Optimization* (Horwood,Chichester, 1988).

[81] H. Ratschek and J. Rokne, *J. Global Optimization* **8** (1992) 1.

[82] R. Kearfott, *Int. Conf. on Interval Methods for Numerical Computations* (Obervolfach, West Germany, March 1990)

[83] L.V. Kolev, *J. Mathematical Physics and Numerical Mathematics* **29** (1989) 223.

[84] J.M. Hasler and P. Verburgh, *IEEE Trans. Circuits Syst.* **31** (1984) 614.

[85] L.V. Kolev, *Electricity* **10** (1991) 47.

[86] G. Bogilov, L. Kolev, V. Mladenov and S. Vladov, *Scientific Session of the Higher Institute for Mech. and Elect. Eng.* (Sofia, Bulgaria, Oct. 1989).

SUBJECT INDEX